A CULTURAL HISTORY OF CHEMISTRY

VOLUME 6

A Cultural History of Chemistry
General Editors: Peter J.T. Morris and Alan J. Rocke

Volume 1
A Cultural History of Chemistry in Antiquity
Edited by Marco Beretta

Volume 2
A Cultural History of Chemistry in the Middle Ages
Edited by Charles Burnett and Sébastien Moureau

Volume 3
A Cultural History of Chemistry in the Early Modern Age
Edited by Bruce T. Moran

Volume 4
A Cultural History of Chemistry in the Eighteenth Century
Edited by Matthew Daniel Eddy and Ursula Klein

Volume 5
A Cultural History of Chemistry in the Nineteenth Century
Edited by Peter J. Ramberg

Volume 6
A Cultural History of Chemistry in the Modern Age
Edited by Peter J.T. Morris

A CULTURAL HISTORY OF CHEMISTRY

IN THE MODERN AGE

VOLUME 6

Edited by Peter J.T. Morris

BLOOMSBURY ACADEMIC
LONDON • NEW YORK • OXFORD • NEW DELHI • SYDNEY

BLOOMSBURY ACADEMIC
Bloomsbury Publishing Plc
50 Bedford Square, London, WC1B 3DP, UK
1385 Broadway, New York, NY 10018, USA
29 Earlsfort Terrace, Dublin 2, Ireland

BLOOMSBURY, BLOOMSBURY ACADEMIC and the Diana logo are trademarks of
Bloomsbury Publishing Plc

First published in Great Britain 2021
Paperback edition published in 2025

Copyright © Bloomsbury Publishing Plc, 2025

Cover design: Rebecca Heselton
Cover image © B&M Noskowski/Getty Images

All rights reserved. No part of this publication may be reproduced or transmitted in any form or by any means, electronic or mechanical, including photocopying, recording, or any information storage or retrieval system, without prior permission in writing from the publishers.

Bloomsbury Publishing Plc does not have any control over, or responsibility for, any third-party websites referred to or in this book. All internet addresses given in this book were correct at the time of going to press. The author and publisher regret any inconvenience caused if addresses have changed or sites have ceased to exist, but can accept no responsibility for any such changes.

A catalogue record for this book is available from the British Library.

A catalog record for this book is available from the Library of Congress.

ISBN: PB: 978-1-3505-5219-7
Pack: 978-1-3505-5229-6
ePUB: 978-1-3502-5157-1
ePDF: 978-1-3502-5156-4

Series: The Cultural Histories Series

Typeset by Integra Software Services Pvt. Ltd.
Printed and bound in Great Britain

To find out more about our authors and books visit www.bloomsbury.com and sign up for our newsletters.

CONTENTS

LIST OF ILLUSTRATIONS — vii
LIST OF TABLES — xi
SERIES PREFACE — xiii

Introduction — 1
Peter J.T. Morris

1 Theory and Concepts: Stability and Transformation in Chemical Problems and Explanation 1914 to the Present — 29
   Mary Jo Nye

2 Practice and Experiment: From Laboratory Research to Teaching and Policy-making — 51
   José Ramón Bertomeu-Sánchez and Antonio García-Belmar

3 Laboratories and Technology: An Era of Transformations — 73
   Peter J.T. Morris

4 Culture and Science: Materials and Methods in Society — 99
   Carsten Reinhardt

5 Society and Environment: The Advance of Women and the International Regulation of Pollution — 123
   Peter Reed

6   Trade and Industry: The Growth, Diversification, and Dissolution
    of a Global Industry   149
    Peter J.T. Morris and Anthony S. Travis

7   Learning and Institutions: Global Developments since 1914   173
    Jeffrey Allan Johnson, Yasu Furukawa, and Lijing Jiang

8   Art and Representation: From the "Mad Scientist" to Poison Gas
    and Chemical Pollution   203
    Joachim Schummer

NOTES   224
BIBLIOGRAPHY   225
LIST OF CONTRIBUTORS   256
INDEX   257

LIST OF ILLUSTRATIONS

0.1 Professor Perkin at the Dyson Perrins laboratory at its opening in 1916. Photograph by Science and Society Picture Library/Getty Images — 2

0.2 Insulin model constructed by Dorothy Crowfoot Hodgkin. Science and Society Picture Library/Getty Images — 6

0.3 A researcher uses a Bruker Ultrashield 500 Plus NMR spectrometer in the laboratory of Johnson & Johnson, in Val-de-Reuil, northwestern France, 2017. Charly Triballeau/AFP/Getty Images — 8

0.4 The supernova remnant M1 (Crab Nebula), which exploded in 1054, harbors the strange chemical species ArH+. Photograph by Heritage Space/Heritage Images/Getty Images — 15

0.5 Ammonia storage tanks for IG Farben at a nitrogen fixation plant, 1930. Photograph by Margaret Bourke-White/The LIFE Images Collection via Getty Images — 17

0.6 Control room, plastics plant, ICI Billingham 1955. Photograph by Walter Nurnberg/SSPL/Getty Images — 21

0.7 A three-dimensional computer model of an oxyhemoglobin protein molecule. The image was generated by Dr. Manuel C. Peitsch at the Glaxo Institute for Molecular Biology in Geneva, Switzerland. Photograph by © CORBIS/Corbis via Getty Images — 24

2.1 Beckman Instruments, "Beckman Infrared Notes: IR4: New Infrared Instrument," 1956. Beckman Historical Collection, Box 17, Folder 38. Courtesy Science History Institute, Philadelphia. https://digital.sciencehistory.org/works/8s45q923r — 55

2.2 Fisher Hirschfelder Taylor Atomic Organic Model Kit. Photograph, 2017. Courtesy Science History Institute, Philadelphia. https://digital.sciencehistory.org/works/1c18dg352   58

2.3 Perkin–Elmer Atomic Absorption Spectrophotometer at an Industrial Hygiene Laboratory in USA, 1970–1979. It was employed to test the lead content of workers' urine. It was part of comprehensive program to ensure compliance with Occupational Safety and Health Administration regulations. Photograph from the Perkin–Elmer-Applera Collection, Box 4. Courtesy Science History Institute, Philadelphia. https://digital.sciencehistory.org/works/9g54xj45n   66

2.4 Johnny Horizon Environmental Test Kit, 1971. Popular chemical test for air and water pollution. It included a magnifying glass, eyedropper, test tubes, tape, measuring cup, filters, and paper packets of additives. Photograph, 2016. Courtesy Science History Institute, Philadelphia. https://digital.sciencehistory.org/works/79407x355   69

3.1 Vitamin B_{12} crystal structure model, 1957–1959. This crystal structure model, made for the x-ray crystallographer Dorothy Crowfoot Hodgkin, shows the structure of Vitamin B_{12}. It was displayed at the Brussels Universal Exhibition in 1958. Photograph by Science and Society Picture Library/Getty Images   78

3.2 Nuclear magnetic resonance (NMR) machine at the Merck & Co. pharmaceutical company, New Jersey, November 2, 1982. Photograph by Barbara Alper/Getty Images   84

3.3 AEI MS9 double-focus mass spectrometer at the University of Sydney. With the A$80,000 [about A$1,000,000 or US$700,000 today] machine are Prof. Charles Shoppee, Head of the Department of Organic Chemistry (on the left) and Dr. Alex Robertson, in charge of the spectrometer. 1966. Photograph by Frank Albert Charles Burke/Fairfax Media via Getty Images   87

3.4 Chief chemist Pierre Beaumier at Mann Laboratories using a GC-MS device, 1988. Photograph by Jeff Goode/Toronto Star via Getty Images   90

3.5 Photograph of a UMIST laboratory *ca.*1980s by Dr. Jonathan P. Miller. Used with permission   94

4.1 US Department of Agriculture poster showing a woman killing an insect with a pesticide, 1943. Hulton Archive/Getty Images   105

LIST OF ILLUSTRATIONS   ix

4.2 1945 illustration of crop dusting by helicopter. Hulton Archive/
     Getty Images   106

4.3 Advertisement celebrating the "age of synthetics," by British
     Celanese Ltd., 1946. Photograph by Culture Club/Getty Images   107

4.4 The illuminated Bayer cross, Leverkusen, ca.1935. Photograph by
     Imagno/Getty Images   109

4.5 Depiction of a couple walking inside chemistry beakers. 1952.
     Illustration by GraphicaArtis/Getty Images   111

5.1 A US Public Health Service worker measures samples of air
     pollution from the American Steel and Wire Company's zinc
     works in Donora during a four-day test. Photograph by Bettmann/
     Getty Images   128

5.2 The British chemist Dorothy Hodgkin showing the model of
     protein molecular structure. Anni Settanta. Photograph by
     Mondadori Portfolio via Getty Images   142

5.3 Number of graduate degrees granted in the United States by year
     and gender, 1950–1999. © American Chemical Society   143

6.1 There was a massive explosion at the Oppau Haber–Bosch factory,
     adjoining the BASF factory in Ludwigshafen, on September 21,
     1921 when a fertilizer silo exploded. Over 560 people were
     killed and the damage to property was estimated to be $7 million.
     Photograph by ullstein bild Dtl. via Getty Images   150

6.2 Soderberg electrode hall in Montecatini's aluminum factory at
     Bolzano, 1959. Montecatini diversified into a large number of
     sectors including light metals. Photograph by Mondadori Portfolio
     via Getty Images   153

6.3 "Cheese at Its Tempting Best – in Cellophane: Cellophane Shows
     What It Protects! Protects What It Shows!" A 1949 advertisement
     for DuPont Cellophane from *The Saturday Evening Post*. Hagley
     Museum and Library with permission   159

6.4 A child crying as she is sprayed with DDT delousing powder at the
     Cecilienschule, Nikolsburger Platz, Wilmersdorf, Berlin, Germany,
     October 1945. Photograph by George Konig/Hulton Archive/Getty
     Images   163

6.5 Synthetic rubber plant in the open air, 1950. Photograph by
     Andreas Feininger/The LIFE Picture Collection via Getty Images   165

LIST OF ILLUSTRATIONS

7.1 In the student laboratory, Sterling Chemical Laboratory, Yale University, New Haven, Connecticut, 1926 (the laboratory was built in 1923). Soapstone table tops supported on metal standards provide working space. Note use of electric lighting and gas for burners. These were graduate students; Yale first admitted women undergraduates in 1969. Photograph by Print Collector/Getty Images — 178

7.2 Sterling Chemical Laboratory, Yale University, New Haven, Connecticut, 1926. Exterior view of the building constructed in 1923. Photograph by Print Collector/Getty Images — 179

7.3 Students in a large chemistry lecture, The University of Iowa, 1930s. Note the table and apparatus for demonstration experiments, the short-form periodic table, and also that the students appear to be all or nearly all male. Frederick W. Kent Collection of Photographs, University of Iowa Libraries, Iowa City, Iowa. Reproduced by kind permission of the University of Iowa Libraries — 180

7.4 The American physical chemist Linus Pauling (1901–1994; Nobel Prize in Chemistry, 1954; Nobel Peace Prize, 1962) presenting a lecture in 1967 using various types of molecular models. The largest ones, on the right, are space-filling molecular models based on quantum principles. Photograph by Universal History Archive/Universal Images Group via Getty Images — 181

7.5 Organic chemistry teaching laboratory, Yenching University, China, 1929. Note that the laboratory appears to be open to both male and female students. Rockefeller Archive Center, RF Records Collection — 193

7.6 The five most productive nations in chemistry, 1996–2017, showing annual totals of publications in selected years. Drawn by the author (Johnson) from data for selected years shown, and for the entire period (SJR 2018). Note that this graph excludes countries that appeared in the top five during some of the selected years, but were not among the top five in cumulative publications — 201

7.7 Using electronic media in chemistry instruction in China, 2017. The instructor is an American chemist, Jay Siegel, Dean of the School of Pharmaceutical Science & Technology of Tianjin University. Tremblay 2017. Reproduced by kind permission of the School of Pharmaceutical Science and Technology, Tianjin University — 201

LIST OF TABLES

5.1 All chemists by employer, highest degree, and gender, 1999–2000 (%). © American Chemical Society — 145

5.2 Demographics of all chemists by gender (%). © American Chemical Society — 146

7.1 Numbers of chemists (in thousands – census figures) in the United States and Germany; memberships in principal professional associations in USA, Germany, and Japan (in thousands) — 175

7.2 German doctorates in chemistry awarded by universities and other academic institutions 1914–1938, with percentage awarded to women — 182

7.3 Academic chemists (excluding postdoctoral assistants) in German higher educational institutions, 1910–1938 — 182

7.4 Assistants and doctoral students in German higher education, 1920–1943 — 183

7.5 United States: Bachelor's degrees and doctorates granted in the natural sciences and chemistry — 183

7.6 Diplom certificates and doctorates awarded in German universities, 1956–1985; bachelor's and doctorates in US universities, 1950–1990 — 187

7.7 Numbers of students majoring in chemical sciences at universities and graduate schools in Japan; female students in parentheses — 190

7.8	Numbers of chemical researchers in Japan (women in parentheses)	191
7.9	Doctorates in German and US universities, 1990–2015, including nationality, ethnicity, and sex	198
7.10	Postdoctoral fellows in US universities 2001–2010, including percentage of foreign and women postdocs (chemistry and biochemistry)	199
7.11	The five most productive nations in chemistry, based on total publications, 1996–2017, ranked by average citations per published document	202

SERIES PREFACE

A Cultural History of Chemistry examines the history of chemistry and its wider contexts from antiquity to the present. The series consists of six chronologically defined volumes, each volume comprising nine essays; these fifty-four contributions were written and/or edited by a total of fifty scholars, of ten different nationalities. Of Bloomsbury's many six-volume *Cultural Histories* currently in print, this is the first in the physical or natural sciences; it is also the first multi-volume history of chemistry to appear since James Riddick Partington's four-volume *History of Chemistry* concluded more than fifty years ago. It is distinguished, among other qualities, by its endeavor to take the subject from antiquity right to the present day.

This is not a conventional history of chemistry, but a first attempt at creating a cultural history of the science. All cultures, including the various branches of natural science, consist of mixed constructs of social, intellectual, and material elements; however, the cultural-historical study of chemistry is still in an early stage of development. We hope that the accounts presented in these volumes will prove useful for students and scholars interested in the subject, and a starting point for those who are striving to create a more fully developed cultural history of chemistry.

Each volume has the same structure: starting with an interpretive overview by the volume editor(s), the eight succeeding chapters explore for each respective era in chemistry its theory and concepts; practice and experiment; laboratories and technology; culture and science; society and environment; trade and industry; learning and institutions; and art and representation. Readers therefore have the option to read multiple chapters in a single volume, thus learning about the cultures of chemistry in a single era; or they may prefer instead to read corresponding chapters across multiple volumes, learning about (e.g.) the art and representations of chemistry through the ages. Though the scope is global, major emphasis is placed on the Western tradition of science and its contexts.

Whether read synchronically or diachronically, in any multiauthor undertaking like this one readers will inevitably notice overlaps and repetitions, conflicting historical interpretations, and (despite the magnitude of the project) occasional gaps in coverage. These are inescapable consequences, but they actually offer advantages to the reader, both in making each chapter closer to self-contained, and in demonstrating the dynamism of the discipline; like science itself, the study of its history is ever contested and incomplete.

Chemistry has been called the "central science," due to its fundamental importance to all the other physical and natural sciences. It is the archetypical science of materials and material productivity, and as such it has always been deeply embedded in human industry, society, arts, and culture, as these volumes richly attest. The editors and authors hope that *A Cultural History of Chemistry* will be of great interest and enjoyment not just to chemists and specialist historians of science, but also to social, economic, intellectual, and cultural historians, as well as to other interested readers.

Peter J.T. Morris and Alan J. Rocke
London (UK) and Cleveland (USA)

Introduction

PETER J.T. MORRIS

CONTINUITY AND CHANGE

At Easter 1916, in the midst of a massive international conflict (sometimes called "the chemists' war") – along with an ill-fated uprising in Dublin – the Dyson Perrins Laboratory opened its doors for the first time in South Parks Road, Oxford. The building was the brainchild of William Henry Perkin, Junior, the new Waynflete Professor of Chemistry at Oxford, who firmly believed in the importance of research in chemical education. (By coincidence, it was exactly sixty years since his father made the first synthetic coal tar dye, mauve, in the East End of London, thereby creating an entirely new branch of the chemical industry.) A year earlier the younger Perkin had introduced the Part II, an extra year of research work added to the three-year undergraduate chemistry degree at Oxford. The "DP" – as it was always called – was the latest reiteration of a style of laboratory construction pioneered by the German chemist August Wilhelm Hofmann in the 1860s. Yet one of the older chemists working in the Chemistry Research Laboratory that opened across the road from the DP in 2004 would have found much of the interior of the building familiar if they had been magically transported back to 1916, not least because they might have done their Part II in the DP. The author recalls his own practical work in the DP in the mid-1970s when an entire class hydrolyzed a sample of an unknown chemical (a chlorophenyl ester) in the open laboratory, thus creating an impressive (and odoriferous) white cloud of the newly liberated chlorophenol below the high ceiling of the laboratory. Certainly, our hypothetical chemist

FIGURE 0.1 Professor Perkin at the Dyson Perrins laboratory at its opening in 1916. Photograph by Science and Society Picture Library/Getty Images.

would have been much more at home in the DP than in an alchemical workshop or even in an early-nineteenth-century laboratory.

Yet in many other ways, chemistry has changed more since 1916 than in any other hundred-year period (Hirota 2016). There are now far more chemists than there were in 1916, and they are less likely to be white European or

American males (although many of them still are). As a group, these chemists produce far more papers and write more books than their predecessors. The actual content of the chemistry they carry out is vastly more sophisticated than that of their nineteenth-century predecessors. This is at least in part because of the much larger number of instruments they use and the much greater power of these instruments to probe compounds and reactions. Above all, the use of computers has completely transformed chemistry in the last forty years. As late as the mid-1970s using a computer as an undergraduate student involved creating a large reel of punched paper tape (or a similar reel of magnetic tape or possibly a box of punched cards), carrying it to the doors of the air-cooled computer room, and handing it over to one of the white-coated experts tending the huge machine. A few days later one would receive a printout of the results on wide green-and-white lined paper in the internal post.

During the twentieth century, chemistry has had a massive impact on society by creating a tidal wave of new materials, chemicals, and pharmaceuticals. The ability of the chemical and pharmaceutical industries to make these products has increased by leaps and bounds (Aftalion 2001; Chandler 2005). An important change has been the shift in starting materials from coal, which was dominant at the beginning of the period, to petroleum and natural gas – petrochemicals – by the mid-1950s (Spitz 1988). At the same time, these industries have had an enormous effect on the environment both directly through its waste products and industrial accidents and perhaps even more significantly through the dispersal of its end products throughout the global ecosphere. Ever-increasing public concern about the impact of chemicals and plastics on the environment has counterbalanced any gratitude for the benefits of chemistry, which are often invisible or taken for granted. Furthermore, the image of chemistry has been damaged by the development of chemical weapons: the first large-scale use of a chemical weapon, namely chlorine gas, took place in Belgium in April 1915 exactly a year before the first chemists moved into the DP (Haber 1986). While it has been argued that chemical weapons can be more humane than the bullet or the explosive shell – the ideal chemical weapon incapacitates the combatant rather than kills – they have always had a bad reputation, for obvious reasons. The use of a chemical agent (a so-called nerve agent) in a botched attempt to kill a former Russian spy in Salisbury, England in March 2018 had a much greater media impact than a shooting would have had. A leading chemist, Fritz Haber, was involved in the development of chemical warfare in 1915, thus linking chemists to the use of chemical weapons. Haber was associated with the chemical firm BASF owing to his development there of the Haber–Bosch process for synthetic ammonia (discussed below). BASF and Bayer merged in 1925 with Hoechst A.G. to create the huge chemical combine IG Farben, which eventually became closely involved with Hitler's aim to use the synthesis of ammonia, oil, and rubber to save foreign exchange and to defeat any future naval blockade of Germany (Hayes 1987).

It is hardly surprising that these events (and others) have had an impact on the public's image of chemistry. Yet it appears that the public do have a positive view of chemists, while often being confused about what they do (Royal Society of Chemistry 2015). Perhaps part of the result for this confusion is that the term chemist is used in the UK for the medical practitioner who is more accurately called a pharmacist. To the ordinary person in the street, the chemist is the person who makes up your medical prescription. So when chemistry or chemists are discussed in this volume, the reader should be aware that the meaning of these terms is strongly dependent on the cultural context in which they are being used.

If we take all the various uses of the term chemistry, what have been the major breakthroughs in chemistry in the twentieth century? Previous centuries had witnessed major changes. The eighteenth century saw the rise and fall of phlogiston and the so-called Chemical Revolution; and the nineteenth century compassed the atomic theory, the periodic table, and the rise of organic chemistry as well as the massive expansion of the chemical industry. While the twentieth century may have lacked a generally accepted "revolution," it certainly witnessed marked changes in the way chemistry was conceived and carried out. In particular, the introduction of the electron as a chemical (as well as physical) entity at the very beginning of our period both transformed chemistry and unified the hitherto largely unrelated branches of inorganic, organic, and physical chemistry (Stranges 1982; Nye 1993). This shift made possible the development of reaction mechanisms (both organic and inorganic), spectroscopy, transition-metal coordination chemistry, electrochemistry, and solid-state chemistry. Scientists' improved understanding of both electronic bonding and the atomic nucleus led to the transformation of chemical instrumentation. The twentieth century also experienced an ongoing expansion of the chemical industry in both scale and scope, and a large-scale evolution in the culture and popular understanding of the science.

Chemical reactions are surely at the heart of chemistry, and our understanding of reactions increased dramatically during the twentieth century. The earlier part of this period saw the development of organic reaction mechanisms, which emerged from a combination of a new understanding of chemical bonding based on electrons and the increasing sophistication of chemical kinetics, especially the measurements of the rates of reactions. The mechanistic interpretation of chemistry was then extended to inorganic chemistry, most notably transition metals (the metals in the middle of the periodic table), whose chemistry is characterized by variable valency (oxidation states) and strongly colored compounds. The spectra of these compounds and their geometry in space were explained by the application of quantum chemistry, in particular a simplification of quantum chemistry called crystal field theory.

Once chemists had gained a solid understanding of ordinary chemical reactions, their focus switched to more extreme cases, notably reactions which occur in a fraction of a second and the reaction between individual molecules, for example by colliding beams of atoms or by using intense flashes of light. This increased understanding of reactions has been underpinned by the development of theories of the bonding between atoms, beginning with theories that simply dealt with the number of electrons in atoms, then models based on simple quantum mechanics, finally to highly sophisticated calculations using computers (Gavroglu and Simôes 2012). By the late twentieth century it was possible to demonstrate chemical bonding and reactions on a screen using computer graphics based on these calculations. This computational chemistry was enthusiastically taken up by the pharmaceutical industry because of its interest in the interaction between drugs and receptor sites in the human body.

If chemical reactions and chemical bonding are central to chemistry, the structures of chemical compounds are just as important, not least because they make it possible to predict how a compound will react. At the beginning of our period there was no way that one could actually "see" a chemical structure, one could only infer it from the compound's reactions (Morris and Travis 1997). The first major breakthrough was the development of x-ray crystallography, which works because x-rays have wavelengths close to lengths of chemical bonds and are thus diffracted by crystals. The technique cannot actually determine the structure directly; it can only determine the positions of atoms in a crystal. For this reason, when it was first introduced by the father-and-son team William and Lawrence Bragg in 1913, it was applied to crystals of inorganic compounds such as sodium chloride (common salt). However, as x-ray crystallography became more sophisticated, it became possible to surmise the structure of organic compounds by the indicated positions of the atoms in their molecules. For instance, Kathleen Lonsdale showed in 1929 that hexamethylbenzene (and by implication benzene itself) was both flat and a regular hexagon. By the 1940s, x-ray crystallographers were able to work out the structure of penicillin before organic chemists could agree on its structure on the basis of its reactions. One of the leading x-ray crystallographers in the penicillin project was Dorothy Crowfoot Hodgkin, who went on to work out the complex structure of Vitamin B_{12} in 1956 before the organic chemists had any success with traditional methods. Awarded the Nobel Prize in 1964, she also worked on the structure of the large hormone insulin.

The chemists working on penicillin also used infrared spectroscopy (based on radiation with a wavelength longer than visible light) and ultraviolet spectroscopy (using radiation with a wavelength shorter than visible light), which also came into general use in the 1940s. However, these techniques

FIGURE 0.2 Insulin model constructed by Dorothy Crowfoot Hodgkin. Science and Society Picture Library/Getty Images.

were only able to show what kinds of bonds a compound had, not how these bonds were arranged in the molecule. The next major breakthrough was the development of nuclear magnetic resonance (NMR) spectroscopy in the 1950s, based on earlier work by physicists in the United States (Reinhardt 2006b). This technique used powerful magnets and radio waves to "tickle" hydrogen atoms in organic compounds to find out how these atoms were arranged in the molecule, and what other parts of the molecule these atoms were near. By itself, the interpretation of NMR spectra of more complex molecules was partly a matter of informed guesswork, but combined with ultraviolet and infrared spectroscopy, the structure of most organic compounds could now be worked out without the use of x-ray crystallography or resorting to the study of the compound's reactions.

Additional help in analyzing a compound's structure was soon forthcoming from another technique borrowed from physics, namely mass spectrometry. The mass spectrometer converts an atom or molecule into a charged particle (an ion) and then shoots it across a magnetic field, which curves its path; the amount by which the path bends depends on the mass of the ion. Francis Aston was concerned only with the issue of isotopes when he built his first mass spectrograph (an early version of the mass spectrometer) in 1919, using it to show that neon and chlorine – two elements with atomic weights that were not close to whole numbers – each had two major isotopes, explaining their odd atomic weights. For many years mass spectrometers were used only to measure the weights of atoms and molecules. Eventually the technique became so sensitive that one could deduce the molecular formula of a compound from the very accurate molecular weight determined by a mass spectrometer.

However, in the 1960s chemists started to use the way that a compound breaks up inside the mass spectrometer – its pattern of fragmentation into multiple charged molecular pieces – to gain additional information about its structure, effectively as an adjunct to NMR spectroscopy (Reinhardt 2006b). Initially both NMR and mass spectrometry were used for relatively small molecules, the larger protein and enzyme molecules being left to the x-ray crystallographers. However, by the end of the twentieth century both NMR spectroscopy and mass spectrometry were being used routinely to study large biological molecules (Reinhardt 2017). Combined with the use of computerized molecular modeling, these techniques revolutionized the study of large proteins around the turn of the twenty-first century and in particular the implications of protein folding for degenerative diseases such as Alzheimer's disease and Parkinson's disease. To sum up, all of these physical methods taken together reduced the timescale for the elucidation of a complex organic compound from a decade (or more) of patient investigation of myriad chemical reactions, to a few hours spent studying the output of these sophisticated and expensive machines.

A completely different approach to "seeing" atoms is the scanning tunneling microscope (STM) which uses an ultra-sharp metal tip to travel over the surface of a metal or a semiconductor (such as silicon) to measure the electrical forces given off by the atoms, and thereby to construct an image of its microstructure. The resolution of this technique is 0.1 nanometers on the horizontal scale and 0.01 nanometers in depth, atoms typically being a few tenths of a nanometer

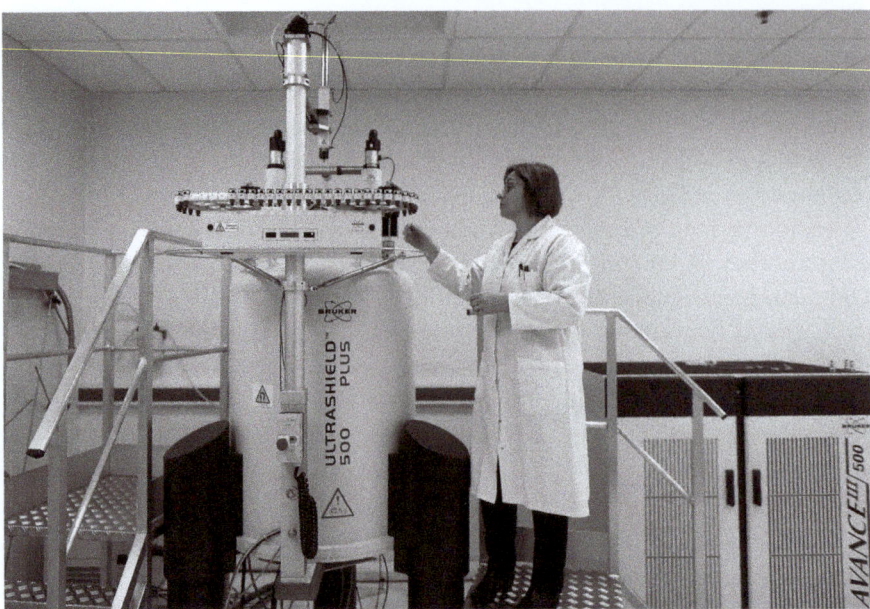

FIGURE 0.3 A researcher uses a Bruker Ultrashield 500 Plus NMR spectrometer in the laboratory of Johnson & Johnson, in Val-de-Reuil, northwestern France, 2017. Charly Triballeau/AFP/Getty Images.

in size. STM was developed by Gerd Binning and Heinrich Rohrer at IBM Laboratories in Zurich in 1981, for which they were awarded the Nobel Prize in 1986. In the same year, Binning with Calvin Quate and Christoph Gerber introduced atomic force microscopy (AFM), which is similar to STM but detects the interatomic forces of the atoms; AFM can be used with insulators as well as metals and semiconductors, thereby greatly extending the scope of the technique.

Another key aspect of chemistry is the analysis of materials and the separation of complex mixtures. Often a chemist will analyze an unknown material to find out what it is. If it is a mixture of compounds, the analyst wants to know what compounds are present and their concentrations. For example, one might want to find out how much iron is contained in an ore, how much DDT there is in a batch of strawberries, or if arsenic was present in an unfortunate victim's dinner. On the other hand, a chemist might want to analyze a known compound, perhaps a product extracted from a plant or a sea organism, to find out the details of its composition. The determination of a chemical formula is usually a relatively straightforward process, but finding out the structure of an organic compound by the application of degradative chemical reactions – a process known as ultimate analysis – was a painstaking activity which could take years or even decades before it was replaced by the instrumental methods discussed in the last paragraph. The 1950s and 1960s saw the development of two methods which greatly increased the sensitivity of analysis of the elements in a material, namely atomic absorption spectroscopy – which was a refinement of the atomic spectroscopy introduced by Robert Bunsen and Gustav Kirchhoff in 1859 – and x-ray fluorescence spectroscopy, which is non-destructive and extremely sensitive.

The twentieth century also saw the development of a very powerful method of separating mixtures, namely chromatography (Ettre 2008). Chromatography has had many variants. The original form of chromatography, which involved passing a solution of a mixture down a column of powdered material, was introduced by the botanist Mikhail Tswett (or Tsvet) in the early 1900s. He called it chromatography (color-writing) because he was using it to separate mixtures of plant pigments which produced colored bands in the column. Tswett died aged forty-seven in 1919, and his technique languished until the English chemist Archer Martin had the idea of loading the powder with a solvent before the separation took place, so that the separation took place in the liquid phase rather than at a liquid–solid interface. He introduced this so-called partition (or liquid–liquid) chromatography with Richard Synge in the early 1940s. Martin and Synge then tackled the problem of detecting the pure compounds as they left the column, rather than relying on the products being colored as Tswett had done. Martin and Synge then applied the technique to filter paper (an idea first attempted by Friedlieb Ferdinand Runge in 1850), thus introducing paper chromatography. Realizing that thin layers of silica on glass plates would be both more rugged and sensitive than filter paper, the American chemist Justus Kirchner developed thin-layer chromatography in 1946.

Collaborating with Anthony James, Martin introduced the very powerful technique of gas (more precisely gas–liquid) chromatography in the early 1950s. This procedure originally used glass tubing coated with silica, later replaced by metal tubes packed with alumina or adsorbent polymers. An inert gas such as nitrogen or hydrogen carries the unknown mixture through the column. Gas chromatography was very useful for mixtures of relatively volatile liquids such as gasoline, which meant that it was taken up with great enthusiasm by the petroleum refining industry. Once again the problem of detection came to the fore, but it was soon solved by the development of two highly sensitive detectors: the flame-ionization detector, devised independently by Ian McWilliam in Australia and Victor Pretorius in South Africa, and a little later the more sensitive electron-capture detector by James Lovelock, who was a colleague of Martin and James. In the long run, however, the most important (and most sensitive) detector for gas chromatography – and a later form of liquid chromatography called HPLC (high-pressure or high-performance liquid chromatography) – was the mass spectrometer. The mass spectrometer was combined with gas chromatography in the mid-1960s, and HPLC-MS was introduced a decade later. Using a mass spectrometer as the detector at the end of a chromatographic separation not only allows the identification of compounds from their precise mass, but also uses a stored library of fragmentation patterns to confirm the identification.

Another important activity in chemistry is synthesis, the making of a compound from simpler compounds (and in theory at least, from the elements). There are two different kinds of organic synthesis. In academic organic chemistry at the beginning of our period, *synthesis* followed the *analysis* of natural products. Hence the prime purpose of this type of organic synthesis up to the 1950s was to confirm the structure of the compound surmised by chemists from the results of degradative reactions. As the chemists could never be certain if their hunch was correct – these degradative reactions sometimes did not take the expected pathway – the final synthesis was a key part of the entire time-consuming procedure. When the process of determining the structure of organic compounds was transformed by physical methods in the 1950s, academic organic chemists had to find a new rationale for organic synthesis. It became a challenge of intellectual and experimental prowess rather like mountaineering, and a testbed for new reactions. There is another type of organic synthesis – "synthesis" is technically incorrect here as it does not arise from analysis, but no alternative term exists – in which completely new compounds are created using simpler compounds, often with the help of catalysts. This type of synthesis is very common in industry, initially in the dyestuffs industry in the late nineteenth century and more recently the pharmaceutical industry, but it is also found in academia.

The space in which all these activities take place, namely the laboratory, has also changed during the period of this volume, not only as a result of the new instrumentation but also due to changing societal concerns, especially new health and safety legislation (Morris 2015). While the expensive, heavy, and bulky instrumentation introduced by the instrumental revolution needs space, they have largely taken over areas formerly occupied by an earlier generation of instrumentation. Spectrometers sit in rooms which were once photographic darkrooms, while the mass spectrometers and NMR machines are lodged in the basements which in older days would have been used for combustion analysis. By contrast, health and safety regulations now dictate that handling of particularly hazardous materials take place in fume hoods (also called fume cupboards), which now dominate laboratories along with "glove boxes" (sealed enclosed spaces equipped with permanently installed long-sleeved gloves with which the chemist manipulates the sensitive or toxic materials within). As it is no longer permitted to eat and drink or even to write up reports in laboratories, glass walls often separate the laboratories from the offices where these activities now take place.

The twentieth century has also seen the development of entirely new subdisciplines. Some of these subdisciplines are explicitly interdisciplinary, such as molecular biology or chemical physics (Reinhardt 2001), but others lie entirely within chemistry (although there is usually a link with other disciplines). A very important and early example of these new disciplines was polymer chemistry, founded by the German chemist Hermann Staudinger in the 1920s. In this period, most chemists assumed that compounds with what seemed to be improbably high molecular weights – rubber and cellulose, for example – were in fact loosely bound aggregates of smaller molecules. By contrast, Staudinger argued from what might be described as a conservative viewpoint that rubber and cellulose consisted of actual molecules that were very large – he called them macromolecules – which behaved in the same way as small molecules. As a result of their x-ray crystallographic studies of natural polymers, Herman Mark and Kurt Meyer gradually accepted Staudinger's thesis. The work of Mark as a physical chemist and Staudinger as an organic chemist brought about the "polymer revolution" (Furukawa 1998). It was not insignificant that Mark worked for the German chemical combine IG Farben during the period he came to accept the macromolecular theory. Two other major contributors to the polymer revolution were Wallace Hume Carothers, who worked at Du Pont, and his young colleague Paul Flory. After the development of the first commercially successful synthetic resin, Bakelite, by the Belgian-American chemist Leo Baekeland in the 1910s, the chemical industry brought out a stream of plastics and synthetic fibers in the 1920s and 1930s, including nylon, polyvinyl chloride (PVC), polyethylene, and synthetic rubber.

One new branch of chemistry that had strong links with physics was solid-state chemistry. It arose in the late 1940s from the study of phenomena on the surface of solids, chiefly with the aim of understanding catalysis on one hand and the development of transistors on the other. A major theoretical breakthrough was the phase diagram, based on Josiah Willard Gibbs' phase rule of 1878, but not introduced until around 1919. Initially used in metallurgy, it became the basis of much of solid-state chemistry, which is concerned with surface reactions and the properties of bulk solids. An early pioneer of solid-state chemistry was Irving Langmuir who was awarded the Nobel Prize in 1932 for his work on surface chemistry. The development of solid-state chemistry was also accelerated by the introduction of x-ray diffraction by the Braggs (see above) and by the introduction of differential thermal analysis, in which the heating (or cooling) of a sample was measured against an inert reference material. This technique was partially displaced by differential scanning calorimetry, introduced in 1964, and eventually by thermogravimetric analysis.

It was from solid-state chemistry, as well as solid-state physics, that the new discipline of materials science arose in the early 1970s (Cahn 2001). One root of this discipline was metallurgy, which existed as a scientific subject since the mid-nineteenth century. This discipline was revolutionized by the application of solid-state physics – in particular x-ray diffraction and the use of quantum theory – since the 1940s. At the same time, the more recent field of macromolecular chemistry (see above) began to apply the methods of physical chemistry and physics to the study of macromolecules, including natural products such as rubber and wood. With hindsight it may seem surprising that two fields such as metallurgy and macromolecular science took so long to find their common ground in the new "materials science," but one should never underestimate the power of disciplinary inertia. Since the 1970s the materials-related sciences have been combined at both the institutional level and intellectually, while each still retains a strong sense of its own identity.

Another new field that was introduced with much hope (and hype) in the 1980s was nanochemistry (Bensaude-Vincent 2009; Mody 2011). At the most basic level it is the concept that one can manipulate and assemble molecules at the atomic (nanometer) level. It soon became a useful umbrella term for techniques such as STM and AFM, and new fields such as "supramolecular" chemistry and the fullerenes. An early example of nanotechnology hype was the spelling out of IBM using thirty-six xenon atoms on a nickel surface in 1989. "Buckminsterfullerene" (named after the American architect Buckminster Fuller) is a molecule composed of sixty carbon atoms in the shape of a soccer football. It had been predicted tongue-in-cheek by the scientific satirist "Daedalus" (David E.H. Jones) in 1966, and more seriously in 1970 by Eiji Osawa, but it was first produced in the laboratory by Robert Curl, Harold Kroto, and Richard Smalley in 1985, for which they were awarded the Nobel Prize in 1996. After a method for the manufacture of gram-sized quantities of C-60 was developed in 1990, there was much excitement about the potential uses of "buckyballs" and its chemical relatives, the fullerenes. Soon afterwards, attention turned to the related material known as carbon nanotubes: long tubes of graphite "chicken-wire" arrangements of carbon atoms which are a few nanometers wide. In 2004, the Russian-British physicists Andre Geim and Konstantin Novoselov produced a monatomic layer of graphite that they called graphene, simply by peeling adhesive tape off a stick of graphite. They were awarded the Nobel Prize in 2010. However, many of the expectations associated with all these new materials remain largely unfulfilled in 2019.

The living organism can carry out numerous complex reactions under mild conditions at room temperature. For over two centuries chemists have admired this ability, while at the same time struggling to reproduce it. During the period covered by this volume, chemists discovered reactions that take place under such benign conditions (for example, the synthesis of tropinone by Robert Robinson in 1917), but they are relatively uncommon. Nature succeeds because it uses the three-dimensional environment to control the reaction through interatomic forces, rather than by the reagents used by the chemist. It was the discovery of the relatively inexpensive "crown ethers" by Charles Pedersen at Du Pont in 1967 that triggered interest in creating a specific three-dimension environment for a given reaction, without having to make large complex proteins similar to the enzymes used by nature. Jean-Marie Lehn made three-dimensional analogues of Pedersen's crown ethers in 1969, the so-called super-molecules, and was able to use them either to distinguish different ions or to carry out chemical reactions to create a specific three-dimensional product. The ultimate aim of this line of research is to create a three-dimensional chemical environment in which different molecules are allowed to enter and combine automatically to produce larger molecules, in the same way that nature uses relays of enzymes.

Today, supramolecular chemistry includes molecular devices and machines, molecular recognition, and processes such as self-assembly and self-organization. Recent work has utilized both non-covalent (e.g. hydrogen bonds) and covalent (metal–ligand interactions) components. Synthetic or artificial enzymes having the characteristics of supramolecular chemistry are also being developed that provide catalysts with rates and selectivity that reach those of natural enzymes. Chemists have also produced polymeric super-molecules. Another approach is to use synthetic zeolites (porous minerals containing aluminum, silicon, and oxygen) with specific pore sizes to influence selectivity and speed up reactions in a similar way. Supramolecular chemistry has evolved from its beginnings in preorganization for the design of molecular receptors that can lead to molecular recognition, catalysis, and transport processes, to systems which can lead to spontaneously generated, well-defined, organized, and functional architectures by self-assembly from their components.

In the nineteenth and early twentieth centuries, there were various forms of medical chemistry, and many chemists worked in hospitals. There was physiological chemistry, hygiene chemistry, pharmacy (and its modern offshoot pharmacology), and biological chemistry. Organic chemistry itself had arisen from the study of the products of the human body, animals, and plants. A good example of this is the isolation of compounds from the human brain by the physician Ludwig Thudichum in the 1870s and 1880s. In the 1920s and 1930s, these overlapping fields left both chemistry and medicine to become biochemistry (Kohler 1982). This was a new discipline, separate from chemistry, in the USA and Britain; however, in Continental Europe, it was still considered to be a subdiscipline of chemistry alongside organic and inorganic chemistry. Biochemistry was strongly supported by the Rockefeller Foundation in the United States, but hindered in Germany by the rigid disciplinary divisions and the power of the professors in German universities. Nonetheless, it is a testament to the strength of the field in Germany that major advances were made there by Otto Warburg and Hans Krebs. Krebs elucidated the metabolically important "urea cycle" in Germany and after he emigrated to Britain in 1933. With the support of the leading British biochemist Frederick Gowland Hopkins, he uncovered the even more significant citric acid (or Krebs) cycle (Holmes 1991, 1993). Despite the success of biochemistry, it was soon overshadowed by the discovery of the structure of DNA by a combination of physicists (Francis Crick and Maurice Wilkins), a biologist (James Watson), and a chemist (Rosalind Franklin), rather than by biochemists (Olby 1974). Thus arose the new field of molecular biology, with its subsequent successes in genetics (Morange 1998). Up to almost the end of the twentieth century, biochemistry and molecular biology appeared to be drifting away from chemistry, but a renewed interest in large biomolecules (for example, the folding of protein chains) by chemists has partly brought these boundary disciplines back under the aegis of chemistry.

Geology and especially mineralogy have always had strong links with chemistry, and the discovery of many elements, notably the rare earths found in the rocks in the mine at Ytterby, near Stockholm. While there are obvious connections between mineralogy and chemistry, the development of the subdiscipline of geochemistry was the result of the interest of certain mineralogists and chemists in the relative abundance of the elements in the Earth's crust on one hand, and the development of x-ray techniques (both spectroscopy and crystallography) on the other (Kragh 2001). The pioneer of modern geochemistry was the Norwegian mineralogist Victor Goldschmidt in the 1920s and 1930s. The technique of radioisotope abundance arises naturally from the concern with elemental abundances, as radioisotopes decay over time and thus allow the dating of rocks and other materials. For instance, the decay of potassium-40 to argon-40 allowed the accurate determination of the age of the Earth's oldest rocks in the 1960s. On an entirely different timescale is the use of radioactive carbon-14 dating to calculate the age of organic materials that were once part of living systems. This technique was developed by the American chemist Willard Libby in the late 1940s and 1950s, for which he was given the Nobel Prize in 1960. A famous example of his work was his 1955 dating of the Isaiah scroll in the Dead Sea Scrolls as being roughly 2,000 years old. While geochemistry was initially much concerned with the abundance of the elements, which was found to be similar in both the Earth's crust and meteorites, it later expanded into other fields such as biological geochemistry (the effect of organisms on the Earth's chemistry) and organic geochemistry (the study of organic compounds in rocks).

Up to the 1920s it was assumed that the stars had the same elemental abundance as the Earth, but it was then shown that the sun and other stars are composed mostly of hydrogen (Kragh 2000; Kragh 2001). Once the "Big Bang" theory was introduced in the 1930s, many scientists assumed that all the elements had been created during the cooling-down stage immediately after the Big Bang. However, astronomer Fred Hoyle (who opposed the Big Bang theory) showed in the late 1940s and early 1950s that the heavier elements were formed in the stars themselves or in supernovae (massive stellar explosions).

This line of research culminated in a famous 1957 paper by Margaret Burbidge, Geoffrey Burbidge, William Fowler, and Hoyle on stellar nucleosynthesis, a paper familiarly known as B²FH. While work continued on the processes involved in the synthesis of the elements in space, attention now turned to the detection of chemical molecules in space, aided by the development of radio astronomy. Like molecular hydrogen (H_2), most of the earlier chemical species discovered in interstellar space were diatomic, and it was assumed that larger molecules would be torn apart by the ultraviolet rays and cosmic rays produced by stars. This assumption was undermined by the detection in space of the tetratomic molecule ammonia (NH_3) by Charles Townes (inventor of the maser) in 1968, and this was quickly followed a year later by the detection of the first organic molecule in space, formaldehyde (CH_2O), by the Green Bank radio telescope in West Virginia. Since then, a "cosmic zoo" of chemical molecules has been detected both in interstellar space and in heavenly bodies such as comets and asteroids, including such exotic species as buckminsterfullerene (C_{60}), benzonitrile (C_7H_5N), and even the "argonium" ion (ArH^+), which was detected in the Crab Nebula, a supernova remnant, in 2013.

FIGURE 0.4 The supernova remnant M1 (Crab Nebula), which exploded in 1054, harbors the strange chemical species ArH^+. Photograph by Heritage Space/Heritage Images/Getty Images.

While research was carried out on high-pressure chemistry in academia, in the first half of the century it was mostly pursued in the chemical industry. It was usually carried out for two different reasons: either to reproduce materials that are made by nature under pressure, such as diamonds, or to use Le Chatelier's principle – the phenomenon that an equilibrium moves in the direction that reduces an imposed constraint – to force two or more gaseous molecules to react to form one molecule, thereby reducing the pressure of the system. The most notable example of a high-pressure industrial synthesis is the Haber–Bosch process, which converts nitrogen and hydrogen into ammonia. It was carried out in the laboratory by Fritz Haber in 1909 and successfully brought to the industrial scale for the German firm BASF by Carl Bosch in 1913; its worldwide development took place in the 1920s and 1930s (Smil 2001; Travis 2018). The Haber–Bosch process led to the industrial synthesis of methanol from carbon monoxide and hydrogen in 1923, and the process to convert coal into liquid hydrocarbons, which was initially developed by Friedrich Bergius before being acquired by BASF (about to become part of IG Farben) in 1925. A completely different form of high-pressure chemistry was developed by Walter Reppe of BASF and IG Farben in the 1920s and 1930s that sought to make organic chemicals from acetylene (and later other compounds) under pressure. A similar attempt by the British firm ICI to make organic chemicals under pressure led by accident to the discovery of low-density polyethylene in 1933. More recently, in 2013, academic scientists used extremely high pressures (around 400,000 times atmospheric pressure) to create exotic chemical species such as sodium trichloride ($NaCl_3$), trisodium chloride (Na_3Cl), and sodium heptachloride ($NaCl_7$). Three years later, using even higher pressures, they were also able to make the bizarre compound disodium helide (Na_2He). Meanwhile, in 2019 attempts continue to make metallic hydrogen in the laboratory under similar high pressures. This form of hydrogen, which behaves as a metal, is believed to exist in the core of the giant planet Jupiter, and was very briefly made by accident at the Lawrence Livermore National Laboratory in California in 1996; attempts to make a sample at the huge pressure of a million atmospheres have been dogged by controversy.

The ammonia from the Haber–Bosch process was mostly converted into fertilizer, at least in peacetime. In the twentieth century, the chemical industry became interested in using its expertise acquired in dyestuff chemistry to create new insecticides to combat insect-borne diseases affecting both humans (for example, typhus) and plants (such as Colorado potato beetles), and to produce herbicides to kill weeds and increase the yields of food plants at a time of great food scarcities (Russell 2001). The best indication of the success of this endeavor is the award of the Nobel Prize to Paul Müller, a chemist at the Swiss company Geigy, who in 1939 discovered the insecticidal properties of DDT, the first synthetic organic insecticide. DDT was heralded as a powerful and

FIGURE 0.5 Ammonia storage tanks for IG Farben at a nitrogen fixation plant, 1930. Photograph by Margaret Bourke-White/The LIFE Images Collection via Getty Images.

remarkably inexpensive broad-spectrum insect-killer, which had the additional benefits of remaining stable on surfaces indefinitely (Kinkela 2011). It was used by the Allies in the latter stages of World War II to dramatically limit disease among troops.

DDT's reign as the queen of pesticides was troubled almost from the start. Many insects rapidly developed resistance to the pesticide, and its broad-spectrum character also eliminated natural enemies of the target species, often resulting in resurgence of the pests. Other environmental effects gradually became evident. Its chemical stability proved to be more of a problem than a benefit, for it was only slowly broken down in nature, and its fat-solubility resulted in its magnification up the food chain. Application of DDT to kill insects in lakes, rivers, and parklands resulted in toxic levels of the substance in apex predators such as raptorial birds. These problems were highlighted by the American biologist Rachel Carson in her famous book *Silent Spring* in 1962. This led to a race against time whereby new pesticides were marketed, but were often later banned; for example, Alar in 1989, and more recently the neonicotinoids. In 2019, the situation regarding the most widely used herbicide glyphosate (Roundup®) remains controversial, with calls for it to be banned. The final outcome is typical of the experience of the chemical industry in the twentieth and early twenty-first centuries: the industry was highly successful in creating new products; the impact of these products was enormous in terms of preventing diseases and making more food available to a rapidly growing world population; but widespread anxiety about their impact on human health and on the environment remain acute.

Early in the century the synthetic dye industry, especially the German firms Bayer and Hoechst, had moved into pharmaceuticals, alongside the existing pharmaceutical companies. IG Farben, the successor of both Bayer and Hoechst, aggressively pursued the development of pharmaceutical drugs under the leadership of Heinrich Hörlein, who introduced the antimalarial Resochin (chloroquine) and the antibacterial Prontosil (sulfamidochrysoïdine) in the 1930s (Lesch 2007). However, Prontosil was soon surpassed by the May & Baker company's sulfapyridine, thanks to May & Baker's links with the French pharmaceutical industry and hence the Institut Pasteur, which had quickly shown that the active element of Prontosil was its sulphanilamide core. The development of the sulfa drugs was then overshadowed by the development of penicillin at Oxford University and then the Northern Regional Research Laboratory of the US Department of Agriculture in the early 1940s. The introduction of penicillin transformed the management of bacterial infections and led to a Nobel Prize in 1945, but was soon dogged by the problem of bacterial resistance, just as DDT was diminished by insect resistance (Bud 2007). Other types of antibiotics were soon found in nature, including the tetracyclines, the cephalosporins, and erythromycin. The problem of bacterial resistance was partly overcome by tinkering with the chemical structure of the original natural products to create semisynthetic antibiotics.

Another breakthrough was a result of the concomitant development of biochemistry and the related field of pharmacology. Discovering the biochemical pathways of a given disease such as high blood pressure enabled scientists to find ways to block them. Coming from the field of veterinary physiology, James Black conceived the idea of blocking the effect of adrenaline on the heart and joined ICI Pharmaceuticals in 1958. There he developed the first beta-blocker drug for high blood pressure, propranolol (Inderal; Quirke 2006). However, to Black's dismay, ICI was not interested in extending this discovery method to the treatment of stomach ulcers. He moved to Smith, Kline & French and developed the anti-ulcer histamine-blocker cimetidine (Tagamet), which was launched in the mid-1970s. Both of these drugs became the best-selling pharmaceuticals in their time (the first so-called "blockbuster" drugs) and Black was awarded the Nobel Prize in 1988.

A different line of development was the development of oral contraceptives in the early 1960s by using semisynthetic steroids to block the body's reproductive system (Marks 2001). After Akira Endo discovered mevastatin as a metabolite of a Penicillium fungus in 1971, the American pharmaceutical company Merck developed Zocor (simvastatin) as a drug to lower blood cholesterol levels in the 1990s. A completely new field of pharmaceutical agents was established with the monoclonal antibodies, which were first discovered in the 1970s and were used as diagnostic tools for infections such as human immunodeficiency virus (HIV) in the 1980s (Marks 2015). The first therapeutic use of monoclonal antibodies was Orthoclone (Muromonab-CD3) introduced by Janssen-Cilag in 1986 to prevent organ rejection in kidney transplants. In 2018, the best-selling pharmaceutical drug in the USA was Humira (adalimumab), introduced by Abbott Laboratories (spun off as Abbvie in 2013) in 2005 as a treatment for rheumatoid arthritis, and subsequently other autoimmune diseases. No less than five of the top-selling drugs in the USA in 2018 were monoclonal antibody pharmaceuticals.

The twentieth century also saw the rise of chemical engineering as a distinct academic discipline and profession. By 1914, several American universities – most notably the Massachusetts Institute of Technology – were offering degrees in chemical engineering. On the professional side, the American Institute of Chemical Engineers was established as early as 1908, showing the extent to which the US was ahead of other countries in this new discipline (Furter 1982). In Britain, chemical engineering was revolutionized by the experience of World War I and the need to set up factories very quickly (Divall and Johnston 2000). Soon after the war, the Institution of Chemical Engineers was founded in 1922. By comparison, in Germany the development of chemical processes remained longer in the hands of chemists rather than chemical engineers, but the situation changed after 1945. DECHEMA (German Society for Chemical Apparatus) was set up in 1926 as a society of chemical apparatus, but duly expanded its remit to cover chemical engineering. With the development of penicillin and later products such as the attempts to make synthetic food in the 1960s, chemical engineering expanded to cover biochemical engineering. Like chemistry, chemical engineering was transformed by the introduction of computers and a growing focus on occupational health and industrial safety. The increased attention to safety was at least partly a response to various industrial accidents including disasters at Flixborough, England (1974), Seveso, Italy (1976), and Bhopal, India (1984), as well as the problems with mercury-laden effluent at Minamata in Japan (1950s), and with vinyl chloride monomer in the PVC industry (1970s; Crone 1986).

One important aspect of chemical engineering in the middle decades of the twentieth century was the development of highly efficient continuous processes, which replaced the older batch processes in which intermediates were moved from one reactor vessel to another in order to carry out the next stage of the process. There was a shift from the major solid chemical materials of the nineteenth century (such as salt, soda, lime, and coal) increasingly to liquid and gaseous products of the petrochemical industry. Furthermore, the chemical industry benefited from breakthroughs in process engineering made by the vast (and wealthy) petroleum refining industry. The introduction of the fluidized bed reactor, which enabled solid powders to be reacted with gases in a continuous process, invented by Fritz Winkler of BASF in 1926, enabled the reaction between powdered coal and hydrogen under pressure in the Bergius coal-to-oil process (Sella 2017). Another important innovation was the development of emulsion technology, which suspended water-insoluble liquids (as most organic chemicals are) in water in the 1920s. The use of emulsions was crucial to the development of the synthetic rubber industry in the interwar period and World War II. Not only did it allow the pumping of liquids from one vessel to another, but the water removed the heat of polymerization. The changeover from batch processes to continuous processes also produced a shift in chemical engineering from a focus on unit processes to transport phenomena.

The introduction of the automatic control of processes – in particular the temperature and pressure of the process – was a vital factor in the development of modern industrial chemistry, especially in the petroleum and petrochemical industry (Bennett 1991; Bennett 2002). Various mechanical controllers for flow regulation had existed since classical times, and by the beginning of our period there were temperature controllers which used thermocouples. The next step forward was the use of air pressure as a regulatory mechanism, but these pneumatic controllers were relatively crude. In 1931, on the basis of an important invention by Clesson E. Mason, the Foxboro company introduced the first so-called pneumatic PI (proportional-integral) controller, which reacted to any deviation from a pre-set value. This breakthrough was followed by the incorporation of electronics in the late 1930s and 1940s, and miniaturization in the 1950s. A further breakthrough was the use of computers for process control in the 1960s, but they were too expensive to replace the existing analogue

FIGURE 0.6 Control room, plastics plant, ICI Billingham 1955. Photograph by Walter Nurnberg/SSPL/Getty Images.

controllers for simple process control. Finally, the introduction of the microchip in the 1970s allowed the production of cheap digital process controllers for the control of individual reactors. By the end of the twentieth century, all processes were controlled from a central computer panel.

The new field of green (or sustainable) chemistry was created in the early 1990s with the support of the US Environmental Protection Agency to bring together several strands of environmentally conscious chemical manufacturing (Roberts 2005). One aspect was prevention of harm, for example the use of environmentally friendly (green) solvents such as ethyl lactate made from maize. The use of ethylene dibromide in the decaffeination of coffee beans has been replaced by the use of carbon dioxide under high pressure. High-pressure (super-critical) carbon dioxide has also replaced traditional solvents in dry cleaning and computer chip manufacture. Another approach was "atom economy," in other words changing traditional manufacturing methods to either achieve nearly 100 percent yields, or to remove undesirable by-products in other ways. One example is the replacement of the traditional multistage synthesis of the widely prescribed statin simvastatin with an enzyme-based synthesis, which takes a simple starting compound and a natural product (lovastatin) to produce simvastatin with a 97 percent yield. A third strand of green chemistry is the use of renewable chemicals, such as chemicals obtained from agricultural products. Plastics have long been made from agricultural products. In the 1930s Henry Ford unsuccessfully sponsored the development of soy-based plastics; ICI originally made polyethylene from fermentation alcohol; and in the 1950s Du Pont made nylon from furfural obtained from oat husks. In the 2010s, chemists were developing plastics made from poly-lactic acid and poly-3-hydroxybutyrate (PHB), both of which can be made from maize, to replace traditional petroleum-based plastics. Several universities now offer courses in green chemistry. While the aims of green chemistry are clearly admirable, they are often difficult to achieve and by 2019 it had not yet had the impact that its supporters have desired.

Green chemistry is different from environmental chemistry, which is the chemical study of the chemistry of the environment, both regarding the absence of contamination and as a result of the impact of human activity. While the subject may seem a new one, it is in fact of long standing, going back to at least the analysis of water supplies and the air in the mid-nineteenth century, when concern about the impact of industrial activities and human waste disposal increased. Effective environmental chemistry relies on the ability to analyze the environment and this means being able to analyze very low levels of substances, regardless of whether these substances should be there (such as oxygen in rivers) or not (such as hydrogen chloride in the air). Hence, mid-nineteenth-century chemists had to develop new analytical methods. By the mid-twentieth century, concerns about the impact of chemicals on the environment were increasing, and biological methods showed that certain new products, most notably the pesticide DDT and its organochlorine relatives, were having a deleterious effect on wildlife, but chemists were unable to measure DDT at such low levels (less than one part in a million). Thanks to the advent of gas chromatography with electron-capture detectors, it became possible by the late 1960s to measure pollutants such as DDT or dioxins (highly toxic contaminants of the herbicides 2,4-D and 2,4,5-T) at the parts per trillion level, thereby showing that these substances were present everywhere, even in the Arctic, at these very low levels (Morris 2002b). Although James Lovelock using a gas chromatograph and his electron-capture detector discovered that CFC (chlorofluorocarbon) refrigerant molecules were floating across the Atlantic from the US to the coast of Ireland in 1967, it was the much older Dobson spectrometer, an ultraviolet spectrometer invented by Gordon Dobson in 1924, in the hands of the scientists of the British Antarctic Survey that detected the famous "Ozone Hole" above Antarctica in 1984 (Gribbin 1993). In the same period, analysis of the ratio of specific elements in acid rain allowed chemists to work out where it was coming from, as they were markers for the fossil fuel being burnt, thereby producing the sulfur and nitrogen oxides responsible for acid rain. More recently, concern has shifted to the levels of carbon dioxide in the atmosphere (anthropogenic global warming). Atmospheric carbon dioxide has been regularly measured since 1958 by Charles Keeling, at an observatory on the Hawaiian volcano Mauna Loa, using the technique of infrared spectroscopy, confirming the steady rise of heat-trapping gases.

Computers have had a massive impact on society since the 1970s and this has been true of chemistry as well. Practically all chemical instrumentation is now computer-driven, which enables automatic loading of samples, storage and processing of the results, and the use of stored libraries of spectra and other data to identify compounds. Modern chemical experimentation would be unthinkable without the ability of computers to control and record experiments, which are often carried out under extreme conditions. As we have seen, computers are also behind the automatic control of industrial processes. Computers are also used to show molecular structures, which is now crucial because the structures of the large biomolecules of interest to many chemists cannot be drawn by hand in the hitherto conventional manner. Computers can also calculate the paths of reactions and carry out reverse engineering of syntheses (called retrosynthesis, pioneered by American chemist Elias Corey, who was awarded the Nobel Prize in 1990). As the number of chemical compounds spiralled upward to the hundreds of millions in the late twentieth century, out of sheer necessity the traditional massive printed reference works initiated by nineteenth-century indexers ("Gmelin" for inorganic chemistry and "Beilstein" for organic chemistry) have been replaced by their online counterparts, as has the venerable periodical index, *Chemical Abstracts*. This has also required the conversion of traditional chemical formulas and names into searchable formats. There is now a large range of computer databases, and by the mid-2010s chemists were beginning to use data mining of these databases to find unexpected connections.

Notwithstanding the stunning achievements of chemists in the twentieth century, we should always bear in mind that most chemists most of the time are not engaged in cutting-edge research. Most chemists are involved in routine work, such as quality analysis or monitoring processes. Others are involved in teaching in secondary schools – although the large majority of their pupils go on to take medical or other scientific degrees rather than chemistry – and in colleges and universities. Some chemists are involved in management, administration, museums, or patent work. Even if they are chemists working in academia or industrial research and development, their contributions to the

FIGURE 0.7 A three-dimensional computer model of an oxyhemoglobin protein molecule. The image was generated by Dr. Manuel C. Peitsch at the Glaxo Institute for Molecular Biology in Geneva, Switzerland. Photograph by © CORBIS/Corbis via Getty Images.

progress of chemistry are likely to be incremental. This is not because they are inept chemists, but because chemistry – like all sciences – moves forward step by step rather than in great leaps. Even when seemingly large breakthroughs are made in chemistry by a single person or a team, they are often the culmination of many smaller contributions made previously by other scientists (Scerri 2016).

It is striking that most governments around the world have sought to increase the number of chemists in their country during our period. This was partly because of the economic benefits that they presumed would accrue as a result of having more chemists working in industry, but also because of the value of chemists as teachers and researchers in the field of medicine. More recently, governments and professional organizations have sought to increase the number of women and ethnic minorities in higher ranks of the chemical professions. Already almost as many women as men are taking first degrees in chemistry in many Western countries, although they were still a small minority of graduates as late as the 1970s. However, the situation varies from country to country. Italy and Portugal have more women than men, while Germany still has more men than women.

The professional organizations of chemistry have expanded in the past century alongside the general expansion of chemistry. The membership of the American Chemical Society has expanded from 7,170 in 1914 to over 151,000 in 2019 (Thackray 1985: table 2.4, 250; American Chemical Society 2019). By contrast, the French chemical society has only trebled its membership from over 1,100 in 1914 to 3,000 in 2017. The Chemical Society, in Britain, founded in 1841, was a learned society based largely on its journal and regular meetings rather than a professional organization. The latter role was played by the Royal Institute of Chemistry (RIC), founded in 1877, which awarded qualifications based on examination as well as having higher professional grades (Russell et al. 1977). The Chemical Society had a membership of about 3,000 in 1914 and the RIC had about half that number. (By contrast, the Society of Chemical Industry founded in 1881 had around 4,000 members.) Clearly there was much overlap between the two organizations, and the possibility of a merger was first raised in 1967. The merger between the Chemical Society and the Royal Institute of Chemistry (along with two smaller organizations, the Faraday Society of Physical Chemists and the Society for Analytical Chemistry) to form the Royal Society of Chemistry (RSC) took place in stages between 1972 and 1980. The membership of the RSC is now over 54,000. Similarly, Germany had two chemical organizations before 1946, the Deutsche Chemische Gesellschaft (DCG, which was modeled on the Chemical Society) founded in 1867, and the Verein deutscher Chemiker (VDCh) for applied chemists founded in 1877 (Maier 2015). In 1915, the DCG had 3,300 members and the VDCh 5,410. In 2018, their successor the Gesellschaft Deutscher Chemiker (GDCh) has 30,000 members. On the international level, the International Association of Chemical Societies had been founded in 1911, but became a victim of World War I. Its successor the International Union of Pure and Applied Chemistry (IUPAC) was established in 1919, although chemists from the Central Powers were initially excluded and chemists from the Soviet Union were also absent. German and Austrian chemists were invited to the IUPAC conference in 1928, after Germany was accepted into the League of Nations in 1926, and these countries (and the Soviet Union) became members of IUPAC in 1931. However, Nazi Germany was barred during World War II, and eventually West and East Germany became members.

Another important feature of the twentieth century has been the global spread of chemistry. At the beginning of the century, although academic chemistry existed in most countries, it was largely concentrated in Europe (notably in

Germany, Britain, France, and Italy) and the USA. During the first half of the twentieth century, leadership of the field moved from Germany to the United States, a result mainly of the policies of the Nazi regime in Germany between 1933 and 1945, but also partly because of the rigidity of the German academic system and the growth of petrochemicals and pharmaceuticals in America. The Soviet Union also became an important player in chemistry, but its achievements were hidden from general view in the West partly by the use of the Russian language (translation services were set up to bring its literature to the attention of Western scientists) and the Cold War. Nevertheless, the half-century between 1950 and 2000 was one of almost unrivaled superiority for the United States. Not only did its chemical and pharmaceutical industry undergo rapid expansion and its chemists make major breakthroughs (collecting Nobel Prizes along the way), but many students and researchers were attracted to America, including a notable "brain drain" from Britain. By the end of the twentieth century, however, a shift in leadership to East Asia was beginning. Japan had been a middling power in chemistry since the 1950s, gaining its first Nobel Prize in chemistry in 1981 (to Kenichi Fukui), but its profile was raised considerably by the award of no less than five Nobel Prizes between 2000 and 2010. In the 1980s and 1990s with the opening up of China there was a steady flow of Chinese students into American universities, but by the 2000s they were beginning to graduate in notable numbers from Chinese universities as well. There is an increasing number of publications from Chinese universities and there is an effort underway in the late 2010s to increase the quality of the chemical research carried out there. While American leadership of the field so far remains almost unchallenged, China in 2019 may be considered to be in a similar position to the United States at the beginning of our period in 1914.

To conclude, both in terms of the number of chemists and the quantity of chemistry being done, chemistry has expanded enormously in the course of the twentieth century, far more than in any other century, at least in purely numerical terms. There was a period in the mid-twentieth century when chemistry seemed to offer a solution to almost any problem, providing wonder drugs, drip-free paints, weed-free gardens, washable suits, and cassette tapes, among so many other things. Yet for all that, the twentieth century has not generally been seen as the "chemical century." To be sure, in certain Western countries, most notably Germany, the United States, and France, there was a kind of chemical era between 1914 and 1945, when the achievements of chemistry in both peace and war appeared to outpace any other field of scientific endeavor. However, since 1945, chemistry as a field has been widely seen as surpassed by physics on the one hand and molecular biology on the other. It is ironic that both the atomic bomb and the study of DNA have overshadowed chemistry, when they both owed much to chemistry. It would also be fair to say that chemistry has perhaps lacked the general appeal of modern astronomy and astrophysics, with its space flights, Big Bang theory, pulsars, black holes, and more recently gravitational waves, or of biology with its concern with evolution, the origin and nature of life, and the manipulation of the genetic inheritance of plants, animals, and humans themselves.

All this considered, some observers argue that chemistry suffered a kind of existential crisis towards the end of the twentieth century (Morris 2008). Some chemistry departments in the UK were closed down or became heritage science or forensic science departments, and many departments in the USA changed their name to "molecular sciences" instead of "chemistry." An increasing number of Nobel Prizes were being awarded to scientists who were clearly not chemists. Some speculated whether chemistry would even survive as a single discipline, or would break up like a scientific Austro-Hungarian Empire into fragments which would be snaffled up by other disciplines. Some parts, it was said, would be absorbed by materials science, others by physics, and a large chunk by biomedicine, thus returning much of chemistry to its medical roots. Yet in the end this did not happen, at least not yet. By casting its net wider, and in particular reabsorbing much of biochemistry, chemistry has survived and has arguably even become larger. The major chemical organizations such as the American Chemical Society and the Royal Society of Chemistry appear to be confident that chemistry will continue to thrive in the twentieth-first century. It would be foolish to predict how chemistry will fare in the next century, but perhaps around 2100 Bloomsbury may commission an additional volume of *A Cultural History of Chemistry* to cover that period.

CHAPTER ONE

Theory and Concepts: *Stability and Transformation in Chemical Problems and Explanation 1914 to the Present*

MARY JO NYE

INTRODUCTION

The starting point for this chapter is 1914, the tragic watershed year of modernity that separated the nineteenth century from World War I and its aftermath.[1] Chemists fought and worked in that war, and they brought chemistry under a new kind of scrutiny and fame, if not sometimes infamy, in what some have called the "chemists' war." Praised in the past for their contributions to science and to industry, chemists now were publicly recognized and often vilified for their role in military operations (Johnson 2010; Freemantle 2014; Friedrich 2017). At the war's end, most chemists returned to their laboratories in university, government, or industry settings, often taking up research from where they had left it in 1914.

A great deal had been achieved in the century before 1914. Chemistry had become a fairly well demarcated discipline by the 1830s, separate from natural philosophy (physics) and natural history (biology, mineralogy, and geology), although still having ties to these fields (Klein and Lefèvre 2007; Klein 2016). During the course of the nineteenth century, organic chemistry became the dominant subdiscipline within chemical science, alongside mineral (or inorganic) chemistry. Physical chemistry began emerging as a significant new field bringing spectroscopy, thermodynamics, and reaction dynamics into chemistry (Ihde 1964; Laidler 1993). A great deal of nineteenth-century chemistry persisted throughout the twentieth century at the core of the chemist's repertoire. Reflecting on that fact, the Harvard physical and theoretical chemist Dudley Herschbach commented in an interview in 2012 (my emphasis) that: "… for problems that are really interesting to *true chemists* they still work more or less the way they did in the nineteenth century and even earlier" (Herschbach 2012, 1559).

Let us identify persistent late-nineteenth-century core chemical concepts. These include the framework of chemical elements, atomic weights, and periodicity of properties; the differentiation of the atom and the molecule; the study of isomers as molecules of the same composition but with different structures and properties; the concepts of chemical affinity or valence, the chemical bond, and the carbon chain; the representation of atoms in a molecule by a three-dimensional structure; the use of "paper tools" such as graphical formulas and representations; and the application of concepts of energy and entropy to chemical problems, such as equilibrium constants and chemical equilibria, extending thermodynamic theory to conditions of absolute zero by 1914 (Levere 1971; Rocke 1993; Klein 2003).

Yet, even though the chemical atom was fundamental to chemical explanation, there was no satisfactory explanation in 1914 for what holds atoms together in a molecule and how atoms pull apart or reassemble when a molecule's chemical or physical environment changes. Electrostatic attraction was not a plausible theory to explain why molecules composed of like atoms, such as hydrogen or chlorine, exist, given that like-charged atoms should repel each other. Nor did simple electrostatic charges explain carbon-to-carbon or carbon-to-hydrogen linkages in hydrocarbon molecules. The challenge to chemists was to solve the problem of affinity, and to supplement the static chemistry of structural formulas and representations with a mechanical or dynamic chemistry of chemical change (Nye 1993).

As discussed in this chapter, the conundrum was resolved in the course of the twentieth century in a new chemistry built on the revolutionary late nineteenth-century discoveries of x-rays, radioactivity, and the electron. "The indivisible atom was shattered. But the atomic theory remained," wrote one chemist in

1907, and the electron would play a revolutionary role as theoretical concept and experimental tool in a new chemical atomic theory (Muir 1907: 377; Nye 1993; Nye 1996). As Noboru Hirota writes in his valuable history of modern and recent chemistry, "While the atom played the leading role in nineteenth century chemistry, ... [t]he understanding of chemical phenomena based on the behavior of electrons was a major focus of twentieth century chemistry" (2016: 227).

Key to what happened in the course of twentieth-century chemistry was reciprocity in development of new theories and new instruments, i.e. the interdependence of conceptual tools and laboratory tools (Morris 2002a; Reinhardt 2006b; Morris 2015). This chapter focuses on concepts and theories, but it necessarily pays some attention to experiments and instruments and to the individuals and research groups responsible for developments that both expanded and obscured the boundaries of what is said to be chemistry. We examine five key domains in which profound conceptual transformations occurred after 1914: the chemical element; the chemical bond; the chemical molecule; reaction mechanisms; and solid states and materials. The chapter concludes with a brief discussion of stability and change in modern chemical concepts and theories, as indicated broadly by trends in awards of Nobel Prizes in chemistry and by chemists' own observations on the epistemological status of their theories.

THE CHEMICAL ELEMENT

Antoine Lavoisier introduced the modern definition of elements in 1789, proposing that the term "element" refers to the endpoint of chemical analysis. By 1914, eighty-six elements had been discovered, some of them filling gaps in the periodic table and others forming a new family group of the so-called noble or chemically inert gases. It only then was becoming clear that atoms are composed of subatomic parts following the discoveries of x-rays, radioactivity, and the electron. These developments led to Ernest Rutherford's experimental confirmation in 1911 that the atom has a positive charge at its center, surrounded at a distance by electrons. Following a stay in Rutherford's laboratory in Manchester, Niels Bohr detailed how the numbers and arrangements of electron rings produce the periodicity of the chemical elements and account for bonding between atoms. In 1913, Henry G.J. Moseley found a regular relationship between shifts in x-ray spectral wavelengths emitted by a series of chemical elements and what came to be called their atomic numbers. A new basis for the periodic table became atomic number rather than atomic weight. Moseley also correctly predicted that there would be ninety-two natural elements up to and including uranium (Heilbron 1974; Nye 1996: 147–71).

As chemists and physicists followed processes of radioactive decay in uranium and other elements, a startling new concept emerged. Two chemical elements could have the same atomic number and the same chemical properties but different atomic weights, a discovery for which Rutherford's former colleague Frederick Soddy coined the term "isotope" in 1912. Following World War I, Francis Aston designed a mass spectrograph that sorted out isotopes on the basis of their differing masses. Chemists also began searching for elements beyond uranium by systematically irradiating elements with neutrons, which were electrically neutral and had been identified by James Chadwick in 1932 as a nuclear companion to the positively charged proton. This trans-uranium search resulted in the unexpected discovery of uranium fission by chemists Otto Hahn and Fritz Strassmann, working with physicist Lise Meitner. The elusive trans-uranium elements were subsequently isolated when uranium was bombarded with neutrons in a Berkeley cyclotron in 1940. Element 93 (neptunium), created by Edwin M. McMillan and Philip H. Abelson, was the first of twenty-four non-natural elements among the 118 chemical elements currently in the periodic table. The heaviest, oganesson (^{118}Og-294), has a half-life of about 0.89×10^{-3} seconds before it decays into livermorium-290 (^{116}Lv) and an alpha particle (Sime 2001; Nye 2003).

Nuclear chemistry gradually became a subdiscipline of chemistry. Nobel Prizes in chemistry for researches on natural and artificial radioactivity and on radioactive and non-radioactive isotopes were awarded to Rutherford, Marie Curie, Soddy, Aston, and Irène Curie and Frédéric Joliot, establishing a stake for chemists in a field also associated with the work of physicists. Beginning in the 1940s, nuclear chemists worked alongside physicists on nuclear weapons. Nuclear chemists also collaborated with biological researchers on more benign uses of radioactivity, for example, as atomic tags in nuclear medicine and in research on reaction mechanisms in photosynthesis, genetic replication, and metabolic pathways (Creager 2013).

The study of the chemical elements, radioactive or not, was part of what was called analytic, inorganic, or mineral chemistry until the mid-twentieth century. It was then, as Jay A. Labinger argues, that the "subject by default" (i.e. what was not organic or physical chemistry) got a new lease on life in the blossoming of organometallic chemistry and in the adoption of electron and bonding theories that combined quantum mechanics with mechanistic studies of kinetics and catalysis, as discussed later in this chapter. Other new theoretical developments after mid-century included the creation of noble gas chemistry, beginning with Neil Bartlett's production in 1962 of xenon hexafluoroplatinate, followed by Howard Claassen's synthesis of xenon tetrafluoride. The development of solid-state chemistry of semiconductors, thin films, and other materials, also to be discussed later in this chapter, rejuvenated inorganic chemistry (Labinger 2013: 2, 49–50).

THE CHEMICAL BOND

The conviction that the noble gases must be chemically inert resulted not only from experimental failures at isolating any noble gas compounds but also from the electron theory of the chemical bond. Bohr had used spectroscopic data and information from the periodic table to theorize that electrons in orbits or shells surrounding an atomic nucleus emit radiation in discrete quanta, or units, of energy as they move from one shell to another. Shells fill up with electrons to limiting numbers of two, eight, eighteen, or in general $2(n^2)$ by losing or gaining electrons, until reaching the configuration of a noble gas. Thus, chemical bonds form or fall apart by the gain or loss of electrons, as proposed without benefit of quantum theory by physical chemists Gilbert N. Lewis in 1916 and Irving Langmuir in 1919. They distinguished between the polar (electrovalent) and nonpolar (covalent) distribution of a pair of electrons in a bond. "Electron rearrangement," said Langmuir in 1921, "is the fundamental cause of chemical action" (Nye 1996: 180).

To this scheme Nevil V. Sidgwick added the "covalent link" or coordination number as a third category of electron bond in his 1927 textbook on *The Electronic Theory of Valency*. Cobalt, for example, has a "primary" valence of three and a secondary valence of six, with cobalt acting as an electron acceptor and each of the groups attached to it as an electron donor (Brock 1993: 573–618).

Bohr had used the early quantum theory to describe the behavior of the electron, and the physicist Arnold Sommerfeld proposed the operation of a relativistic effect whereby the electron's mass changes as its velocity changes during its transit of an elliptical orbit, a hypothesis which was confirmed spectroscopically. Further conceptual complications included the orbit's orientation in space and, in the mid-1920s, the notion of electron spin. In 1923 the physicist Louis de Broglie proposed the theory that an electron is accompanied by a "pilot" wave. Erwin Schrödinger in 1926 reformulated de Broglie's work into a theory of the electron as an intense concentration of waves in a very small space, with the square of the calculated amplitude of the wave a measure of electron intensity or density. Werner Heisenberg preferred a less-realist or visual mathematical formulation, while Max Born argued that Schrödinger's equation should be interpreted statistically to indicate the probability of finding electron particles at a particular location in space and time (Segrè 1980; Gavroglu 2000).

Chemists for the most part did not have the mathematical training and skills to engage in the quantum revolution in physics. They became more directly interested, however, after two different treatments in 1927 of the chemical bond in the hydrogen molecule, the problem with which Bohr had begun in 1911. Working together in Zurich, the physicists Walter Heitler and Fritz London extended to the diatomic hydrogen molecule the quantum mechanical theory that Heisenberg had recently applied to the energy of the coupled system in the two-electron helium atom. Using the hypothesis of an electron potential field that is centered on the individual atomic nucleus, Heitler and London deduced well-known chemical facts such as the existence of the hydrogen molecule and the nonexistence of a diatomic helium molecule or ion. In contrast, in Göttingen, Friedrich Hund proposed that the electron field, or the probability of finding an electron in a region within a molecule, is the result of the electron's interaction with all the atomic nuclei and electrons in the molecule.

The Heitler–London atomic orbital strategy was adopted in the USA by the physicist John C. Slater and by the chemist Linus Pauling, who had been learning quantum wave mechanics in Europe. It became known as the valence bond theory. The contrasting Hund strategy was adopted by the American chemist Robert Mulliken and the German physicist Erich Hückel, whose brother Walther was a prominent organic chemist. Their approach was dubbed the molecular orbital theory. Because it is a wave function that can be visualized as the distribution of electron density in space, the orbital is quite different from Bohr's early electron orbit. The valence bond theory was immediately congenial to most chemists in its visualization of the molecule as a composite of atoms. Molecular orbital theory was less appealing initially to chemists because it regarded each molecule as an indivisible unit (Gavroglu and Simões 2012: 9–129). In a 2003 discussion of the ongoing rivalry between the two theories, the theoretical chemist Sason Shaik noted that Erich Hückel had explained a molecular orbital structure by analogy to a vibrating string, where the motion is delocalized through the whole string. "A vibrating string! Can you imagine a more alien picture to chemistry than that?" (Hoffmann et al. 2003: 752.)

From the 1930s to the early 1960s, the valence bond approach moved from one success to another, particularly in the hands of Pauling and his research group at the California Institute of Technology, as laid out in research publications and in Pauling's widely influential textbooks for both advanced and general chemistry (Nye 2000b). In the early 1930s Pauling and Slater each worked out theories of the mixing or "hybridization" of electron energy levels in order to explain the equivalence of carbon's four valence electrons and different orientation of bonds in space, for example, in methane, ammonia, and water. Pauling, with George Wheland, further began to apply the concept of resonance to molecular systems of alternating single and double bonds, such as benzene, pushing forward a theory that recognized conjugated bonds (delocalized electrons) as neither really a single or double bond but as something in between, an idea earlier expressed by Christopher Ingold as a "mesomer" and by Fritz Arndt as a *Zwitterion* (Park 1999). Quantum wave mechanics allowed the calculation of the relative contribution of each resonance form to a structure, with the relative weights of the contributions calculated from chemical and physical measurements. The correct resonance hybrid structure is the one with the lowest energy value (Brush 1999; Gavroglu and Simões 2012). The valence bond theory also proved useful in the study of chemical dynamics, a reaction's transition state, and the electronic theory of metals, as discussed later in this chapter.

Partly as a consequence of increasing interest in the structures and reactions of large conjugated biological molecules such as glycoproteins and nucleoproteins, many chemists increasingly found the valence bond theory inadequate and less intuitively linked to the molecules they now wanted to model. Mulliken and Michael Dewar in the USA and John Lennard-Jones, Charles Coulson, and H. Christopher Longuet-Higgins in the UK were among those who developed useful approximations to generate molecular orbitals from linear combination of atomic orbitals so that large molecules could be treated successfully by the molecular orbital approach. Postwar computer technologies made calculations possible that had been unwieldy if not impossible in earlier decades. In 1952 Kenichi Fukui at Kyoto University offered an explanation of Ingold's so-called electrophilic (electron-seeking) and nucleophilic (nucleus-seeking) substitution reactions using calculations for the highest occupied molecular orbital and the lowest unoccupied molecular orbital. In 1965 Robert Burns Woodward and Roald Hoffmann explained a series of stereoselective reactions and rearrangements on the basis of conservation of orbital symmetry, for example in an orbital's rotation about an axis. They made risky predictions that turned out to be unexpected and correct (Brush 1999: 269–70, 285–6; Hirota 2016: 506–8).

In their history of quantum chemistry, Kostas Gavroglu and Ana Simões note different characteristics of quantum mechanics in chemistry, distinguishing quantum chemistry as physics mainly in Germany, as chemistry mainly in the USA, as applied mathematics in the UK, and as computer programming in Japan, Sweden, the USA, the UK, France, and Germany (2012). They argue that chemists – especially organic chemists – initially resisted the mathematics of quantum chemistry in favor of the familiar visual imagery of atoms forming molecules through bonds, which now were understood as electron bonds. Physical chemists initially found that they could predict intermediates and transition states more easily with valence bond theory and semi-empirical calculations, while molecular theorists initially aimed to perfect so-called *ab initio* calculations that do not incorporate empirical data into their approximations. As molecular orbital theory prospered after the 1960s, with greater use of semi-empirical methods, which do incorporate empirical data, a new worry emerged that mathematical chemistry was taking chemistry too far away from traditional chemical conceptions such as oxidation–reduction (for example, gain or loss of oxygen), the functional group (a specific group of atoms or bonds within a molecule, for example, the carboxyl group –COOH), and molecular structure.

Some feared that chemists were being encouraged to rely on computer simulations, which could be misleading, rather than on laboratory experiments. In no area was this worry stronger than in organic chemistry, where concepts of the three-dimensional structure, size and shape, and enormous variation in conceivable structure seemed to defy mathematical reductionism in favor of pragmatic and visual realism (Weininger 1984; Hoffmann 1988: 1597; Austen 2016).

THE CHEMICAL MOLECULE

In the course of the twentieth century, it became feasible to understand the structure of very large molecules and to synthesize ever larger and more complex molecules. The number of different kinds of molecules increased exponentially as a result of the synthesis of previously unknown compounds that sometimes could be found in nature, but sometimes not. What had been conceptualized and represented in the nineteenth century as a basic skeletal structure of letters and dashes, or balls and sticks, became an elaborate architecture of chains, rings, fused rings, helices, cages, sheets, and other shapes in space. The external contours and shape of a molecule, the angles in space of its parts, its symmetry or asymmetry, internal strains, intramolecular rotations, and flexibility all became features of an increasingly realistic picture for a substance, its possible behavior, and its functions in living organisms as well as in the laboratory.

Mid- and late-nineteenth-century chemists, notably including Louis Pasteur, August Kekulé, Joseph Achille Le Bel, and Jacobus Henricus van 't Hoff, had theorized an architecture of hydrocarbon molecules built on carbon atoms that link together in open or closed chains – perhaps in a spiral according to Pasteur – and of carbon valences directed outwards in three dimensions, as in the so-called carbon tetrahedron (Rocke 1985; Ramberg 2003; Rocke 2010). A cascade of hypotheses and predictions followed, for example that five- and six-membered rings are more stable than smaller or larger rings because there is less distortion (or "strain") of valences from the symmetrical 109-degree angles of the tetrahedron. The proposal that a molecule might "pucker" into "bathtub" ("boat") or "chair" isomeric structures was confirmed in the 1930s by Odd Hassel, who concluded that the chair form is the more stable *steric conformation*, a term introduced by Walter Haworth in 1929 (Ihde 1964: 621–31; Brock 1993: 573; Nye 1996: 133–8).

In the 1920s, the question still was open as to the limits to molecular size. Some of the debate centered on the chemistry of colloidal substances, which include proteins, cellulose, rubber, and starch. Wolfgang Ostwald and some other German chemists favored the theory that colloidal substances are physical aggregates of small molecules held together by weak intermolecular forces, similar to the secondary valences of Sidgwick's coordination chemistry. In contrast, Hermann Staudinger used traditional "wet chemistry" and Hermann Mark used x-ray and electron diffraction to argue that many colloidal substances are made up of long-chain "macromolecules" of very high molecular weight. Theodor Svedberg confirmed this theory by measurements of molecular weights using a rapidly spinning ultracentrifuge in which particles sort as they settle to the bottom of a cylinder or test tube depending on their weight. Mark and his colleague Kurt Meyer insisted on the flexibility of macromolecular structures or polymers and hypothesized that the elasticity of rubber is due to the spiral molecular shapes. The macromolecule became a staple of twentieth-century chemical theory along with methods of polymerization for producing not only known materials, but also entirely new ones such as synthetic fibers and thermoplastics (Furukawa 1998). By 2017 the commercial polymer industry approached $500 billion in global annual sales and the polymer market sector in the USA was growing more rapidly than the gross domestic product (Lodge 2017: 10).

Partly under the influence of Mark, Pauling was among those who began building up pictures of the way that large molecules are structured, proposing the idea that weak attraction between an electronegative atom, such as oxygen, and an electropositive hydrogen atom that is bonded to another electronegative atom could determine the shape of macromolecules (a mechanism called hydrogen bonding). In 1950, Pauling and his coworkers published a key paper for modern chemical and biochemical theory when they used their theories to propose a single-coiled spiral model (the *alpha*-helix) for the protein keratin, modeling the protein by using both simple paper folding techniques and three-dimensional wooden or plastic materials. Pauling also suggested a triple helix structure, which fits collagen, in which two identical chains entwine with an additional one that differs slightly in its chemical composition. The triple helix made an unfortunate reappearance in Pauling and Robert Corey's proposal for the structure of deoxyribonucleic acid (DNA), the molecule that some biologists were beginning to think might play a major role in genetics. James Watson and Francis Crick's double helix structure, relying on Pauling's modeling approach and Rosalind Franklin's x-ray photographs of hydrated DNA, was published in spring 1953 and won the day. Chemistry and biology now were linked together conceptually more firmly than ever in an expanding biochemistry field, and in what came to be called molecular biology or chemical biology (Olby 1974; Hager 1995; Strasser 2006).

The determination of structure and the synthesis of organic and organometallic molecules, many of them of biological significance, became one of the most prominent features of twentieth-century chemistry. In 1945 Dorothy Crowfoot Hodgkin's research group became the first to correctly determine the structure of penicillin, following up this success with Vitamin B_{12} in 1956 and insulin in 1969. Like Pauling, and like Crick and Watson, Hodgkin used physical models not too different from ball-and-stick models, but she and her group also were modeling three-dimensional contours of electron density with help from whatever was the best computing technology at the time. Like Hodgkin, Max Perutz used the replacement of a lighter-weight metal with a heavier one in a molecule in order to find the structure of hemoglobin in 1953, and John Kendrew announced the structure of myoglobin in 1962 (Nye 2016).

Once structures were known, synthesis of natural products could be achieved both as confirmation of the structure and as replacement for the natural product itself. In collaboration with Albert Eschenmoser and his research groups at the ETH in Zurich, Woodward and his teams at Harvard achieved a total synthesis of Vitamin B_{12} in 1972. This followed earlier syntheses of quinine, cholesterol, cortisone, and other compounds, including tetracycline, by Woodward and his Harvard research groups. His chemistry marked a shift in synthetic organic chemistry from relying mainly on traditional methods of analysis followed by synthesis, characteristic, for example, of Robert Robinson, and instead employed detailed knowledge of new concepts from physical organic chemistry. Woodward's tools came to include not only the so-called Woodward–Hoffmann rules, developed with his younger colleague Hoffmann from their theory of orbital symmetry in 1965, but also Woodward's earlier rules for correlation of ultraviolet spectroscopy maxima in conjugated systems with the amount and type of substitution on the system. Woodward's colleague Elias J. Corey similarly used precise physical data in the method that he called *retrosynthesis* for building up a target compound from simpler structures, aided by computer analysis (Tarbell 1976: 121–2; 536–7; Seeman 2017; Hepler-Smith 2018).

Some of the most exciting new molecular structures bridged organic and inorganic chemistry. One was an iron compound that Woodward and his collaborators found to have a novel sandwich structure with an iron atom sandwiched between layers of cyclopentadienyl rings. Many organometallic compounds came under new study, using revived interest in varied types of bonding (Seeman 2014: 714; Gortler and Weininger 2017). Many of the compounds studied had biological importance, for example, the iron-centered porphyrin heme, which is composed of an iron atom centered within a flat ring of four heterocyclic groups. Charles Pedersen in 1962 proposed another enclosure structure, which he called a crown ether, that results from a large ring of ether groups capturing, or "hosting," a sodium or other metal ion and forming a complex as a result of coordination between the enclosing ring's oxygen atoms and the captured metal cation. Jean-Marie Lehn in Strasbourg and Donald Cram at UCLA independently expanded notions of guest–host bonding with the creation of cage-shaped molecules in which an inorganic or organic cation or anion could be isolated, so that they could study the guest's properties in a stable host environment. Some of these molecules mimicked the behavior of enzymes in living cells (Hirota 2016: 539–45).

A kind of cage without a guest was produced in 1985 through the laser evaporation of graphite, producing a stable spheroidal molecule of sixty carbon atoms. Christened buckminsterfullerene in homage to the architect Buckminister Fuller's design of geodesic domes, the molecule was recognized as a previously unknown elemental form, or allotrope, of carbon. The fullerene as a hollow spheroidal or cylindrical molecule became a fundamentally new concept in chemistry. Held together by single and double bonds arranged into twenty hexagons and twelve pentagons, buckminsterfullerene is a kind of molecular soccer ball representative of a new class of molecules with beautiful symmetry and surprising properties such as shock resistance and superconductivity (Bensaude-Vincent and Stengers 1996: 262; Levere 2001).

REACTION MECHANISMS

The explanation and prediction of reactions that might be expected from molecules in contact with each other and the synthesis thereby of new molecules are at the heart of modern chemistry. Late-nineteenth-century chemists, drawing upon the work of physicists and upon empirical laboratory chemistry, had gradually created a macroscopic thermodynamics that made possible reliable predictions of the course of chemical reactions from heats of reaction and chemical bond strengths. The introduction into chemistry of graphical formulas, visual representations, and three-dimensional models of chemical molecules quickly led to proposals for precise reaction mechanisms and possible new reactions. However, there were many warnings against hypotheses that might be mere fictions, including hypotheses about internal vibrations or motions within the molecule (Nye 2014).

Many chemists objected to speculations about hypothetical reaction intermediates (fleeting molecular forms halfway between reactants and products) that could not be isolated or empirically verified. Yet, a hypothetical mechanism was a valuable tool if it resulted in correct predictions, even if the correct outcome did not guarantee the validity of the mechanism. By the 1930s, theories of reaction mechanisms and reaction intermediates began to be applied widely for predicting substitutions and for creating new compounds. These theories also became increasingly important in the fields of catalysis and surface chemistry.

Victor Grignard and Paul Sabatier were among early-twentieth-century organic chemists who engaged in speculations about reaction intermediates. Grignard established a method to build up carbon-to-carbon linkages using powdered magnesium, and Sabatier and his collaborator Jean-Baptiste Senderens developed a process for the hydrogenation of unsaturated compounds over finely divided nickel. Both proposed the fleeting existence of unstable intermediates, a hypothesis confirmed in the 1920s for Sabatier's intermediate nickel hydride by

Wilhelm Schlenk and Theodor Weichselfelder using x-ray diffraction. Sabatier hypothesized that the formation and decomposition of transient intermediates must correspond to a diminution of free energy, but he did not take the step, nor did Grignard, of applying the electron concept of the chemical bond to the problem of the reaction mechanism (Nye 2014).

While resistance to electronic theories of reaction mechanisms was strong among organic chemists, especially in France, and also in Germany, a different attitude emerged in Great Britain and the USA. Joseph J. Thomson was among British scientists who had a strong interest in a unified theory of matter for physics and chemistry, and he expressed that confidence in his book *The Electron in Chemistry* published in 1923. His theory of the particulate nature of the electron included a portrayal of directional tubes of force associated with bonding electrons, resulting in polarities within the molecule, even in an organic molecule. Some chemists began using directional or arrow formulas showing incipient charges within the molecule, including Howard Lucas at Caltech, Martin Lowry in Cambridge, and Fritz Arndt in Breslau. After they heard a talk by Langmuir in 1921, Arthur Lapworth and Robert Robinson jointly proposed that the development of slight internal polarization within molecules is the crucial step in many chemical reactions, visualizing the effect, which Johannes Thiele earlier had attributed to development of partial valencies within a molecule, as the movement of electron pairs. Hans Meerwein offered a similar proposal to the one by Lapworth and Robinson in 1922 (Weininger 2018).

By the late 1920s and 1930s Christopher Ingold with his wife Edith Ingold proposed the terms "electrophilic" and "nucleophilic" for predicting dispositions at reaction sites. They used the notation of δ^+, δ^- for partial or fleeting positive and negative charges that appear at the instant of internal electronic redistribution. Ingold, however, like Robinson, had no interest in introducing quantum mechanics into his highly visual theories of electronic reaction mechanisms (Gortler 1985; Saltzmann 1986; Nye 1993: 163–223). Among Ingold's proposed ionic intermediaries were transient trivalent carbocations, such as $(CH_3)_2CH^+$, and in the 1960s George Olah produced stable carbocations at low temperatures and studied their structure and stability using high-resolution nuclear magnetic resonance (Weininger 2000).

Stephen J. Weininger has described the delayed acceptance by German chemists of electronic reaction mechanism theories and quantum chemistry, with the result that German organic chemists became avid researchers in these fields only after the 1960s, building at that time upon a new familiarity with American textbooks and laboratories as a result of postwar exchanges (Weininger 2018). In his history of inorganic chemistry, Labinger found a similar German resistance to Anglo-American mechanistic theories, with influential inorganic or mineral chemists focusing on synthesis and characterization of new compounds or what Walter Hieber in 1955 called "real chemistry" (2013: 30). Weininger argues that the organizational structure in German universities kept teaching and research programs separate for organic, inorganic, and physical chemistry, with physical chemists the least numerous and least influential among German chemists for the first decades of the twentieth century (2018).

One area of inorganic chemistry that became increasingly important in chemical science and a fruitful ground for studies of reaction mechanisms was surface chemistry and catalysis. A good example is the Haber synthesis of ammonia from nitrogen and hydrogen gases, originally using a surface catalyst of osmium powder in 1909. One of the great successes in a new era of surface chemistry was clarification of the exact molecular mechanism of the ammonia synthesis. In the 1970s Gerhard Ertl and his research group, then at Munich, showed that nitrogen molecules dissociate on an iron catalyst surface to produce surface-bound nitrogen atoms that react in stages with hydrogen atoms produced by the dissociative adsorption of molecular hydrogen. The reaction intermediates NH and NH_2 precede the ammonia (NH_3) that finally escapes into the gas phase (Ertl 2008; Hirota 2016: 511–14).

In the same period of the 1970s, Mario Molina and Sherwood Rowland postulated the destruction of stratospheric ozone by the breakdown of ozone into oxygen. The chemical reaction was catalyzed by free chlorine and fluorine atoms generated by the action of ultraviolet radiation on atmospheric chlorofluorocarbons, which were being used globally in coolants and aerosol sprays. A chain reaction mechanism was proposed that regenerated the halogen atoms for further ozone destruction, and Paul Crutzen found that the mechanism includes surface catalysis on ice crystals. These investigations constituted major conceptual breakthroughs for understanding catalytic processes and for studying atmospheric and environmental chemistry (Rowland and Molina 1994; Hirota 2016: 565–6).

An important key to understanding general reaction mechanisms and intermediate states lay in combining thermodynamics and quantum chemistry. This step came in 1931. According to Hirota, "the biggest contribution to chemical reaction rate theory ... in the first half of the twentieth century" was the transition state theory of the 1930s (2016: 270). This advance emerged from investigations on simple gas systems, such as sodium vapor and chlorine, or hydrogen with chlorine or bromine, investigations that also led to collision and chain-reaction theories of gas reactions as developed by Cyril Hinshelwood and Max Bodenstein.

Michael Polanyi was among those who originally developed the transition state theory. When Henry Eyring arrived from Wisconsin to work with Polanyi in Berlin, Polanyi suggested that they combine some of Fritz London's results on exchange energies in the chemical combination of three- and four-atom systems with the thermodynamics of activation energies in order to study chemical activation and reaction mechanisms. The result in 1931 was a now iconic graphical potential energy surface that describes the energies of the reacting atoms and the energy of the activation complex or transition state (Glasstone et al. 1941: 1–13; Nye 2012: 118–30). By arriving at a plausible description of the transition state, chemists could make rational decisions about the optimum reaction conditions that would bring about their desired result. Bretislav Friedrich writes that this work made 1931 the *annus mirabilis* of theoretical chemistry (2016: 5382).

Direct experimental corroboration of the transition state came in the early 1960s with methods of flash photolysis developed by physical chemists George Porter and Ronald Norrish to track short-lived chemical species. Ahmed Zewail used laser pulses in the late 1980s to directly observe a transition state of about 50 femtoseconds (1 fs is 10^{-15} s) in the decomposition reaction of iodine cyanide: ICN → I + CN (Rawls 2000: 35). Polanyi's son John C. Polanyi obtained the first infrared chemiluminescent spectroscopic evidence for generalized transition stages in the reaction of a fluorine atom with diatomic sodium. He shared the 1986 Nobel Prize with Dudley Herschbach and Yuan T. Lee, who used crossed molecular beams to study reaction mechanisms of simple vapor and gas reactions (Herschbach 2017). In these experiments, a beam or stream of gas particles moves in a single direction at very low pressure and density before colliding with another molecular beam.

The study of reaction mechanisms at the other end of the temperature spectrum became equally important by the 1990s, when physical chemists and molecular physicists set out to study reactions of "cold molecules" at temperatures near absolute zero, using a combination of molecular beams with laser cooling in electric, magnetic, radiofrequency, microwave, and optical fields. For simple systems such as $H_2^+ + H_2 \rightarrow H_3^+ + H$, the approach, interaction, intermediate states, and final products of the atoms and molecules could be monitored. Among the results have been new insights into chemical reaction dynamics and what has been called the "exploration of emergent collective phenomena in interacting many-body systems, which represents one of the central challenges in science" (Doyle et al. 2016; Bohn et al. 2017: 1002).

SOLID STATES AND MATERIALS

The physicist Philip W. Anderson called "emergence" the "fundamental philosophical insight" of twentieth century science. What he had in mind was a new attitude among physicists who were investigating many-body systems. Behaviors were being discovered that could not be predicted or understood on the basis only of the fundamental laws of elementary particles and wave mechanics (Anderson 1972; Anderson 2011: 90; Martin 2015). In a simple example, Anderson noted that a single atom of gold is not by itself shiny, yellow, and electrically conducting because metallicity is a property meaningful only for a macroscopic sample. Chemists had recognized the problem of emergence for a long time, with Pierre Duhem speaking eloquently to this subject at the turn of the twentieth century in describing how it is that a newly created compound substance does not have the properties of its original ingredients (Bensaude-Vincent 1998; Bensaude-Vincent and Simon 2008: 122–8). Speaking as a physical chemist and chemical physicist, Michael Polanyi reflected in 1936, "Just link up two or three of the atoms of physics, and their behavior becomes so complex as to be beyond the range of exactitude" (1936: 234; Harré 2013: xvii; Manafu 2013).

Physicists' insights into emergent properties owed much to their new interest in solid-state "materials" previously identified with mineral and metal chemistry or with the chemistry of fibers and crystals. New conceptual and disciplinary categories arose after the 1950s, crisscrossing physics, chemistry, biology, and engineering. Scientific fields emerged for "nanomaterials," "biomaterials," and "green materials," in which the word "material" carried the connotation of multi-molecular or multi-atomic substances. Thus, physicists and chemists found themselves working on complex systems both at the very large scale of oceans, atmospheres, and the global environment, but also at the microscopic scale of nanometers (Bensaude-Vincent 2018).

By the 1990s, approximately one-third of American physicists identified themselves as condensed matter physicists (Kohn 1999: S77). These scientists, like some chemists, studied the behaviors and properties of crystals, metals, polymers, colloids, gels, and glasses, with attention not only to quantum mechanical explanation of electron behavior, but to pattern formations, fractures, dislocations, and the flow of materials. Among chemists, fields of interdisciplinary research now included not only molecular biology and chemical physics, but also the materials science of "solid-state chemistry" (Bensaude-Vincent and Simon 2008; Teissier 2014; Martin and Janssen 2015).

Generally speaking, there have been two different quantum mechanical treatments of solids. Ionic crystals have usually been studied via valence bond theory, but metals required the molecular orbital theory, which in the case of metals is energy band theory. Beginning in the 1950s, the physicist and quantum chemist Per-Olov Löwdin showed that band theory could also be applied to semiconductors and insulators, and that it explains the difference in electrical conductivity of metals, semiconductors, and insulators by the size of gaps between allowed and forbidden energy transitions (Anderson 2011: 123). The restriction of superconductivity to metals at temperatures below 30 °K ended with discoveries beginning in 1986 of superconductivity in ceramic materials at relatively high temperatures up to 160 °K, as both physicists and chemists tried to develop different kinds of materials with little electrical resistance well above absolute zero. The idea of organic superconductors occurred to

Woodward in the late 1970s, and Klaus Bechgaard achieved this marvel in 1979 in Copenhagen with the organic complex tetramethyltetraselenafulvalene (TMTSF)$_2$PF$_6$ (Seeman 2014: 6; Hirota 2016: 557).

What came to be called "nanoscience" also produced a major transformation in both chemistry and physics. Recognition of the existence of carbon nanotubes dates back to publications beginning in 1952 when the physical chemists Leonid Viktorovich Radushkevich and V.M. Lukyanovich published an image of carbon tubes 50 nanometers in diameter (where 1 nanometer is the length of three gold atoms). In 1991, using a transmission electron microscope, the physicist Sumio Iijima identified carbon nanotubes (CNTs) in the detritus of fullerenes. These CNTs have radii of a few nanometers and are constructed from networks of carbon hexagons. Depending on how a sheet is rolled, CNTs exhibit metallic or semiconducting behavior, and they have mechanical properties of lightness, tensile strength, and elasticity. They are structurally equivalent to monatomically thick graphite films rolled into cylinders. These films, or graphene sheets, are an allotropic form of carbon, an infinitely large aromatic molecule that first was separated from a backing material by peeling sheets from a graphite crystal in 2004 (Hirota 2016: 545–6). Among the pioneers in this work, MIT physicist Mildred Dresselhaus became known as the "queen of carbon" (Angier 2017).

Norio Taniguchi coined the term "nanotechnology" in 1974 in his investigation of semiconductor processes and thin-film deposition. The study of the deposition of single-molecule thick films, including organic molecules, on a metallic or semiconducting surface became a leading field of research at the intersection of physics and chemistry, with precedents going back to the work of Irving Langmuir, Michael Polanyi, and others in the 1920s and 1930s. Late-twentieth-century novelties included the discovery of the self-assembly of molecules into monolayers on surfaces (SAMS), a field in which John Polanyi has figured (Harikumar et al. 2011). The theoretical investigation of nanoscale materials provided unusual opportunity for study of properties that result directly from quantum mechanical effects in materials at the nanoscale rather than in the dimensions of the macroscopic world (Marcovich and Shinn 2014: 5–6; Bensaude-Vincent et al. 2017; Channel 2017).

By the early twenty-first century, many chemists, particularly theoretical chemists, were moving back and forth across the spectrum from the nanoscale to the macroscale as their province of study. On the one hand, new algorithms in computational quantum chemistry were being developed to study many-body problems with simple "quantum computers" that store and process information as entangled quantum bits (or qubits) rather than as states of "0" or "1," potentially vastly expanding the computer's storage of information and predictive capabilities in the hands of quantum chemists. On the other hand, the systems often studied by these same chemists extended to the wider world of inorganic and biomaterials and of energy sources that interact in complex systems at both the scale of the organism and the biosphere (Johnston 2010).

A typical example of such current research can be found in the work of Garnet Chan's group at Caltech. In an interview in 2016, Chan explained that:

> I'm interested in going from the very simple equations of quantum mechanics – which are the fundamental equations of nature, the most basic equations we know about the world – to the actual behavior of molecules and materials and real matter that we can touch around us. It's a discipline that involves finding computer algorithms that allow us to simulate these equations, at least approximately.
>
> (Clavin 2016)

Chan's research problems include computer modeling to find materials that could exhibit high-temperature superconductivity and his group's investigation of the biological mechanism by which enzymes fix nitrogen, expanding the "frontier where the tools are being developed to study the most complex problems of biology and materials" (Clavin 2016).

STABILITY AND CHANGE, 1914 TO THE PRESENT

As demonstrated in this chapter, during the course of the twentieth century the theories, instruments, and communities of the cultures of chemistry and physics became more tightly bound together in a reciprocal process in which physics was also more like chemistry and both intersected more closely than in the past with biology and medicine. By way of seeing how chemical science after 1914 retained key theories and concepts from the past and how it incorporated novelty and change, the awarding of Nobel Prizes is an informative, if imperfect and partial, gauge that registers these developments. To be sure, the awards are subject to the vagaries of nominating procedures and of personal prejudices, rivalries, and agendas among Swedish Academy of Sciences members and the many external award nominators, but that is the case for all social processes (Friedman 2001). What the Nobel Prizes do tend to demonstrate is the persistence in recent chemistry of core chemical concepts dating back to the nineteenth century, the emergence of new concepts and theories, and the erosion of boundaries between traditional science disciplines.

It often has been noted that some of the first chemistry Nobel awards went to chemists identified with the then new field of physical chemistry – van 't Hoff in 1901, Svante Arrhenius in 1903, and Wilhelm Ostwald in 1909 – but there were no more awards for chemists who commonly have been thought of as "physical chemists" until Walther Nernst in 1920. Early chemistry Nobel awards for work on radioactivity and isotopes recognized their revolutionary significance for understanding the elements, atomic weights, and periodic table. The radioactivity and isotope Nobel awards also indicate the sometimes difficult distinction between chemistry and physics, given that some of the Nobel chemistry prize recipients – such as Rutherford, Aston, and Frédéric Joliot – thought of themselves as physicists. The early electron, despite its implications for chemical valence and bonding theory, received little mention in Nobel chemistry awards until the 1936 chemistry prize to physicist Peter Debye for his experimental investigations on dipole moments and the diffraction of x-rays and electrons in gases. A Nobel chemistry breakthrough for the electron came only with the 1954 award to Linus Pauling for the electron theory of the chemical bond and its applications in explaining molecular structure.

From 1901 to 2016, the fields of organic chemistry (twenty-seven prizes) and biochemistry (twenty-seven prizes) received approximately 50 percent of the chemistry Nobel Prizes, with a marked increase of recognition in biochemistry after 1950. By 2016, physical chemistry (twenty-four prizes, thirty-six laureates) had garnered about 25 percent of the awards, with an additional seven prizes (eleven laureates) to "theoretical chemistry," beginning with Pauling's 1954 award (Borman 2016). When Robinson received the Nobel Chemistry Prize in 1947, it was not for his theory of electron reaction mechanisms, but for his "researches on plant products of biological importance." With a nod to the past, Robinson was praised "as a student of molecular architecture" who pursued practices "emerging from Kekulé and Couper" (Fredga 1964).

In the 1950s, the Nobel awards began to reflect transformations in research problems and conceptual frameworks that were driven both by earlier and newer developments in electron theory and studies of reaction mechanisms. Physical instrumentation proliferated in laboratories, including the introduction of a succession of improved electronic computers. Standard research equipment, often shared among different disciplinary laboratories in university or industrial settings, began to include instruments for x-ray and electron diffraction, atomic and molecular spectroscopy, electrophoresis, nuclear magnetic resonance spectroscopy and, after the 1980s, scanning tunneling and atomic force microscopy which allowed scientists not only to "see" atoms and molecules, but also to manipulate them. The 2017 Chemistry Prize went to researchers who developed cryo-electron microscopy that not only sees biomolecules, but also freezes them in mid-movement in order to visualize their functions. While historians often have emphasized the resistance of chemists, particularly organic chemists, to physical instruments, Gavroglu and Simões suggest that resistance mostly was directed at mathematics rather than at physical concepts and techniques (2012).

X-ray and electron diffraction methods revolutionized the understanding first of salts and simple inorganic substances and then of protein structures and other biologically important molecules. Beginning with the 1958 award to Frederick Sanger for protein work that included the structure of insulin, and strongly influenced by the 1962 Nobel award in Physiology or Medicine to Watson, Crick, and Wilkins for the structure of DNA, an increasing number of Chemistry and Physiology or Medicine prizes recognized research on molecular structures and drew chemists' attention to investigating relationships between molecular structure and biological function. William Brock suggests that one effect of the de-glamorization of classical structure determinations by ever more sophisticated instrumentation was to push organic chemists toward questions about function. This was a theoretical field in which Pauling had been a pioneer in the 1940s in his investigations of mechanisms that account for how abnormally shaped hemoglobin triggers sickle cell anemia and how the structure of protein antibodies explains their neutralization of pathogens (Brock 1993: 631–2; Hager 1998).

As the different kinds of instruments and their commercial availability expanded, the necessity of sharing them often brought together scientists trained in different disciplines but who found themselves needing the same equipment to address a problem. These problems ranged across a wide spectrum of temperatures, pressures, and phases of matter at which properties and behaviors of atoms and molecules escape disciplinary boundaries. In the Nobel Prize category of Physiology or Medicine, the dominant field became genetics after 1950, with citations in genetics or molecular biology recognizing investigations on molecular structure and manipulation of proteins, viruses, nucleic acids and other materials of organic and biochemistry. In the Physics Nobel awards, particle physics and quantum electrodynamics became fields newly recognized after 1950, but so, too, did condensed matter research after the early 1960s (Nobelprize.org).

If quantum physicists around 1930 had thought that chemistry could be reduced to physics, they found themselves at the dawn of the twenty-first century conceding that they had come to some of the same conclusions as chemists and biologists about the non-reducibility of molecular behavior to the mathematics of quantum electrodynamics. Pauling, in studying theoretical physics under the influence of Arnold Sommerfeld, had spoken in the 1920s of the goal to work out "a complete topology of the interior of the atom and, beyond this, a system of mathematical chemistry." By the 1930s, however, after his structural and functional studies of both inorganic and organic molecules, Pauling was more circumspect: "I have become somewhat more cautious ... there is more to chemistry than an understanding of general principles. The chemist is also, perhaps even more, interested in the characteristics of individual substances – that is, of individual molecules" (Nye 2000a: 478, 484).

A similar theme of individuality rather than generality can be found in the chemical and philosophical essays of Roald Hoffmann, whose work in quantum and theoretical chemistry has mirrored, or helped to create, some of the trends in recent modern chemistry. His own researches moved from organic molecules to inorganic molecules, and then to solids and surfaces (2012b: 81). Hoffmann insists on realism, based in experimental work, as the dominant epistemology in chemistry. The central activity of chemistry, he argues, is the identification and creation of molecules, whether to make something new or to confirm a structure or to test a prediction. Chemistry uses and combines what Hoffmann calls a vertical understanding in the classical reductionist sense, but also a horizontal understanding that is constituted by concepts, definitions, and symbolic structures at the same level of complexity as its object. These horizontally placed ideas, such as aromaticity, the acid–base concept, and the functional group, mostly are not reducible to physics. Chemistry, Hoffmann writes, moves upscale, horizontally as well as vertically, as it climbs "ladders of complexity, creates new molecules and emergent phenomena" (Hoffmann 2012a: 30).

The history of modern chemistry demonstrates that complementarity of mechanisms and representations are natural to chemistry and not at all concessions of failure. Alternative and superficially inconsistent theories coexist and contribute to a richer understanding of chemical phenomena, for example in the different visual representations of the electronic structure of benzene or in the rival mathematics of valence bond, molecular orbital, and the newer density functional theory, which began to be developed in the early 1960s by Walter Kohn. The theories or representations that are chosen depend on the question asked and the convenience of applying one framework or another. Herschbach compares the special epistemology of chemistry to an impressionistic painting:

> A physicist tends to stand too close to chemistry, looking to reduce things to first principles. The old-time biologist wanted to stand too far away to avoid getting swamped in too much molecular detail … But the chemist's intermediate domain, where you see the impressionistic beauty emerge, is … where you blend intuition and rigor.
>
> (2012: 1559)

In this view, the chemical scientist is practicing an art and a craft that is more tightly tied to observation, logic, and prediction than a painter or sculptor, but the resulting creation, whether an immaterial theory or a material substance, is a thing of beauty. In the course of the twentieth century, the electron joined the chemical atom and the chemical molecule as a core concept and tool of a chemical scientific culture that explains the structure of the molecule and its behavior both mathematically and visually. In what might seem an unexpected twist, the use of physical instruments, such as atomic force microscopy, which facilitate building up materials one atom at a time, rather than only analyzing and resynthesizing them, has strengthened rather than reduced the persistent realism that undergirds what might be called the chemical middle view.

CHAPTER TWO

Practice and Experiment: *From Laboratory Research to Teaching and Policy-making*

JOSÉ RAMÓN BERTOMEU-SÁNCHEZ AND
ANTONIO GARCÍA-BELMAR

INTRODUCTION

In master narratives, the history of twentieth-century chemistry has been characterized by the dramatic growth of the discipline, alongside revolutionary theories and experimental breakthroughs. The traditional focus by historians of chemistry on theory has been replaced in the last few decades by a greater concern with experimental practices, material culture, and instruments. Most of the current studies on experimental practice show that practical techniques are combined with interpretive tools to understand the results and reveal the rhetoric employed in the publications of chemists. Three decades of analysis of laboratory notebooks have shown that the published accounts barely describe everyday experimental practices in laboratories and the "fine structure of scientific creativity" (Holmes 2004). Moreover, recent works have moved from history centered on discoveries and innovations to a history of circulations, social uses, and cultural practices. The "geographical turn" and the growing interests on previously scarcely explored sites of chemistry have introduced new spaces, actors, and problems to current narratives.

To start, let us take into account recent scholarship on the history of laboratories. The "classical laboratory," which was well-established by the 1870s, dominated our scene, but the number and diversity of experimental spaces expanded in the twentieth century: university, industrial, pharmaceutical, and government laboratories spread all over the world with different features and peculiarities, hybrid situations, and in-between laboratories were far from exceptional (Morris 2015). Moreover, chemical laboratories overlapped with other disciplines: experimental spaces such as physical and biological laboratories followed the spectacular changes in the boundaries of chemistry and its neighboring areas (Reinhardt 2001; Bertomeu Sánchez et al. 2008). A broad range of instruments and experimental practices circulated across the contact zones and encouraged exchanges and hybridizations. They promoted, for instance, connections from chemistry to mathematics and technology, and set the basis for new research fields, in some cases even changing the borders of disciplines. At the same time, methods and practices were transformed in their migration through different research areas, social institutions, and geographical sites (Reinhardt 2001; Kikuchi 2013). Local practices were forced into unbalanced exchanges with modern chemistry technologies prompted by transnational corporations and governments. And yet, even from subaltern positions, citizens and activists were able to contest and resist processes of globalization at the local level (Mukharji 2016).

The range of issues relating to the history of both experimental and theoretical practices in twentieth-century chemistry is wide and varied. In his study on the impact of physical methods in chemistry, Carsten Reinhardt distinguished "five dimensions of method-making science": (a) the making of new methods in the laboratory; (b) the adaptation of chemical concepts into model systems; (c) the processes of standardization and teaching; (d) the university–industry nexus; and (e) the required social organization (2006b). The following discussion is organized around these issues, including also the previously mentioned ones. We will examine four groups of practices: experimental practices; practices of theory; practices of teaching and popularization; and standardization and regulatory activities. These practices are related to several categories of chemistry sites: university and industrial laboratories, scientific societies and libraries, classrooms and mass media, and governmental laboratories and regulatory offices. Needless to say, all these places are interconnected by frequent and unequal exchanges of people, objects, concepts, and data. The same applies to the different practices and methods discussed in the following sections.

INSTRUMENTAL PRACTICES

Most historians accept the existence of an "instrumental revolution" in the middle of the twentieth century (see Chapter 3 in this volume), which was characterized by the advent of new physical methods in both analytical and organic chemistry, in particular infrared spectroscopy and nuclear magnetic resonance. These changes came with an increasing role of synthesis in organic chemistry and the arrival of computers in quantum chemistry. New experimental and theoretical practices were introduced, thus marking the transformation from classical to modern chemistry (Morris 2002a). However, as mentioned in other chapters, many substantial changes took place in the material culture of chemistry since the "glassware revolution" of the early nineteenth century (Jackson 2015). While the use of the senses (colors, smells, flavors) remained as a fundamental resource for chemists (Reinhardt 2014; Bertomeu-Sánchez 2015; Kiechle 2017), new physical instruments entered the laboratory in the second half of the nineteenth century: spectroscopes (Hentschel 2002), saccharimeters (Pohl and Baumgartinger, 2005; Warner 2007), and colorimeters (Garrigós et al. 2006). They introduced new connections between chemistry, industry, and medicine and new forms of visual culture, sometimes difficult to be managed (Hentschel 2014).

Saccharimeters introduced new practices in the analysis of glucose in human blood and the control of quality in the sugar industry, including the making of a new scales and international standards, resulting from complex negotiations and metrological conflict between industry, instrument makers, producers, and chemists. They were adopted in a broad range of industrial and academic activities, sometimes at the service of commercial and imperialist purposes (Singerman 2017). Since the dramatic discoveries of new elements by Robert Bunsen and Gustav Kirchhoff around 1860, early visible spectroscopy was introduced in areas such as mineral analysis (James 1983; James 1988). During the second half of the nineteenth-century, analysis of visible spectra enabled the discovery of new elements (helium), established the basis of cosmochemistry, and sparked the imagination of chemical cosmologists (Kragh 2000). At the turn of the century, the introduction of new ultramicroscopes would soon impart a new dynamic to colloidal chemistry, a mushrooming area related to both theoretical and practical issues, from the understanding of the origins of life and the visualization of the "experimental atom" to the advancement of chemical industries (Cahan 1997; Bigg 2014). The advent of ultracentrifuge and electrophoresis apparatus changed the nature of research on colloids during the 1920s and 1930s (Ede 2007b). Another pioneering area for the introduction of new physical instruments was nuclear chemistry and the early industry of radium, which employed electrometers, ionization chambers, and x-ray and electron-diffraction devices (Pestre 1997).

Rooted in the earlier history of chemical practice, the new methods produced by the "instrumental revolution" emerged from complex processes of interaction between experimental practice and theoretical knowledge, connecting instruments and technological practices in very complex ways. According to Carl Djerassi, the main techniques related to the transformation of organic chemistry in the second half of the twentieth century were ultraviolet and infrared spectroscopy, nuclear magnetic resonance spectroscopy, and mass spectrometry. "The principles behind these techniques were discovered by physicists," Carl Djerassi recalled, "who also developed the first instruments … but it was the chemists interested in structure elucidation, rather than in synthesis, who first applied these new physical tools to the solution of organic chemical problems" (2015: 199). These techniques not only changed research areas in academic chemistry but also revolutionized other research areas, chemical industries, medical practice, and governmental agencies.

Many technologies coming from physics entered the chemical laboratory in the mid-twentieth century. Infrared spectroscopy was transformed as one of the main laboratory tools between the 1950s and 1960s. With its roots in nineteenth-century spectroscopy, the transit into chemical laboratories was promoted by the studies of William Coblentz at the National Bureau of Standards (NBS) in Washington. He undertook the spectral identification of many organic compounds and provided experimental evidence of the link between the molecular structure and the spectral lines. He offered evidence that certain absorption frequencies remain constant for given molecular groupings, thus laying the basis for a wide variety of analytical uses of infrared spectroscopy in the first decades of the twentieth century, but it was not adopted by chemical laboratories until many decades later. It was employed in the 1930s in the oil industry due to the application to the identification and separation of octane isomers. Substantial research was performed at the American Petroleum Institute by Edward Washburn and Frank W. Rose. On the eve of World War II, infrared technologies were firmly established as an analytical tool in major chemical and oil companies, but were scarcely employed by academic chemists, who relied on traditional analysis for identifying organic compounds. The war changed the scenario under the programs for improving fuel production, petroleum refining, and the making of synthetic rubber. By the mid-1940s, infrared spectroscopy had become a standardized technique for analyzing complex organic mixtures, which was commonly used in many industrial companies and federal laboratories in the United States and Great Britain. After the war, and profiting from the previous developments and experience, new marketing strategies from the main instrument manufacturers promoted the use of infrared spectroscopy in the academic world, from Perkin–Elmer summer schools to providing large numbers of instruments free of charge (Rabkin 1987).

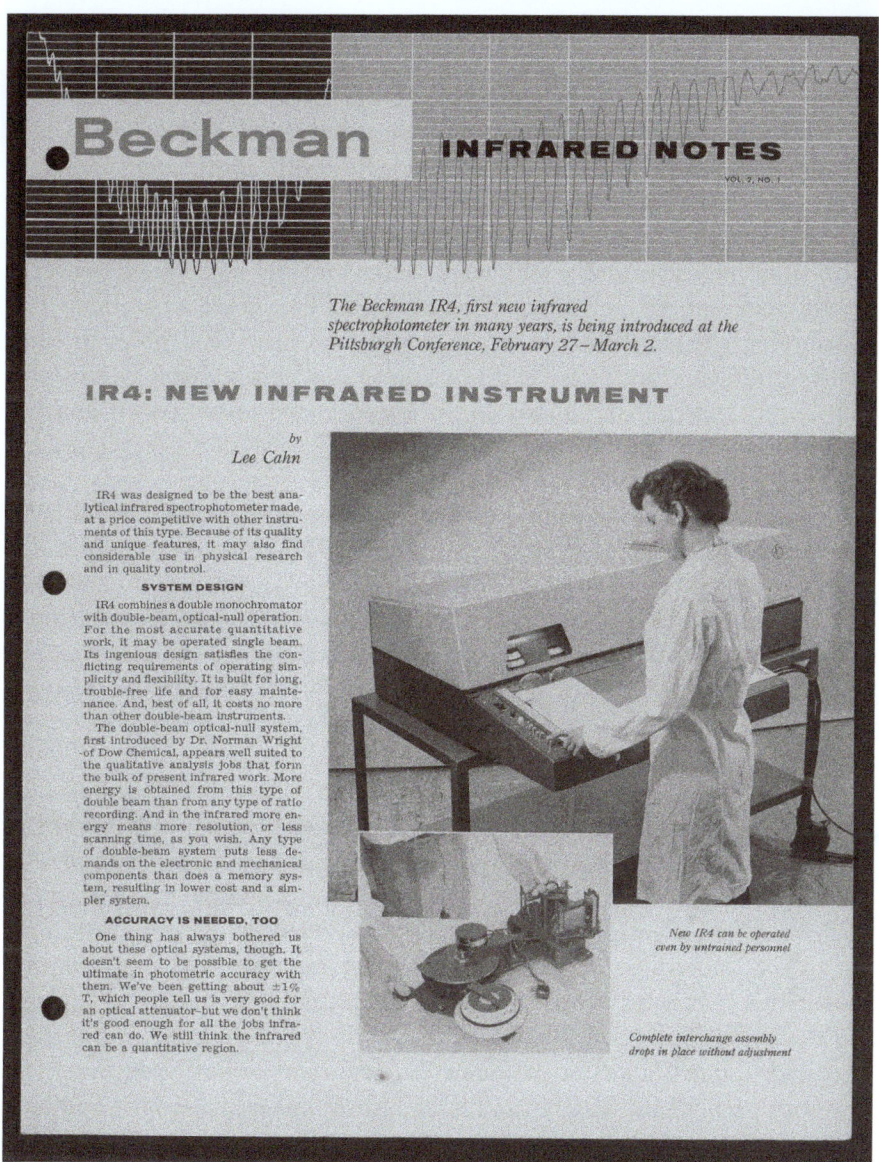

FIGURE 2.1 Beckman Instruments, "Beckman Infrared Notes: IR4: New Infrared Instrument," 1956. Beckman Historical Collection, Box 17, Folder 38. Courtesy Science History Institute, Philadelphia. https://digital.sciencehistory.org/works/8s45q923r.

In Germany, the spread of infrared spectroscopy was shaped in some respects by the anti-intellectual trends in the Nazi government, along with the relative lack of demand for these technologies in chemistry-based industries. In contrast, the fascist Italian policies helped to stimulate the research program of Giovanni Battista Bonino at the University of Bologna. With the help of infrared and Raman spectroscopy, Bonino and his colleagues studied a large number of organic aromatic compounds and applied group theory to the study the symmetry of molecules, thus producing the research style around which quantum chemistry developed in Italy during the 1940s (Karachalios 2001).

The previous example shows how the introduction and accommodation of new practices into chemical laboratories were shaped by the interplay between universities, industry, and government, along with the circumstances of war and the different national styles of research. Another group of methods that transformed chemical practice in academies, industries, and hospitals were nuclear magnetic resonance (NMR) technologies. Using technology rooted in military radar research during the World War II, Herbert Gutowsky (University of Illinois) adapted the method to the concepts and needs of organic chemistry. In the 1950s, he introduced the new NMR spectrometers, involving the experimental control of chemical effects and the connection between physical phenomena and chemical concepts. By adapting NMR to chemical research interests, Gutowsky and his research group made possible the first uses of NMR in the study of the structure of organic compounds. Thanks to the creative role of users (expanding its applications and methodologies) and the producers (the commercial practices of industries such as Varian Associates), NMR was transformed into a routine practice in organic chemistry by the 1960s (Reinhardt 2004).

Further uses were introduced thanks to the work on basic NMR methodology by the physical chemist Richard E. Ernst in close collaboration with instrument manufacturers. NMR was employed in the study of large molecules in connection with the models provided by traditional structural chemistry. In all these developments, the new technology involved the making of research groups and leaderships, new operative concepts in physical organic chemistry, cooperation of scientists with instrument builders, and new practices of teaching and textbooks (Reinhardt 2006b). While NMR became routine in chemistry laboratories, one of its direct descendants (magnetic resonance imaging, MRI) spread in medicine in the 1970s, first to identify cancerous tissues, and later becoming one of the leading diagnostic technologies. NMR spectrometers were also largely crucial in the oil industry in the 1990s for petroleum and gas exploration, providing data on rock porosity and permeability.

Electron paramagnetic resonance (EPR), also called electron spin resonance (ESR), was a similar method based on electromagnetic radiation. It was introduced after World War II thanks to the work of Yevgeny Zavoisky in the Soviet Union. First employed for studies in quantum chemistry, the new technology was subsequently used for studying metal complexes and organic radicals. In the 1980s, EPR proved to be crucial for the development of new materials and polymers because it is sensitive at the right space and timescales for delivering precious information on the structure and dynamics of polymeric systems. By the beginning of the twenty-first century, the method was employed to study a broad range of issues in biochemistry, medicine, and material science: free radicals, transition ions, semiconductors, etc. (Weil and Bolton, 2007).

Mass spectrometry was also another important technology in the instrumental revolution. Its origins can be traced to key experiments in physics which established the basis of future experimental arrangements employed in nuclear physics (for instance, measurement of isotopes and their stability). However, its first use in chemistry can be found in industry during World War II. The first attempts to connect it to chromatographic techniques were made in the following years. It started to be used in the chemical and oil industry and fully entered the organic chemistry university laboratories during the 1960s. Several authors from academic and industry departments provided the basis for the "chemistry of the instrument": Fred W. McLafferty (who worked at Dow Chemical, Purdue University, and Cornell University), Klaus Biemann (MIT), and Carl Djerassi (Stanford University). They provided key concepts for interpretation of data, particularly a consistent view of reaction mechanisms. By the 1970s, mass spectrometry was turned into one of the most sensitive analytical techniques for complex organic substances, of great interest for both academic and industrial chemistry as well as for environmental pollution monitoring agencies. Its uses expanded in subsequent years from astrochemistry to climatology, archeology, and drug abuse diagnosis (Gedeon 2006; Reinhardt 2006b).

These examples show how new instruments could be crucial for the development of new fields, not only for their potential and sometimes unexpected uses, but also for practical, organizational, and ideological reasons. For instance, it is hard to envisage the expanding field of nanotechnology without the advent of scanning tunneling microscopes during the 1980s. The new technology not only provided new methods for research but also encouraged the emergence of an "instrumental community," that is, a group of academic and industrial people committed to its production, improvement, use, and promotion (Mody 2011). The new instruments were transformed into the key technology for manufacturing molecules "from the bottom up," so they played also a symbolic role in the making of the social imaginaries of the nanotechnological world, along with the visionary work of Eric K. Drexler (Maestrutti 2011).

INTERACTION BETWEEN INSTRUMENTAL PRACTICE AND THEORY

Nanotechnology is a good example of the interplay between theory and practice in the development of new research fields. When dealing with the practices of chemistry, historians of science have focused on experimental activities in laboratories. In contrast, the innovations of theoreticians are frequently attributed to immaterial placeless "eureka moments," abstract thought, and insights of the minds of geniuses. Recent historical studies have adopted a more symmetrical approach between practice and theory, confirming that the "practices of theory" are related to embodied skills, tacit knowledge, and particular material cultures, from pen and ink to typewriters, blackboards, and computers (Galison and Warwick 1998; Kaiser 2005). In these recent studies, theoretical work is regarded less as a "cerebral, contemplative, and introspective accomplishment" and seen more as "the deployment of practical and embodied skills that have to be learned, developed, and actively communicated" (Warwick 2003: 26). Theoreticians employ many "paper tools" in making calculations, models, or theories. In many aspects, these paper tools are comparable to laboratory instruments. They can be based on hidden (or "black-boxed") intellectual assumptions, and their rules of construction and combination can largely shape their intellectual productions and representations (Klein 2001).

Needless to say, theoretical and experimental practices overlapped in the work of most chemists. In some cases, the boundaries are far from being clear. Take, for example, the study of reaction intermediates, which flourished during the first half of the twentieth century combining research on thermodynamics, kinetics,

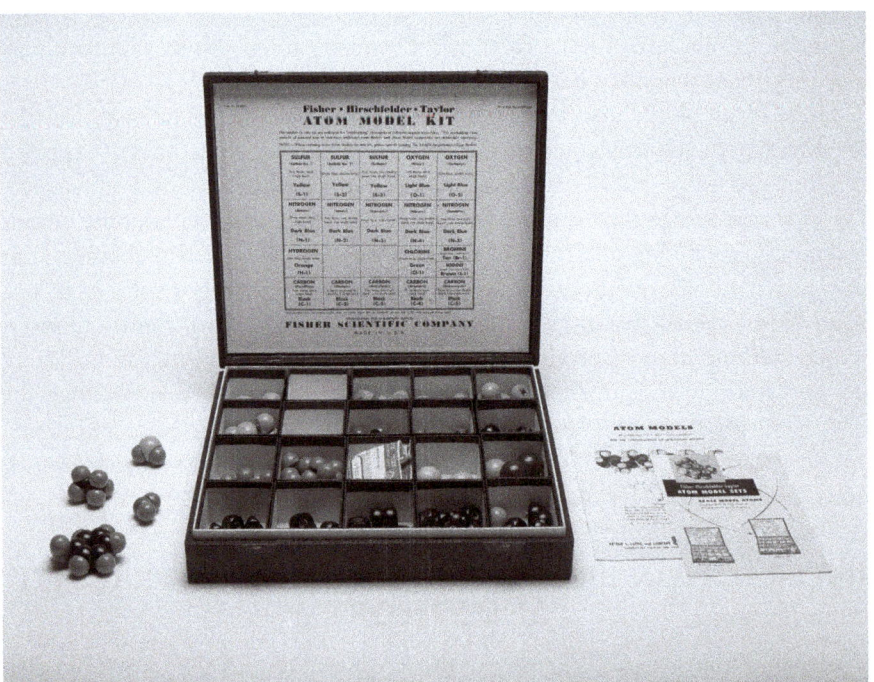

FIGURE 2.2 Fisher Hirschfelder Taylor Atomic Organic Model Kit. Photograph, 2017. Courtesy Science History Institute, Philadelphia. https://digital.sciencehistory.org/works/1c18dg352.

and quantum chemistry models. There was a range of reaction intermediates extending from those that could be easily isolated as material compounds, to those that could be only be indirectly observed. In many cases it was up to each author to decide whether the activated complexes were empirical facts or theoretical interpretations, that is, "states of matter or states of mind" (Nye 2014).

The main paper tools employed by twentieth-century chemists were the broad range of formulas introduced in inorganic and organic chemistry during the first third of nineteenth century (Klein 2003). However, the forms, meanings and uses of these formulas substantially changed. For instance, by the 1870s, the new structural formula offered a picture of the spatial arrangement of the atoms and prompted the emergence of the new field of "stereochemistry," providing not only new explanations of key concepts such as isomerism but also hints for new laboratory research in synthetic organic chemistry (Ramberg 2003). Moreover, structural formulas were employed in the making of the new chemical terminology, which became an important tool for new practices of recovering information in databases such as *Beilsteins Handbuch* and *Chemical Abstracts* (García-Belmar and Bertomeu-Sánchez 1999; Hepler-Smith 2015).

By the beginning of the twentieth century, with the advent of the electronic interpretation of chemical bonding, new paper tools and forms of representation emerged. One of the most enduring representations was the shared-electron-pair bond, introduced by Gilbert N. Lewis in the prequantum era. In classical structural chemistry, the chemical bond was represented as a simple line, a line which became the center of attention for new chemical investigations in the early twentieth century. Lewis' successful picture of the chemical bond as a shared pair of electrons was a fundamental paper tool for a new generations of mechanistic chemists, offered a reliable interpretation of organic reactions mechanisms, and eased the way for the application of quantum mechanics to the chemical bond (Kohler 1971; Kohler 1975). Atomic and bond models and representations became a fruitful space of interaction between teaching and research, experimental and theoretical investigations, and even art and science. Initially conceived as a graphic teaching device to illustrate the empirical rule of eight, Lewis' cubic atoms soon became a serious theory of atomic structure, while artists such as Kenneth Snelson imagined the atom's portraits inspired by Alfred L. Parson's magneton and cubic octet model (Heartney 2013).

Three-dimensional molecular physical models were another outstanding example of interactions between teaching and research practices, theory and empirical data, as well as technical abilities and funding availability. Linus Pauling's space-filling molecular models proved to be extremely fruitful tools for experimental and theoretical discoveries, for the renovation of science pedagogy methods, and for the creation of a new positive image of chemists as "magician of molecules" (Nye 2001). In this sense, one can say that molecular models have defined and redefined, rather than represented, molecules. As chemistry tools were transformed into objets d'art, three-dimensional molecular models were placed in popular scientific iconography under the scrutinizing gaze of their designers and users. Following their look, these chemical sculptures have helped to shape our "direct sensory experience of something that by definition can neither be touched nor seen" (Francoeur 1997: 32).

Although not so powerful in the popular imagination, numerical methods and computer calculations were also crucial to the work of chemists, particularly during the second half of the century and in the subfield of quantum chemistry (see Chapter 1 in this volume). The different institutional settings encouraged, constrained, or limited the exchanges between physical, chemical, and mathematical methods. National styles of research and the structure of chemical communities played a role. For instance, the physically oriented approaches of German physicists proved to be successful in the first decades of the century, providing most of the pioneering works in the field. By the second half of the century, these methods were superseded by the "chemically oriented" and "pragmatic" approach of American scientists such as Robert S. Mulliken and Linus Pauling in conjunction with the works of their British colleagues (Lennard-Jones, Hartree, Coulson) who enlarged the domain of applied mathematics, so as to include quantum chemistry in the decades after World War II. The advent of electronic computers in the 1960s changed the connections between calculations and experiments in the field of quantum chemistry. New semi-empirical methods, virtual experiments, and visual models emerged and allowed productive exchanges among the different research styles in the area (Gavroglu and Simões 2012).

TEACHING PRACTICES

Like theoretical practices, teaching practices have been scarcely explored by historians of twentieth-century chemistry. More historical studies are still needed to grasp how teaching and learning content and practices of a school subject such as chemistry were shaped in the twentieth century by institutional, social, and pedagogical factors, as John L. Rudolph claimed more than ten years ago (2005a: 343). For now, we have to content ourselves with the references included in general histories of science education, most of them local histories focused on the USA and UK (DeBoer 1991; Rudolph 2002), or with the reports and retrospective narrations provided by chemistry educators taking stock of the past with an eye on the future (Kornhauser 1980; Johnstone 1993; Orna 2015). Chemistry was taught in the first half of the twentieth century using places, objects, and methods largely created and consolidated during the previous century (Brock 1975; Brock 1993). We will review some of these long-term features and then introduce some new elements of twentieth-century pedagogical practices.

In the nineteenth century, the incorporation of chemistry into technical, university, and secondary education curricula posed an important pedagogical challenge. Lecture demonstrations were the most common response. They were performed by professors and their assistants in front of large audiences in auditoriums (García-Belmar 2006). With slight variations, this teaching method was followed in chemistry lecture halls of technical schools, university faculties, and secondary schools throughout Europe and America. The collections of scientific instruments preserved in still-active secondary schools created in the mid-nineteenth century are testimony to the continuity of this way of teaching chemistry well into the twentieth century (Simón and Cuenca 2012). However, the limitations of this teaching practice soon became apparent. Several ways of transforming the laboratory into a teaching space were tried throughout Europe, the most famous being the teaching laboratory set up by Justus Liebig in Giessen in the 1840s. Despite the undeniable influence that Liebig's laboratory had among his contemporaries, the coexistence of similar laboratories elsewhere suggests that this didactic innovation was a response to formative concerns shared by many other teachers in Europe and America since the beginning of the nineteenth century (Holmes 1989).

In addition to lecture demonstrations and laboratory practical instruction, chemistry textbooks were the other important nineteenth-century legacy. This new publishing genre underwent important transformations throughout the century, which affected teaching and learning practices (García-Belmar et al. 2005). New printing techniques allowed the transfer of the images from the end plates to the body of the text, and represented with ever greater detail and realism the chemical phenomena and experiments described. Moreover, stoichiometric diagrams that became pencil-and-paper analytical "instruments" offered students the possibility of interpreting and even predicting phenomena visualized through images (Bensaude-Vincent et al. 2003). Following in the wake of physics textbooks such as that of Adolphe Ganot, chemistry textbooks of the second half of the nineteenth century incorporated collections of problems and practical exercises to be solved with paper and pencil (Simon 2011).

The "heuristic method" advocated by Edward Armstrong at the end of the nineteenth century was perhaps one of the last proposals for a novel chemical pedagogy with its own adjective. Armstrong advocated the need to devise ways of teaching chemistry centered on the learner rather than the teacher. He strongly criticized the teaching of chemistry based on the memorization of data and formulas, which prepared students to answer questions in an examination and not to think and solve practical problems. He developed a teaching and learning method in which students were placed in the position of the scientist and guided in problem solving by applying the "scientific method." From his position at the City and Guilds of London Institute, an institution financed by London businessmen and aimed at the technical training of workers and artisans, Armstrong advocated a "methodological use of knowledge" in which the student was confronted with the phenomena and its properties, before the teacher gave a name and an explanation for them. The heuristic method was one factor in an exponential growth in the number of chemical laboratories installed in British schools (Brock 1973: 34–5).

Classrooms and laboratories, lectures and experiments, textbooks and notebooks remained key components of chemistry teaching practices in the twentieth century. However, along with these lines of continuity, the objects, spaces, and methods that were used to teach chemistry in the nineteenth century were reconfigured to produce new teaching and learning practices adapted to the new contexts, audiences, and interests. Chemistry teaching was aimed at two different audiences: pupils in primary and secondary schools, and future chemists in colleges and universities. Their backgrounds, interests, and expectations largely shaped different teaching practices and school disciplines related to chemistry. The popularization of chemistry in the new mass media also created its own audiences and practices.

General science programs and new forms of popularization were different ways to tackle two of the most debated issues in the second half of the twentieth century in many countries: the decline in interest in chemistry, and the degradation of its popular image. Many authors thought that some of the causes of disaffection were the methods of teaching and the curricula, in particular the descriptive and encyclopedic disciplinary approaches found in chemistry textbooks. As the influence of professional educators in science curricula increased, new pedagogical proposals emerged aiming at dissolving the rigid frontiers of academic disciplines in the school environment. For instance, at the beginning of the twentieth century a group of professional educators in Chicago launched a general science curriculum that broke with disciplinary boundaries and offered a program designed to provide new students with an appreciation of the value of science in modern society and the skills to apply scientific thinking in their daily lives. In contrast to the descriptive chemistry programs, teachers who adopted the general science curricula relied on problem-based methodologies, aiming to transmit the value of scientific problem solving in daily life (Rudolph 2005a; Rudolph 2005b).

Interest in the methods of science extended beyond the schools. Home laboratories and chemistry sets were popular during the interwar years. In the USA, companies like Gilbert and Chemcraft produced chemistry sets aimed at middle-class boys. In contrast with eighteenth- and nineteenth-century demonstrations in salons, which male and female adults and children attended together, the new experiments were conceived as individual childhood investigations without adult guidance. They promoted the emergence of chemistry clubs and the organization of "magic chemistry" shows performed by children. They created a culture for curiosity, based on experimental facts, hands-on learning, and "innocent science," which was still present in the science centers created during the 1970s, such as the Exploratorium in San Francisco, under the direction of Frank Oppenheimer (Onion 2016).

Since the late 1950s, important chemistry curriculum reforms appeared in very different national contexts. The trauma created in American society by the launch of the Soviet Sputnik satellite in the middle of the Cold War has usually been considered as the trigger for this renewed interest of governments, academics, and chemical educators in the USA. Major historical narratives describe measures adopted in the USA to confront the crisis of vocations and the insufficiency of scientific training among young people, which was regarded as one of the main sources of the perceived loss of American technological leadership. However, a more global examination reveals a very different picture: between 1960 and 1984 more than ninety new chemistry-education projects were set up in forty-five countries (Waddington 1984: 45–85). Thus, the causes of the reform of teaching practices seem to be more complex and varied than are often depicted: perceptions by key protagonists on the obsolete character of chemical curricula, increasing criticism of problem-solving methodologies, the advent of new ideas of science education, changes in university student populations, new forms of recruitment and training for teachers, and so forth, all played a role. Perhaps the main difference regarding the case of the United States was the enormous amount of money invested in funding chemistry-education projects, and the international circulation of the didactic materials produced (Waddington and Heikkinen 2015).

Under the auspices of the National Science Foundation, some of the most influential chemistry education projects of the second half of the twentieth century emerged in the United States. Two well-known examples are the Chemical Bond Approach (CBA) Project, and the Chemical Education Material (CHEM) Study Project. On the other side of the Atlantic, the Nuffield Foundation funded similar school chemistry projects for the production of O-level and A-level chemistry materials in the UK. Apart from the differences between the curriculum projects on either of the Atlantic, they all shared a similar conviction: interpretations should emerge and be tested by observations, so chemistry instruction must be laboratory-orientated and evidence-based, intended to present chemistry as systems of enquiry rather than stable bodies of knowledge (Fast 1963). Focus was shifted from studying substances and properties, to investigating reactions and processes, in this way confronting students with issues governing chemical changes. The new didactic strategy also involved substantial changes in the role of teachers, who abandoned their stance of being the "sage on the stage," instead becoming guides alongside students. Evaluation practices also underwent significant changes. Especially significant was the possibility of using textbooks in the examinations, which were designed to evaluate the student's abilities in the resolution of problems with the help of information sources, rather than memorizing the data collected in them (Orna 2015).

Promoters of these programs took into account the importance of international diffusion of these ideas, and they designed a strategy for translations and adaptations to each national context, as well as the training of a new generation of chemical educators. CHEM Study materials were one of the most successful in the global educational market. The international circulation of these programs in developing countries was possible thanks to the support given by UNESCO to the translation, adaptation, and training of chemistry educators in several Asian and African countries (Kornhauser et al. 1980; Simon 2019).

In the 1980s a new strategy was attempted to stimulate student interest by connecting chemistry and society. The authors of the new context-based projects tried to expand high school chemistry topics beyond conventional general chemistry curricula in order to cover social problems related to environmental chemistry, nuclear energy, or biochemistry. They wanted to encourage the students' interest while promoting new practices of teaching and learning. Influential examples are *Chemistry in the Community* (ChemCom) or *Chemistry in Context* (CiC), from the USA, and *Salters Chemistry*, from the UK (King 2012).

The perspective of two centuries of chemistry teaching reveals the survival of some basic tensions still fueling the debate around the role of chemistry, in both the training of scientists and the instruction of citizens. The relationship between the contours, contents, and methods of academic disciplines and school disciplines; the distinction between training for future scientists versus training for future citizens, with a growing relevance of the latter; and the role of laboratory and classrooms as teaching and learning places, are some of them. Attracting the interest of the students by bringing chemistry teaching to the needs of daily life is another constant regarding pedagogical projects of the twentieth century. The interests, the needs, and the aspects of chemistry that could serve as a bridge are questions to which very different answers were given, and which remain largely open still today.

INVISIBLE PRACTICES

The last group of practices related to chemistry discussed here are the most invisible ones. They are employed in everyday and routine practices, not only in academic laboratories but even more in governmental offices, quality control departments, municipal and customs laboratories. Many of them are intended to create uniformities across time, space, and cultures. They are employed to regulate both academic research and social life: processes, instrument designs, terminology, measurements, quality control, risks assessment, etc. The making, circulation, and application of standards is connected to particular groups and practices, and are frequently shaped by practices of accommodation, change, or resistance (Timmermans and Epstein 2010; Schaffer 2015).

One example of such standardization is the metric system. It was devised at the end of the eighteenth century and developed during the nineteenth century, and has culminated in the establishment of the International System of Units, which was adopted by the most of the international scientific community during the second half of the twentieth century. Although this history is reasonably well studied, there are many less well-explored related topics regarding how the metric system was employed in chemistry and other fields connected to production and consumption, in which alternative metrological systems coexisted. The analysis of errors was another important area that helped shaped the development of the practice of chemistry in the twentieth century, both in research laboratories and classrooms (Olesko 1991).

The development of chemical terminology and nomenclatures standardized language practices in chemistry, including the important activities of information gathering and recovering, which expanded in the twentieth century with the emergence of large chemical information systems, such as the new versions of *Beilsteins Handbuch* and the *Chemical Abstracts*. The editors of these projects significantly influenced the development of chemical terminology in the twentieth century, sometimes suggesting new conventions or rules which were adopted by the International Union of Pure and Applied Chemistry (IUPAC), and sometimes developing their own system (such as the Chemical Abstracts number) tailored for the purposes of that particular information systems (Verkade 1985; Hepler-Smith 2015). The use of national and global languages in meetings and publications led to the transition from the multilingual world of nineteenth-century chemistry to a kind of imperialist predominance of English at the end of the twentieth century, involving different communicative practices and works of translation according to the linguistic area (Gordin 2015). Like other standardized practices, new chemical terms coexisted with older chemical names (from alchemy to the late-eighteenth-century reform), terms coined by other academic fields (such as pharmacy or medicine), and popular and commercial expressions for chemical products (García-Belmar and Bertomeu-Sánchez 1999).

Standards based on science also circulated outside laboratories and academies. During the twentieth century, they became the basis of many regulations in food quality, pharmaceuticals, occupational health, and environmental law. Many of these standards were largely based on chemistry. One example is nutritional and dietary standards. They were conceived in terms of quantity of biochemical nutrients required in the diet during the era of quantifying the science of nutrition, which started in the mid-nineteenth century due to the work of Justus Liebig and others. The methods of determining these requirements were constantly negotiated and changing, as well as the standards themselves, and

FIGURE 2.3 Perkin–Elmer Atomic Absorption Spectrophotometer at an Industrial Hygiene Laboratory in USA, 1970–1979. It was employed to test the lead content of workers' urine. It was part of comprehensive program to ensure compliance with Occupational Safety and Health Administration regulations. Photograph from the Perkin–Elmer–Applera Collection, Box 4. Courtesy Science History Institute, Philadelphia. https://digital.sciencehistory.org/works/9g54xj45n.

eventually the quantitative reductionist approach was contested in the second half of the twentieth century (Neswald et al. 2017).

Nutritional standards were employed in food labeling, which was introduced by manufacturers as a form of publicity in the nineteenth century. New standardized forms of labeling were promoted by the growth of packaged foods whose content and quality could not be easily assessed by consumers. Food labels turned out to be a conflict-laden area of encounter between chemists and nutritionists, governmental offices, consumer associations, and the food industry. These conflicts affected the amount and nature of the information to be included, the different aims and targeted audiences, and the prescriptive character of the regulations. The recent "informational turn" in food labeling has introduced a new scenario. Changes in labels produced important transformations in practices of production, distribution, and consumption related to different chemicals, from food additives to pesticides and pharmaceuticals (Frohlich 2017).

Standardization of drugs also involved a large number of new practices of drug quality control in the pharmaceutical industry, hospitals, and government laboratories. Different ways of regulating drugs (professional, administrative, industrial, public, juridical) coexisted during the twentieth century, mobilizing different values, purposes, regulatory tools, decision-makers, and forms of evidence, from chemical analysis and animal experiments to different types of statistically controlled clinical trials (Gaudillière and Hess 2013). These different forms of evidence attained different uses during the century, sometimes under the pressure of public controversies surrounding animal or even human experimentation (for instance, the infamous Tuskegee syphilis experiment), or scandals such as the thalidomide crisis (Jones 1993; Stephens and Brynner 2001; Guerrini 2003).

The multiplicity of interests, actors, methods, and administrative tools can also be described in the different ways of regulating toxic chemical substances during the twentieth century. Many of the early regulations that were established at the national level concerned the circulation of toxic substances or workplace hazards. One of the first attempts at an international agreement was developed in 1919 by the International Labour Organization (ILO) regarding lead poisoning for women and children working in the chemical industry. During the following years, the ILO also established recommendations concerning white phosphorus and white lead (Lönngren 1992). Further regulations introduced new threshold limit values (TLV) for hundreds of chemicals, the figures for which were in constant revision and negotiation, frequently fueling conflicts among victims, polluters, and governments (Bruno et al. 2011; Sellers and Melling 2012).

Assuming a threshold below which the substance was unlikely to affect human health, limit values were adopted as regulatory concepts with the necessary ambiguity for connecting different social spheres and for fulfilling specific functions in each social context (Reinhardt 2012). Thanks to these features, the approach was adopted in the second half of the century and encouraged new laboratory research, tests, methods, and regulations. At the turn of the twenty-first century, it was increasingly challenged by the advent of new products and problems (for instance, endocrine-disrupting chemicals), which could hardly be handled under this framework (Langston 2010; Vogel 2013; Boudia and Jas 2014). Moreover, white lead, asbestos, and many other products show the limits of the implementation of these regulations based on threshold limit values for chemical in workplaces and homes. By combining practices of concealment, deceit, and denial, industry managers could circumvent or minimize the regulations for decades in spite of the growing concerns from labor unions, public health doctors, and governmental agencies (Markowitz and Rosner 2002; Oreskes and Conway 2010; Markowitz and Rosner 2013).

During the second half of the twentieth century, the "chemical intensification" of world economies was not followed by international agreements such as those relating to radioactive products. The proliferation of regulations did not assure the control of toxic risks, but the making of "structural production of ignorance on the chemicals and their hazards, along with the ill-protection of human populations and the environment" (Jas 2014). The attempts to control harmful effects of chemicals is linked to scientific research on chemical toxicity: analytical tests for detection, animal and in vitro experiments, and epidemiology. With their roots in nineteenth-century legal medicine and toxicology, many of the new assays emerged when facing specific problems and dramatic accidents, but were applied to a large number of new substances. For instance, the elixir sulfanilamide tragedy in the USA in 1937 set the stage for the development of one of the most important twentieth-century frameworks regarding toxicity: the median lethal dose for 50 percent of the tested population (LD_{50}), which was adopted by the USA's Food and Drug Administration and in many other national and international regulations during the second half of the twentieth century. Moreover, the analysis of dose–response curves involved the use of a large number of statistical methods, chemical tests, and animal experiments which were introduced in the everyday activity of governmental laboratories (Davies 2014).

Like the Wasserman test for syphilis studied by Ludwik Fleck, the practice of these new tests involved the use of embodied skills, mathematical analysis, and a certain "chemical touch," in many cases connected to specialized groups (Fleck 1979). Unexpected results and puzzling problems emerged. The balance of false positives and false negatives was frequently a source of controversy among the stakeholders, before the test became an everyday practice in customs, municipal laboratories, or regulatory agencies. How new results were connected to regulation and decision-making is another relevant issue. New experimental practices could provide decision-makers with plausible justifications for taking authoritative action. In other cases, however, an expert report can be rejected or made invisible, or questioned by new research. The relative value of the different types of evidence was also an issue fraught with controversy. For instance, the new short-term tests for mutagens were contrasted in the 1970s with the long-term bioassays and epidemiology in assessing risks for cancer. In different ways, the new tests were connected with different classificatory practices regarding carcinogenesis or the estimation of safe thresholds and the subsequent regulations at the international or national level (Brickman et al. 1985).

Many groups and organizations contributed to making and reshaping new chemical tests on food quality, pharmaceutical effects, and toxic risks. International organizations related to health, industry, and commerce played an important role. When the International Program of Chemical Security became operational in the early 1980s under the aegis of the United Nations, one of the main purposes was to provide guidelines on "methodology for exposure measurement, risk assessment, toxicity and epidemiological studies." Several groups worked on standardized tests for predicting mutagenic and carcinogenic potential, methods for evaluating neurotoxicity or behavioral toxicity, and the effects of chemicals on reproduction. The OECD published several books on test guidelines and "principles of good laboratory practice," along with recommendations concerning from data confidentiality to new systematic research on existing chemicals (Lönngren 1992). Moreover, activists and

FIGURE 2.4 Johnny Horizon Environmental Test Kit, 1971. Popular chemical test for air and water pollution. It included a magnifying glass, eyedropper, test tubes, tape, measuring cup, filters, and paper packets of additives. Photograph, 2016. Courtesy Science History Institute, Philadelphia. https://digital.sciencehistory.org/works/79407x355.

citizenships groups also developed alternative practices for collecting information concerning cancer epidemiology related to chemicals or employing new simple tests for gathering information regarding pollution and contamination in homes, workplaces, or the environment. These data were frequently contested by other forms of proof, particularly from experts connected to governments, industries, and other stakeholders (Brown 1992; Allen 2003).

As in other practices reviewed in the chapter, old tests and methods, coming from both inside and outside the chemical community, coexisted with new approaches, instruments, and methodologies. For instance, in the field of milk quality control, nineteenth-century methods (hydrometers, volumetric analysis, butyrometers) were combined with different paper tools (mathematical analysis) and fast commercial methods (in which speed rather than precision was the key), along with traditional testing based on the senses (color, flavor). By the middle of the twentieth century, new tests included polarimetric apparatus, different quantitative methods for nitrogen analysis, lactomers and conductivity measurements, freezing points, and centrifugal and sediment analysis. They were employed by a broad range of purposes in governmental laboratories for food control, courtrooms dealing with litigations related to milk adulteration, and also in making national and international regulations. The advent of the "instrumental revolution" in the 1960s took place in this complex scenario along with a displacement of the focus of testing interests from fat and proteins to vitamins, enzymes, and other micronutrients (Atkins 2010).

A final example of the coexistence of old and new practices in twentieth-century chemistry is offered by the persistence of qualitative group analysis for inorganic substances. Its main features were established during the second half of the nineteenth century due to the work of Heinrich Rose, Carl Remigius Fresenius, and many other chemists (Szabadvary 1966). Analytical methods involved the use of test-tube glassware, standardized color reagents, and considerable tacit knowledge concerning nuances of colors and flavors (Homburg 1999; Bertomeu-Sánchez 2013; Jackson 2015). Group analysis involved the use of the dangerous and distasteful hydrogen sulfide, obtained from Kipp's apparatus and employed in the first phases of the separation as a result of the low solubility of most metallic sulfides. As result of the "instrumental revolution," group analysis was progressively abandoned in research laboratories after World War II, but it remained as a common practice in teaching laboratories until the end of the twentieth century (Baird 1993).

Of course, when a global perspective is adopted, many local changes and accommodations can be found in these general trends. For instance, toxic hydrogen sulfide gas was replaced by carbonates in many Spanish universities, thanks to the works of Siro Arribas Jimeno, an influential analytical chemist with good connections to the Spanish authoritarian regime. Using similar traditional analytical techniques, two American connoisseurs of Italian coffee cups (Don and Fran Wallace) developed in the late 1980s a test kit for detecting lead in ceramic glazes and distributed it worldwide, attracting public attention to the problem "in a way that armies of learned experts could not" (Wedeen 1993). Meanwhile, new cutting-edge technologies (such as x-ray fluorescence measurements in bones) remained contested and faced serious resistance to employment in regulatory policies concerning toxic risks.

CONCLUSIONS

In this chapter, we examined experimental and theoretical practices in twentieth-century chemistry, taking into account a broad range of issues, from academic research productivity and pedagogical efforts to the economic impact of chemistry on world economies and its cultural impact in mass media and social imaginaries. Our examples confirm that many twentieth-century chemical practices were highly relevant for making and spreading new concepts, models, measurements, curricula, standards, and regulations. They also proved to be important in the making of new disciplines, university–industry nexuses, and research groups. In other words, these practices were crucial to "to gain knowledge of the natural world by transforming it, along with the society in which they were embedded" (Reinhardt 2018).

In this changing and intertwined context, assessing the relative importance of experimental and theoretical practices is an elusive task for historians. An experimental technique (volumetric analysis), an instrument (scanning tunneling microscopes), a model (G.N. Lewis' molecules), or a mathematical method (the Hartree–Fock approximation), any of these can be regarded as crucial in terms of discipline building, teaching innovations, or research programs in certain academic areas. However, they could be completely irrelevant in other subdisciplines and might hardly affect regular practices of chemistry in industry or agriculture. By contrast, other technologies (such as gas chromatographs, pH meters, or mass spectrometers) could easily travel from industrial to academic laboratories and back. Moreover, they could be also transformed into classroom demonstrations or adapted for the purposes of regulatory practices or quality control in customs laboratories – or the other way around. New teaching practices encouraged by large pedagogical projects such as Nuffield or CHEM Study shaped the image of chemistry for thousands of students from Portugal to Japan. And yet, the significance of the new instruments and practices could be very different depending on the geographical contexts. In their creative circulations, techniques and practices were enlarged, transformed, or neglected in certain environments, while potentially contributing to substantial changes in the social, cultural, and-natural worlds in which they were introduced.

With these issues in mind, the pursuit of a general periodization is nugatory, particularly when narratives focused on inventions and novelties are replaced by the history of circulations and social uses. From this perspective, social processes of accommodation, resistance, and persistence are brought to the fore and, when taking into account different contexts, the resulting picture is complex. To be sure, the advent of new physical methods (the so-called "instrumental revolution") introduced substantial changes in chemical practices and can be regarded as a milestone in the general development of twentieth-century chemistry. However, many traditional chemical techniques (from blowing glass to the uses of the senses) also largely endured in the everyday life of regular chemists. The productive role of old and new technologies also varied according to the different research areas, educational environments, or regulatory practices throughout the twentieth century.

New versions of old chemical tests (such as those developed by the Wallaces) coexisted in time with cutting-edge research technologies and new practices in fields such as nanotechnology, high-pressure chemistry, and ultra-fast reactions, which emerged at the end of the twentieth century and expanded into the twenty-first century. These old and new practices evolved and interacted in different academic settings and social spaces along with the dramatic disciplinary changes in chemistry and related areas. These changes have been so dramatic that historians have wondered whether "what chemists do" at the beginning of the twenty-first century was "actually chemistry insofar as it would have been recognized by chemists living 50 or 100 years ago as being chemistry" (Morris 2008). Dealing with these issues will be a challenging task for historians in the future.

CHAPTER THREE

Laboratories and Technology: *An Era of Transformations*

PETER J.T. MORRIS

FROM STASIS TO TRANSFORMATION

For three decades after 1914, the material culture of chemistry altered comparatively little. A chemist trained in the late nineteenth century would still have found the typical academic (or industrial) laboratory and the instruments it contained very familiar. However, between 1945 and 1965 the chemical laboratory, the instruments it contained, and the overall practice of chemistry were all transformed. This transformation was a consequence of developments within physics, which were then transferred into chemistry by physicists and chemists, some of whom were working in industry or for instrument manufacturers, resulting in the fashioning of completely new chemical techniques. While the pace of change then slackened somewhat, from the 1990s onwards the laboratory itself was divided into a set of sharply defined working areas. This chapter will concentrate on this "instrumental revolution" and the later transformation of the laboratory, while not neglecting the period before 1945. This account not only describes the introduction of new techniques and their impact on chemical practice, but also the changing entrepreneurial practices which brought new firms into the field of instrument-making. Novel instruments and different kinds of manufacturers also meant new networks of influence and information exchange between chemists and manufacturers, such that the relationship between the chemist and the instrument maker was fundamentally altered. These changes took place at a time when the United States was both scientifically and economically dominant, especially in the nascent field of electronics that underpinned this new instrumentation. It is thus not surprising that the instrumental revolution also assisted the rapid rise of the United States as the global leader in chemistry, replacing the war-battered countries of Europe, especially Germany.

WHERE CHEMISTRY TAKES PLACE

The first two chapters in this volume have looked at what chemists do, namely the new theories they developed and their experimental achievements. Developments in theory and experimental practice clearly changed the culture of chemistry during the twentieth century. Yet surely the places where chemistry is carried out must also have a strong influence on its culture: chemistry is shaped by its surroundings and apparatus in addition to what chemists think and what chemists do. If chemistry is the science that creates its own object (according to Marcellin Berthelot's famous phrase), the places where chemistry happens also help to create chemists. Conversely, the prevailing culture of chemistry must surely influence how its sites and apparatus are created and developed. Thus, exploring the laboratories where chemistry was done and the apparatus used by chemists can be an important way to trace how the culture of chemistry has changed during the century after 1914.

Chemistry has always been carried out in many places, even in domestic kitchens and garden sheds. Nonetheless, by 1914 the laboratory had become both the most likely place and certainly the most iconic. As its Latin root indicates, a laboratory is simply another name for a workshop. However during the nineteenth century the laboratory had left its humble origins behind and had become a highly specialized room with a typical layout and specific furnishings (Morris 2015). By the late nineteenth century, laboratories were rarely freestanding, but were gathered together – at least in academia – with other spaces such as lecture rooms, libraries, museums, and storerooms, to create "chemical palaces." The standard format for a laboratory at the beginning of our period was two or more rows of laboratory benches with washbasins at their ends and surmounted by bottle racks. Round the edge of the room, often at the ends, there would be several fume cupboards supplied with air extractors that were used for particularly dangerous experiments. Because many experiments were carried out on the open benches, including those which would now be considered hazardous (for example, the use of highly toxic hydrogen sulfide), the whole room had a ventilation system that would regularly change the air in the room. From the 1850s onwards, most laboratory benches were supplied with town gas, water, and sometimes even steam. The availability of these utilities greatly increased the capabilities of the worker at the bench, who was now able to heat a reaction mixture with a Bunsen burner, or produce a modest vacuum to assist filtration using a water aspirator.

By 1914, electricity was beginning to make its mark in the laboratory. Current from a central battery had been available since the 1850s, but it was the advent of AC electric power (mains electricity) from the 1890s onward which eventually transformed the laboratory. However, the adoption of mains electricity was slow, and for many years the slide projector in the lecture hall was the heaviest user of electricity; the University of Cambridge still had gas-lighting in its laboratories in the early 1940s. Laboratories were most conspicuous in university chemistry departments, but they also existed in industry, most prominently in the chemical and pharmaceutical industries, but also in the food industry, the steel industry, and the railways (Hudson and Russell 2012). Furthermore, most schools had laboratories by 1914 (Brock 2017). There were 1,165 school laboratories in England recognized by the new Board of Education in 1903, largely thanks to the efforts of the photographic chemist William de W. Abney at the Science and Art Department, which had just become part of the Board of Education (Morris 2004).

CHEMICAL PRACTICE IN THE INTERWAR PERIOD

The culture of experimental chemistry in the years between 1914 and 1945 remained largely the same as it had been at the end of the nineteenth century. In the standard chemical laboratory there were hardly any new instruments, although some of the existing ones – notably balances – became more accurate. Even electrically powered pH meters and melting-point apparatus were rarely present in academic laboratories until after World War II.

In the absence of physical instrumentation, there were some attempts made to use physical properties to work out the structure of organic compounds. One was the "parachor" introduced by the English chemist Samuel Sugden in 1924 (Sugden 1924). This unit was determined by measuring the surface tension, viscosity, and molecular weight of the compound, all of which then were combined using a complicated equation to generate the final figure. The utility of this otherwise meaningless number lay in the fact that Sugden argued it could also be calculated by adding up the constants calculated on the basis of other parachors to specific atoms or chemical groups. Hence a parachor could – at least in theory – be used to find out the structure of a compound. The key difficulty was that the sum produced by adding up the values for equally feasible alternative combinations of the different groups yielded a similar parachor for the whole compound. Not surprisingly, the use of the parachor rapidly declined when more reliable methods for determining the structure of organic compounds arrived in the laboratory after 1940. The young American chemist Robert Burns Woodward used then-novel ultraviolet spectroscopy in 1941 to work out the structure of certain intensively studied steroids, employing the so-called Woodward rules, which in effect were another system of parameters added together (Benfey and Morris 2001: 43–56; Slater 2002). These efforts demonstrate the sheer desperation of chemists to find quicker and easier methods of working out the molecular structure of organic compounds, a task which was taking years or sometimes even decades using purely chemical methods.

There was, however, one physical method in this period which could be used to determine the structure of both inorganic and organic compounds, and it led to one of the most striking instrumental innovations affecting chemistry after 1914. X-ray crystallography was first used after its discovery by the German physicist Max von Laue in 1912 to study the arrangements of ions in crystal structures (Ewald 1962; Authier 2015). The technique uses the scattering of the x-rays (called diffraction) to work out the distances between the atoms and their arrangements within this crystal structure (Glazer 2016). Historically, the scattered x-rays were captured on photographic film, like a medical x-ray, although electronic detectors are now usually employed. The phenomenon relies on the coincidental fact that the spacing of crystal lattices is similar in size to the wavelength of x-rays, namely around one-billionth of a meter (a nanometer).

While it was very useful to know the crystal structures of inorganic compounds and minerals, the real prize lay in using this new technique to determine the structure of complex organic compounds. This was an inherently different process from inorganic crystals that do not have a molecular structure, but rather consist of an assembly of ions (and sometimes small molecules such as water) in a structural array. It was a leap of faith that the method might work with the molecular structure of large molecules (Bragg 1975). Nor was the process at all easy. Many organic compounds do not form well-defined crystals, at least not at temperatures that were convenient for the x-ray crystallographer. However, chemists had spent nearly a century making well-defined crystals by converting organic compounds into crystalline derivatives in order to purify them. X-ray crystallographers could use similar methods to obtain suitable crystals.

Lawrence Bragg, at Cambridge and the Royal Institution in London – where his father William Bragg was also based – made the first x-ray determination of the structures of naphthalene and anthracene in 1921 (Hunter 2004; Jenkin 2008; Glazer and Thomson 2016). This was followed by the determination of the structure of hexamethylbenzene by the Irish x-ray crystallographer Kathleen Lonsdale in 1928 (Wilson 2015). One of the first complex organic compounds to have its structure determined in this way was the yeast compound ergosterol in 1932 by another Irish x-ray crystallographer, John Desmond ("Sage") Bernal, who had moved to Cambridge from the Royal Institution (Brown 2005).

However, it was not easy to analyze the patterns generated by the diffraction of these organic compounds. The x-ray diffraction pattern reflects the density of the electrons around atoms rather than the positions of the atoms themselves. The hydrogen atom has only one electron, and hence its electron density is low, so it was difficult to spot the positions of the hydrogen atoms in the data. This problem was partly solved by the Scottish x-ray crystallographer John M. Robertson while he was based at the Royal Institution in the 1930s. By chance he was carrying out an x-ray determination of a new type of dye invented by Scottish Dyes Ltd. (part of ICI), the phthalocyanines. These dyes contained a large organic compound which surrounds a central metal atom (just as chlorophyll contains magnesium at its center). Realizing that the presence of such a heavy atom simplified the analysis of the diffraction pattern, he sought to generalize this method. However, most compounds do not naturally contain such an atom. Robertson then had the idea of inserting a suitable atom into these compounds, which did not affect their overall crystal structure (Law 1973). With the help of Robertson's techniques, along with contemporaneous improvements in calculation methods, Dorothy Crowfoot (later Hodgkin) at Oxford used x-ray crystallography in the 1940s and 1950s to deduce the structures of penicillin and Vitamin B_{12}, both of which had been problematic for organic chemists (Ferry 1998).

Nevertheless, x-ray crystallography could not provide a complete solution for the longstanding problem of determining the structure of organic compounds. Not all organic compounds can form suitable crystals, or are able to accept a heavy atom. X-ray crystallography involved time-consuming calculations in the days before computers and could not give rapid results. Various mathematical techniques and physical aids were developed to assist with the interpretation of the initial data. Among these were the Patterson function developed by the x-ray crystallographer Arthur Patterson in 1935, and paper strips used in the calculations, especially the Lipson–Beevers strips introduced by Arnold Beevers and Henry Lipson in 1936 (Bernal 1968). Modern x-ray crystallography is underpinned by so-called "direct methods," for which the American x-ray crystallographers Jerome Karle and Herbert Hauptman were awarded the Nobel Prize in 1985 (Glazer 2016). The introduction of computers in the 1950s and 1960s transformed x-ray crystallography, beginning with the structure of Vitamin B_{12} (Frenz 1988). This breakthrough was then followed by the determination of the structure of the protein myoglobin by John Kendrew at Cambridge in 1958, hemoglobin by Max Perutz also at Cambridge in 1959, and the enzyme lysozyme by David Phillips at the Royal Institution in 1965 (Olby 1985;

FIGURE 3.1 Vitamin B_{12} crystal structure model, 1957–1959. This crystal structure model, made for the x-ray crystallographer Dorothy Crowfoot Hodgkin, shows the structure of Vitamin B_{12}. It was displayed at the Brussels Universal Exhibition in 1958. Photograph by Science and Society Picture Library/Getty Images.

Johnson 1998c; Perutz 2007). Yet, as we will see, it required another wave of advances brought about by electronics that revolutionized structure determinations in the chemistry laboratory.

Much of the early x-ray crystallographic apparatus up to and beyond World War II was made in the laboratory: Kathleen Lonsdale and her group even used empty National Dried Milk tins as makeshift cameras. However, thanks to the demand for medical x-ray apparatus from hospitals, the large electrical machinery firms, notably Philips, Siemens, and US General Electric, also moved into the manufacture of x-ray crystallographic apparatus by the 1930s. The German company STOE (which unusually came from the optical crystallography field) and the Dutch firm of Enraf-Nonius (absorbed by Bruker in 2001) had entered the sector by the 1960s.

THE UNCHANGING LABORATORY

There is little to say about the development of chemical laboratories in this period, as they changed remarkably little between the late nineteenth century and the 1950s. A good illustration of the slow rate of change is the adjacent organic chemistry and physical chemistry laboratories in South Parks Road, Oxford (Morris 2015). The organic chemistry laboratory building was constructed for the new Waynflete Professor of Chemistry, William Henry Perkin, Jr. The son of the inventor of mauve dye, Perkin valued chemical research and had already erected a new laboratory building in Manchester in collaboration with the famous laboratory architect Alfred Waterhouse (Morrell 1993). The new building was funded by Charles William Dyson Perrins, whose family had made their fortune from Lea & Perrins Worcester Sauce. The Dyson Perrins building (always known as the DP) was designed in a Queen Anne style, typical of the period, by Alfred Waterhouse's son, Paul. The internal arrangement was the then-standard design of wooden benches and bottle racks, along with wooden floors and fume cupboards.

Two decades later the automobile magnate Lord Nuffield agreed to fund a new physical chemistry laboratory to replace the joint Balliol–Trinity college laboratory, where Cyril Hinshelwood was carrying out world-class research in chemical kinetics under rather primitive conditions. Despite Nuffield's generosity, the laboratory was not named after him, unlike the social sciences college he funded at the same time. When it opened in 1941, twenty-five years after the DP, the rather austere neoclassical exterior was very different, but the internal fittings were similar. The main dissimilarities were a result of the differences between the two subdisciplines, rather than the passage of time. For example, the benchtops lacked bottle racks, in order to leave more room for the bulky glassware and equipment used in physical chemistry. The workshops were also an important feature of the Physical Chemistry Laboratory (usually called the PCL until 1994, when it merged with Theoretical Chemistry, to become the PTCL). The PCL also obtained the first mass spectrometer made by the British firm of Metropolitan-Vickers in 1949. This huge instrument, which ran on vacuum tubes, was temperamental, partly because the PCL chemists did not realize that vacuum tubes work best when they are left on all the time (Danby 1991). Yet for all its problems, this beast was the forerunner of a transformation that was about to take place in chemistry.

THE INSTRUMENTAL REVOLUTION OF THE 1950S AND 1960S

After World War II and especially during the 1950s there was a radical shift in the culture of experimental chemistry (Baird 1993; Morris 2002a; Reinhardt 2006b). This was part of the wider impact of the development of electronics, which touched the lives of ordinary people in the form of television and transistor radios. To be sure, chemists remained chemists, and they continued to do much the same bench chemistry as before. However, the introduction of electronic instruments employing techniques hitherto little used by chemists changed the relationship between them and their equipment in the material realm, as well as the relationship between chemists, instrument manufacturers, and physicists in the human world. The term "instrumental revolution" was first used in 1962 in the context of chemical analysis (Bottle 1962: 82). However, D. Stanley Tarbell and Ann Tracy Tarbell (1986) were the first historians to discuss the instrumental revolution in organic chemistry in detail under that term.

The instrumental revolution spawned many new instruments and techniques, and it is a matter of debate exactly which techniques can be said to belong to it (Baird 1993; de Galan 2003). For example, atomic absorption spectroscopy using hot flames to detect tiny concentrations of the chemical elements can be traced back to the work of Robert Bunsen and Gustav Kirchhoff in the 1860s (McGucken 1969), but it only took off in the 1950s when new techniques and new instrumentation were brought into the field (de Galan 2012). In this chapter I will focus on five techniques that had a major impact across the whole of chemistry, and on organic chemistry in particular. These were ultraviolet and infrared spectroscopy, which use electromagnetic radiation (invisible light) in the regions on either side of the visible light spectrum; nuclear magnetic resonance spectroscopy, which is based on the behavior of atomic nuclei; mass spectrometry, which measures the masses of pieces of molecules; and chromatography, which separates pure compounds from mixtures.

This new instrumentation practically destroyed an entire specialty within organic chemistry, namely the determination of the structure of organic compounds by chemical means (Morris and Travis 1997). While this shift in chemical practice greatly reduced the time needed to determine chemical structures, it also completely changed the culture of organic chemistry (Slater 2001). Hitherto, organic chemistry had been based on the two pillars of analysis (structure determination by chemical means) and synthesis (the creation of compounds by chemical means). The two sides reinforced each other.

Degradative analysis yielded intermediate compounds that could be useful stepping stones in the synthesis, and alerted chemists to possible complications when the synthesis was attempted, such as a rearrangement in a compound's structure. As the analysis of a compound could not determine the structure beyond doubt because of these potential rearrangements, the subsequent synthesis was necessary in order to confirm a compound's structure. Thanks to the new physical instrumentation, chemical analysis was now redundant and chemists had to find a new rationale for organic synthesis. Furthermore, the analytical side had thrown up unexpected reactions and new compounds that became part of the knowledge base of organic chemistry. This fount of new chemical discoveries arising from this degradative analysis dried up as a result of the instrumental revolution.

Even the very meaning of an organic compound's structure was transformed as a result of this shift. Up to the 1950s, a molecular structure was a coded registry of its known chemistry, even if the determination was as yet incomplete. Each bond and spatial arrangement of the atoms in a structure represented a step in its chemical analysis. Once structure determination became a physical process, the structure now became simply the symbolic representation of physical data, not chemical knowledge. To be sure, in keeping with the earlier technique of x-ray crystallography, the physical data could not automatically and unambiguously be transformed into a structure. It had to be converted into such a structure by a chemist, and these structure proposals – especially in the early days of the new instrumentation – were not always infallible.

SHEDDING LIGHT ON CHEMICAL STRUCTURES

Ultraviolet spectroscopy examines the light with a shorter wavelength (i.e. higher frequency and energy) than visible light. By contrast, infrared spectroscopy examines the light that has a longer wavelength (hence is less energetic) than visible light. An atom or a compound can either emit light or absorb light. Whereas the older visible spectroscopy employs both absorption and emission spectra, infrared and ultraviolet spectroscopy involve only the absorption of light. In simple terms, a spectrophotometer generates infrared (or ultraviolet) light, spreads out this light to form a spectrum (like the familiar colored spectrum in the case of visible light), and then passes the light beam through a sample of the compound, which may be in solution. The compound absorbs some portions of this light, and this reduction in the amount of light is measured by a light meter (a photometer). The theory of this method was well understood by the 1930s, but what was lacking was an automatic method of recording the results. A chemist patient enough to use either technique had to make slow and laborious measurements across the spectrum, using the so-called point-to-point method.

The demand for the rapid analysis of chemical mixtures in World War II, for example in the American petroleum industry and the related synthetic rubber industry, led to the production of instruments that could make these measurements without human intervention, and conveniently display the resulting spectrum on a roll of graph paper. Both techniques gave rather limited information about the bonds in a chemical compound, but they were inexpensive once the necessary spectrophotometers were mass-produced, and easy for a chemist to interpret. As a result, both ultraviolet and infrared spectroscopy were well established in chemical laboratories by the mid-1950s. At the most basic level, ultraviolet spectroscopy was employed to identify carbon–carbon double (and triple) bonds, whereas infrared spectroscopy was more useful for finding specific chemical groupings such as the carbonyl or hydroxyl group. One important application of the two techniques, especially infrared spectroscopy, was "fingerprinting," in which the spectrum of an unknown compound was compared with a collection or "library" of the spectra of known substances in order to identify the compound.

Signals from the Atomic Nucleus

The next technique to be adopted, nuclear magnetic resonance (NMR), was more complex, and required a major cultural shift in order to gain acceptance. The NMR phenomenon was first discovered in 1946 independently by two physicists, Felix Bloch of Stanford University and Edward Purcell at Harvard University. Any form of spectroscopy that uses light, whether ultraviolet, visible, or infrared, is based on the behavior of electrons that move around the central atomic nucleus. The NMR phenomenon, on the other hand, results from the behavior of nuclei in a strong magnetic field interacting with an electromagnetic radiofrequency pulse; the nucleus can behave as if it contains a small bar magnet with a particular resonance frequency. Just as a compound exhibits absorption bands in infrared light as a result of the vibrations of its chemical bonds, it can also produce signals that reveal the environment of its central nucleus. This environment is affected by nearby atoms and the exact relationship between the target atom and its neighbors. As it happens, only a few elements have nuclei that behave as suitable internal magnets, but fortunately two of them are commonly found in organic chemistry, namely ordinary hydrogen nuclei (protons), and a less-common isotope of carbon, carbon-13.

For the average chemist in the late 1940s, the physics of NMR was very complicated, and at first the chemical implications of this phenomenon were not well understood (Reinhardt 2006b). However, by 1950 it was clear that the position of a proton in an NMR spectrum was affected by its location in a compound, thus making NMR a possible technique for determining the structure of organic compounds. The first full NMR spectrum of a chemical compound (ethanol) was made by Martin Packard, a physicist at Stanford, in 1951. The next step was the realization that there was an interaction between the "magnets" of the hydrogen atoms on neighboring carbon atoms, what is called spin–spin coupling. The discovery of this phenomenon was made soon afterwards independently by Warren Proctor and Fu Chun Yu at Stanford, and by a pioneer in the chemical use of NMR, Herbert Gutowsky, his colleague David McCall, and the physicist Charles Slichter (a former student of Purcell) at the University of Illinois. Gutowsky's group then became an important center for the development of NMR spectroscopy in chemistry.

Although the basic technique had now been established, it still had to be worked out in detail. However, the complicated and expensive apparatus that was needed to generate the NMR spectrum had to be produced and marketed. Just as importantly, initially skeptical (even bewildered) chemists had to be persuaded to take it up. It was a major cultural shift. Earlier techniques had involved relatively straightforward apparatus which could be used directly by chemists given a brief introduction to the instrument, and the results were easily comprehended by them. Such techniques – older ones such as polarimetry or new ones like infrared spectroscopy – were merely adjuncts to the chemist's work, and most chemists did not specialize in them.

NMR was different. The equipment came straight out of the physics laboratory, and it was completely unlike any existing chemical instruments. It was extremely expensive and had to be handled carefully and expertly. It could not be located in the laboratory itself, needing a special room with a well-trained technician in attendance. Furthermore, the output from the equipment was not readily understood, and required careful analysis by chemists trained in NMR spectroscopy. Some chemists (and physicists) even specialized in NMR spectroscopy, a subdiscipline located between physics, physical chemistry, and organic chemistry. These features of NMR spectroscopy were not wholly novel, as x-ray crystallography was similar in these respects, but x-ray crystallography was also set apart from mainstream chemistry. The advantage of NMR over x-ray crystallography is that it can capture the dynamic behavior of the molecule, including liquid substances and solid substances in solution, in a way that the essentially static x-ray crystallography cannot. This aspect of NMR became important as molecular biologists and chemists increasingly turned to the study of large biomolecules, especially proteins, in the late twentieth century (Reinhardt 2017).

The development of NMR brought the instrument manufacturer to the forefront of the instrumental revolution. Hitherto, the manufacture of chemical instruments had been a specialized trade, dominated by small businesses. One of the few firms that moved into these new areas was the British instrument maker Adam Hilger, a firm that was strong in ultraviolet spectroscopy (Bigg 2002). Having moved into the field early in the twentieth century, the company produced massive instruments cased in mahogany and brass that were still

FIGURE 3.2 Nuclear magnetic resonance (NMR) machine at the Merck & Co. pharmaceutical company, New Jersey, November 2, 1982. Photograph by Barbara Alper/Getty Images.

sold in the 1950s and used up to at least the early 1980s. Another important but small British firm was Unicam Instruments founded by Sidney Stubbens, a former foreman at Cambridge Instruments, in 1934. It was taken over by W.G. Pye & Co., which had expertise from making radio parts as well as optical systems. The two firms were managed independently of each other, although they were both located in Cambridge.

The major firms in early postwar spectroscopy were both American, and their involvement with spectroscopy arose largely as a result of World War II. The Perkin–Elmer corporation started out as an optics firm founded by two amateur astronomers, Charles Elmer and Richard Perkin, in 1937. When war broke out, they initially made military optics such as gunsights (as Pye had done in World War I), but with the pressing need to develop spectroscopy for the war effort, the firm was asked by the American Cyanamid Company, which was literally next door, to develop an infrared spectrophotometer, which became the model 12 (Travis 2002b). The rival firm of Arnold O. Beckman (later Beckman Instruments) had started life in 1935 as National Technical Laboratories to make pH meters to measure acidity. As early as 1941 Beckman was already developing its "DU Model" ultraviolet spectrophotometer, and was commissioned by Shell at the beginning of World War II to produce an infrared spectrometer for use in the petroleum industry. Thanks to Beckman's earlier use of electronics for the pH meter and the brilliance of Shell's Robert Brattain who had developed the prototype, the IR-1 was superior to Perkin–Elmer's model 12, which was grounded in prewar optics. However, as part of the wartime synthetic rubber program, the IR-1 was shrouded in secrecy, allowing Perkin–Elmer to gain a commercial advantage (Rabkin 1987).

The manufacture of NMR machines was very different. Crucially, there were no optics involved, thus giving no advantage to firms such as Perkin–Elmer or Hilger; nor was there any overlap with earlier chemical instrumentation. Having no prior connection with chemistry, the NMR machine was the outcome of nuclear physics on one hand, and the new electronics industry on the other. It was thus logical that the industry should be set up near the birthplace of NMR in Stanford, in what is now called Silicon Valley in central California (Morris 2015). The firm of Varian Associates was founded in 1948 by two brothers, Russell and Sigurd Varian, in San Carlos, about eight miles north of Stanford University, with the support of several members of the university faculty (Lenoir and Lecuyer 1995; Freeman and Morris 2015). Initially the firm was set up to make the klystron – a type of radio amplifier invented by the Varians in 1937 – for military use, but the Varians were also keen to develop NMR. The NMR expertise was supplied by Martin Packard, a former student of Bloch. The firm produced its first commercial NMR machine in 1952, but it was so expensive that only large firms such as Shell Development at nearby Emeryville or Du

Pont in Delaware could afford it. A year later, Varian became one of the first businesses to move into the Stanford University industrial park.

When NMR first became a chemical technique around 1953, the young Varian company was faced with two daunting tasks. On one hand it had to develop a machine that was small enough and inexpensive enough to be used by academic chemical laboratories. On the other hand, it had to persuade chemists who knew little or nothing about nuclear physics that NMR spectroscopy was easy to use and represented the future. Like a new religion, NMR (and Varian) needed a missionary, letters to the faithful, and a gospel. The missionary was chemist James Shoolery, who went round chemistry laboratories in America and Europe with Packard to convince chemists to buy Varian's machines, in return for exceptionally generous technical support. An element of this support was the Varian newsletter that announced new developments along with information about the technique. The gospel came later, when organic chemist Jack Roberts published *Nuclear Magnetic Resonance* in 1959, which sold 8,000 copies despite a then-high price of $7.50. Roberts was typical of the converts to NMR: he was interested in the relatively new field of physical organic chemistry, and he was comparatively young, just thirty-two in 1950 when he was first inducted into the mysteries of the new technique by Richard Ogg of Stanford University (Reinhardt 2006c). As Roberts later remarked, "It was clear there were applications to chemistry, even if I didn't understand what they were" (1990: 151). Further enthused by the work done on NMR at Du Pont, Roberts then set about converting fellow organic chemists, such as William Johnson at Wisconsin in 1957. Both Roberts and Johnson then moved to California: Roberts went to Caltech in 1952, where he persuaded Linus Pauling to buy a Varian machine in 1955, and Johnson went to Stanford in 1960. Introduced in 1961, the compact, robust, and relatively economical Varian model A60 finally brought NMR into most academic laboratories. Nonetheless, the new instrument was still kept out of the reach of ordinary chemists; it was housed in its own room, and samples had to be handed over to a technician who carefully tended the awe-inspiring machine.

Breaking Up the Molecule

NMR spectroscopy was not the only invader of the chemical laboratory from physics. Mass spectrometry had been invented at Cambridge by Francis Aston, a student of the discoverer of the electron, Joseph J. Thomson, in 1919 (Remane 1987; Grayson 2002). In essence, mass spectrometry allows the mass of a positively charged ion (an atom, a molecule, or a piece of a molecule with one or more electrons removed) to be measured by passing it through strong magnetic and electric fields. However, as chemists had other, cheaper methods of determining the atomic weight of compounds, this technique did not seem to offer any advantage. The method was transformed in accuracy by the development of the modern mass spectrometer through the work of Alfred Nier at the University of Minnesota in the 1930s (de Laeter and Kurz 2006). The Consolidated Engineering Company, later named Consolidated Electrodynamics Corporation (CEC), then started building these advanced machines for the petroleum industry (Meyerson 1986; Grayson 2002).

In these years the value of mass spectrometry for organic chemistry still seemed limited. The situation was transformed in the mid-1950s when a few organic chemists saw that this increased accuracy allowed them to determine the chemical formula of a compound from the precise mass of its ion (Reinhardt 2002; Reinhardt 2006a). Furthermore, they realized that an organic compound would break up inside the machine, producing a pattern (a "mass spectrum") of fragments from which it was possible to work out portions of the structure of the compound. Deducing the structure of compounds from their fragments was not a wholly new idea. The degradative analysis used by organic chemists up to the 1960s also broke up large molecules into smaller compounds that were easier to study, thus determining their structure. But this earlier kind of molecular demolition used chemical reagents. The most difficult step was to reassemble

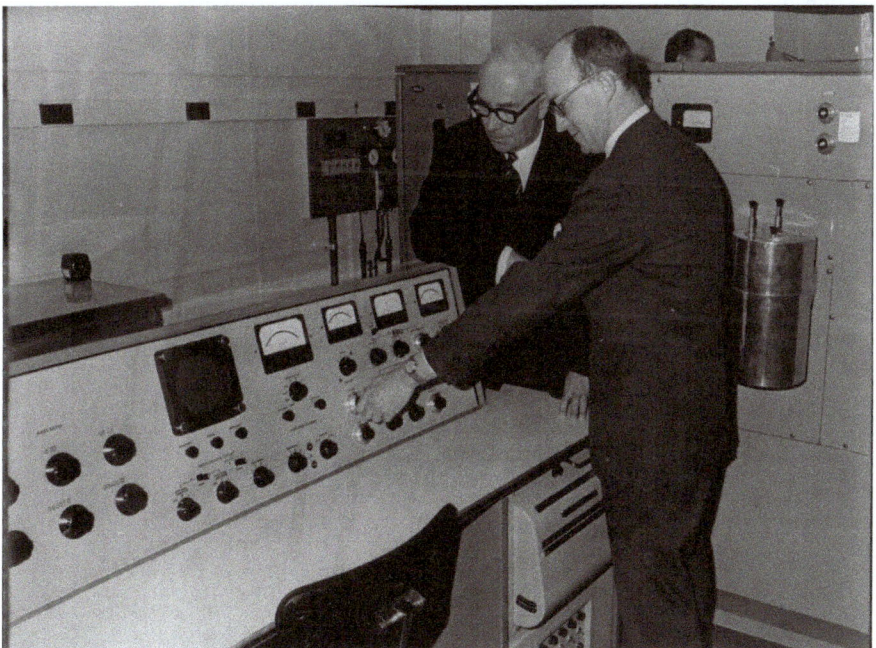

FIGURE 3.3 AEI MS9 double-focus mass spectrometer at the University of Sydney. With the A$80,000 [about A$1,000,000 or US$700,000 today] machine are Prof. Charles Shoppee, Head of the Department of Organic Chemistry (on the left) and Dr. Alex Robertson, in charge of the spectrometer. 1966. Photograph by Frank Albert Charles Burke/Fairfax Media via Getty Images.

the structure of the original molecule from its fragments, regardless of whether they had been produced in a chemical flask or in the mass spectrometer.

The new application of mass spectrometry to organic chemistry was vigorously taken up by Klaus Biemann, an Austrian organic chemist who had moved over to analytical chemistry after he joined MIT in 1955. He applied mass spectrometry to peptides – which has remained an important topic in mass spectrometry (along with the larger proteins) – and to the complex compounds called alkaloids. In developing this technique he was supported by Woodward (an MIT graduate) at neighboring Harvard University, who supplied him with interesting compounds.

Carl Djerassi was a restless chemist who was always seeking new ways of exploring organic chemistry (1990; 1992). He had already developed the technique of optical rotatory dispersion, which uses the spectrum of plane-polarized light to determine the structure of certain compounds that can exist as mirror-image isomers (so-called chiral isomers). However, this technique provided only limited structural information. Djerassi was aware of the work on mass spectrometry by the organic chemist Ivor Reed at Glasgow University, and decided to use the technique to examine the complex organic compounds – such as alkaloids – which he was already studying. To work out how these compounds broke up in the mass spectrometer, he labeled specific parts of the compound by replacing the normal hydrogen and carbon atoms with their heavier counterparts deuterium (hydrogen-2) and carbon-13.

The Art of Separation

These new techniques (with the exception of NMR) could be used to quickly detect and analyze small amounts of material. This meant that they could be used with another important new procedure called gas chromatography (GC). For centuries chemists had struggled to separate mixtures of compounds into their pure components. For liquids, distillation was useful, but the heat could also break up delicate compounds. Nineteenth-century chemists had developed crystallization to a fine art, but it could only be used with compounds that were solid at or just below room temperature and that could be dissolved in a suitable solvent. The burgeoning petrochemical and pharmaceutical industries were desperate for a method which could rapidly remove small amounts of the compound of choice from a much larger body of material, whether it be in a batch or a continuous stream. Scientists working in the discipline of biochemistry also needed to be able to purify tiny samples.

The brilliant but eccentric English chemist Archer Martin had developed two separation techniques (Ettre 2001; Ettre 2008). To separate relatively large amounts of material in solution, with the help of Richard Synge in the early 1940s he converted the basic concept of chromatography (begun in the work on separating plant pigments – hence "chromato"-graphy – by the

Russian–Italian botanist Mikhail Tswett in the early 1900s) into the much more powerful technique of liquid–liquid chromatography, which, despite its name, uses a column of powdered silica in a glass tube. He then had the idea of using filter paper instead of silica, which was not only cheaper, but also made it possible to carry out the separation in one direction, then turn the paper by 90 degrees to attain additional separation of the mixture. Finally at the beginning of the 1950s, Martin teamed up with Anthony James, his colleague at the National Institute of Medical Research at Mill Hill, just north of London, to develop yet another form of chromatography, which he had predicted in one of his papers with Synge (Ettre 1977; Morris 2002b). Instead of using a solid column of silica powder, gas chromatography used a narrow glass tube coated with silica gel on the inside and an inert gas such as nitrogen or hydrogen as the carrier of the material, rather than a solvent. It worked best in separating mixtures of gases and liquids with a low boiling point, but less-volatile liquids could be heated.

Such was the explosive demand for this technique that gas chromatography developed rapidly during the 1950s and early 1960s (Ettre and Zlatkis 1979; Gerontas 2014). While chemical and petroleum companies were enthusiastic users of all the new instrumentation, not least because they had the funds to buy them, researchers at the petroleum firms British Petroleum and Shell played an important role in the further development of gas chromatography. Chemists quickly invented new instruments to detect tiny amounts of compounds as they emerged from the glass tube, most notably the flame-ionization detector (FID) and the electron-capture detector (ECD). But it was also possible to use infrared spectroscopy to detect the compounds, and employ the fingerprint technique (using a data library stored in the instrument) to identify them. The original form of infrared spectroscopy was too slow to be used in this way, but the new and much faster technique of Fourier-transform infrared spectroscopy (FTIR) was introduced in the 1960s.

Although the FID and the ECD were very good, from the 1980s onwards chemists increasingly switched to a new machine, a combination of the gas chromatograph and the mass spectrometer. This "hyphenated" GC-MS allowed the compounds separated by the GC to be quickly identified in the MS by their exact mass and their fragmentation pattern. Gas chromatography fundamentally changed not only organic (and inorganic) chemistry, but also the new field of environmental science. While Rachel Carson was writing *Silent Spring* at the end of the 1950s, there was no way of positively detecting the pesticide DDT at levels below one part in a million using chemical or physical methods, yet it was clear to Carson as a biologist that birds of prey were being affected by even lower levels of DDT in their environment. Within three years of the publication of *Silent Spring* in late 1962, thanks to the ECD, it became possible to detect DDT at levels of parts per billion and by the mid-1970s at levels of parts per trillion.

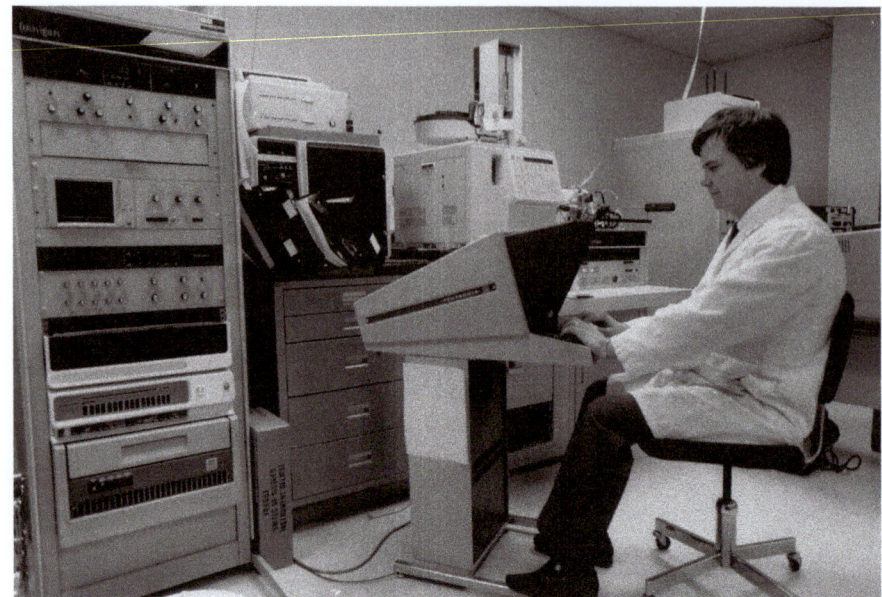

FIGURE 3.4 Chief chemist Pierre Beaumier at Mann Laboratories using a GC-MS device, 1988. Photograph by Jeff Goode/Toronto Star via Getty Image.

Yet even gas chromatography was not without its drawbacks. As in the much older case of distillation, it did not work for heat-sensitive compounds or for non-volatile liquids or solids. This was a particular problem in pharmaceutical chemistry and biochemistry. What was needed was a method that combined the best features of liquid–liquid chromatography with gas chromatography. The solution was to use liquid chromatography under pressure, a technique which was initially called high-pressure liquid chromatography, later high-performance liquid chromatography, HPLC (and high-priced liquid chromatography by some of the more cynical chemists; Ettre and Zlatkis 1979; Ettre 2008). Several groups of chemists worked independently on the development of this method in the late 1950s and 1960s, including the eminent biochemist Stanford Moore (with William Stein) at the Rockefeller Institute, the industrial chemist Lloyd Snyder at Union Oil, Csaba Horváth, a Hungarian chemist at Yale University, and the Austrian chemist Josef Huber at Eindhoven University. Huber was also aided by Archer Martin, who had predicted HPLC as well as GC in his key 1941 paper on liquid–liquid chromatography.

As in the case of Varian and NMR, one firm took the lead in its commercialization. Waters Associates had previously taken up the similar technique of gel permeation chromatography, developed in 1962 by John Moore at Dow Chemicals, so it was relatively easy for Waters to move into

HPLC (McDonald 2008). The founder of the firm, James Waters, persuaded Woodward in 1972 to allow his use of a Waters machine during his famous synthesis of Vitamin B_{12} to appear in his company's publicity materials. Varian also entered the field, as it was already making Aerograph gas chromatographs. However, Waters soon dominated HPLC and continued to do so for many years. By contrast, although the Perkin–Elmer company also entered the field, no single firm managed to dominate gas chromatography, partly because it was relatively easy to construct a basic gas chromatograph in-house. By the end of the 1970s, the speed and sensitivity of HPLC was comparable to gas chromatography.

MADE IN AMERICA

One important result of this wave of innovation in chemical instrumentation was to hand the USA a major advantage in academic (and industrial) chemistry. This was partly because much of this innovation took place in the United States, although gas chromatography was partly a British development and Harold Thompson at Oxford University had been a pioneer of infrared spectroscopy. Crucially, most of the firms who developed and commercialized the equipment, including Beckman Instruments, Perkin–Elmer, Varian, and Consolidated Electrodynamics, were also American. There were two reasons for this. Some academic researchers moved into industry, and the leading academics worked closely with the firms, favoring geographical closeness between the two groups. Furthermore, Britain and Germany's historic advantage in traditional chemical instrumentation was lost because the new equipment was completely different from the earlier refractometers or polarimeters. In particular, the new instruments required advanced electronics, a field in which the United States was more advanced in the late 1940s than Britain or war-damaged Germany. These advances stemmed largely from the pioneering research at Bell Telephone Laboratories, and MIT's radiation laboratory in the 1940s. The new American firms had benefitted both from the financial benefits of wartime contracts and the expertise gained from wartime projects, such as for the Beckman IR-1. It is not insignificant that these wartime projects (such as the American radar project and the synthetic rubber program) brought academic and industrial scientists together. Perkin–Elmer was on the east coast, in Connecticut, but CEC, Beckman, and Varian were all based in California. Both Harold Washburn of CEC and Arnold Beckman were graduates of Caltech in Pasadena. Russell Varian graduated from Stanford University, which was also where Purcell had his laboratory. Herbert Hoover Jr., the founder of CEC and the son of the US president, had also been educated at Stanford. Both Beckman and Varian were involved (along with another chemist, Gordon Moore) in the later development of Silicon Valley around San Jose, some sixteen miles from Stanford (Thackray and Myers 2000; Thackray et al. 2015).

As the leading firms were American, this meant that their products were priced in dollars, which was a major problem for cash-strapped European universities who were unable to obtain dollars because of currency restrictions. The new instrument firms looked first and foremost to the American market, and worked closely with American academic chemists, giving them both advice and access to the latest models. The instrument makers then benefited from the feedback from these chemists and the publicity from world-class chemists using their instruments. These instruments were hugely expensive, which was less of a barrier on that side of the Atlantic because academic science was well supported in postwar America. Initially this funding came from industry and the military (Leslie 1994). After the wartime cash had dried up, money came from the research funding arms of the US armed forces, and then from the National Science Foundation (founded in 1950). Additional funding, especially for buildings and equipment, came from private philanthropists such as John Stauffer. During the 1950s and 1960s, the National Institutes of Health increased its support for chemistry because it was seen as an important sector for the advancement of medicine; Stanford particularly benefited from this funding. US government funding increased exponentially after the "Sputnik crisis" of 1957, when Americans became alarmed that their country was falling behind the Soviet Union in science and technology. While chemists seeking the new instrumentation still had to win their argument for this funding, their position was in stark contrast to their impoverished colleagues in Europe, who often had to make their own apparatus from ex-army surplus parts.

LABORATORY DESIGN BETWEEN 1945 AND 1980

In what way did these enormous changes in chemical instrumentation affect the laboratory building itself? Perhaps surprisingly, the short answer is "very little." The classical chemical laboratory hardly altered, partly because it was a very robust design that was capable of absorbing such changes, and partly because of the expense and upheaval of building a wholly new laboratory. The new apparatus, exotic as it might have been, did not require new utilities; the electricity and gases needed had been available in standard laboratories for several decades. NMR machines and mass spectrometers had to have their own rooms, but the buildings that housed the classical laboratories already had specialized rooms for visible spectroscopy and combustion analysis. A room formerly used for polarimetry or a chemical museum may well have been converted for these newcomers. To be sure, the heavy mass spectrometer needed a basement room. However, most chemistry departments already had basements which hitherto had been used for storage and existing heavy equipment such as electrical generators or boilers.

So well suited was the classical laboratory design for novel apparatus that it was largely retained when new laboratories were built in the 1960s and 1970s. Of course, given the passage of time, these new laboratories were not carbon copies of the late-nineteenth-century laboratories. The materials used for the laboratory were more rugged and stain-resistant than the older stained wood or stone. In the 1960s heat-resistant ceramic materials were introduced for bench tops, avoiding the need to use asbestos mats. The new science of ergonomic planning favored smaller laboratories and peninsula benches at a right angle to the wall, rather than island benches in two rows in the middle of the room. The need for good lighting moved the benches to the window wall. The fume cupboards moved from the window wall (where they had been placed so the air could be vented to the outside) to the inside wall (venting through piping to a central flue), and there were more of them. The overall effect was to create a brighter and more intimate space.

A classic example of a laboratory building in this period was the Stauffer One laboratory building at Stanford University (Morris 2015). This building was constructed because the energetic provost Fred Terman attracted two leading chemists to Stanford's hitherto lacklustre chemistry department at the beginning of the 1960s with the promise of new laboratories, which already formed part of his ambitious plan to boost the standing of chemistry at Stanford (Gilmour 2004). Both of these chemists – William Johnson and Carl Djerassi – were enthusiastic users of the new technology, and it was clear that the building would have to be designed to house spectropolarimeters (for optical rotatory dispersion), NMR machines, and mass spectrometers. The building was designed by the university's architect, Birge Clark, rather than an architect who specialized in laboratory construction. The external appearance of the building was in keeping with Clark's work on the revolutionary Stanford Industrial Park. Rather than a grandiose statement of neoclassical architecture or the Romanesque "Rundbogenstil" favored by many earlier German laboratory buildings, Stauffer One was a straightforward mid-twentieth-century office building, a style that now populates many university campuses and research parks. Even the very name of Stauffer One was purely functional. The building had been generously funded by John Stauffer of the Stauffer Chemical Company and his niece Mitzi Briggs, and it was planned to be the first of two or three such buildings (which would naturally be called Stauffer Two and Stauffer Three). Despite his generalist background, Clark was a good listener and understood what Johnson and Djerassi wanted (Hutchinson 1977). In particular, he made the building as flexible as possible and even constructed a basement at Terman's insistence, although there was no immediate need for one. The result internally was an outer ring of relatively small laboratories with six central windowless rooms, which contained the advanced equipment such as the spectropolarimeter. The very similar Stauffer Two was constructed in 1963 after the arrival of two future

Nobel laureates Paul Flory (Nobel Prize 1973) and Henry Taube (Nobel Prize 1983). The heavy equipment, especially the new mass spectrometers, was placed in the basement laboratories between the two Stauffer buildings.

It is interesting to compare the Stauffer buildings in the United States with the chemistry building designed by Robert Haszeldine, head of chemistry at UMIST (University of Manchester Institute of Science and Technology, but invariably known by its acronym) from 1957 (School of Chemistry, University of Manchester 2015). One striking difference was the tower block construction, in contrast to the two-storey Stauffer building. The construction of these tower blocks for both housing and offices was common in Britain (and elsewhere) in the 1960s, and was at least partly a result of the scarcity of land compared with the United States. Other tower block laboratories of this period in the UK included the neighboring University of Salford, Imperial College, and the biochemistry building in Oxford. An interesting feature of the UMIST building was the construction of a research tower and a separate teaching block connected by a bridge and an underpass. As Britain's leading fluorine chemist, Haszeldine devoted the top floor to fluorine production and electrochemical fluoridations, presumably with the idea of venting any noxious escape of fluorine or fluorine compounds to the roof (as well having nice views of Manchester). Internally the fittings were similar to Stanford, but not of as high a standard. A photograph of a UMIST laboratory in this building shows a small room typically housing

FIGURE 3.5 Photograph of a UMIST laboratory *ca.*1980s by Dr. Jonathan P. Miller. Used with permission.

five or six researchers with a central bench at right angles to the window, and with a traditional bench running along the window in contrast to Stanford. At both Stanford and UMIST, the traditional wooden bottle rack above the bench has been replaced – at Stanford by tubular steel racking which was both cheaper and could be removed if necessary, and at UMIST by metal racking at the window and by a wooden bookcase on the bench, which presumably could also be removed. This shows an increasing desire for flexibility in laboratories to allow for a greater variety of experimental set-ups.

THE SHAPE OF THINGS TO COME

The development of chemical instruments and laboratories in the period after 1965 was one of increasing sophistication and incremental change, contrasting with the upheaval of the 1940s and 1950s. New techniques were introduced into both NMR and mass spectrometry that have greatly improved their value, but completely new methods have not appeared. An important breakthrough in NMR was the introduction of powerful superconducting magnets. At first the hegemony of Varian appeared to be unaffected; the firm marketed the first commercial instrument to use a superconducting magnet, the HR 220, in 1964. However, a new European competitor appeared on the scene. The German firm of Bruker, founded by the physicist Günther Laukien in 1960, began to collaborate with the former Varian chemist Richard Ernst, who had moved back to the Swiss Federal Polytechnic (ETH) in Zurich in 1968 (Reinhardt 2006b; Ernst 2010). Ernst used his earlier experience at Varian to develop the concept of pulsed NMR, and then adopted a powerful mathematical technique called Fourier transform (FT). Bruker brought out an NMR machine in 1969, followed a year later by a superconducting Fourier transform model. Varian was now the underdog, and sought to regain lost ground by marketing the XL 200 superconducting FT machine in 1978. The firm was taken over by Agilent Technologies in 2010, but was closed down four years later (Freeman and Morris 2015). Another key figure in the development of new NMR techniques was Ray Freeman, who like Ernst worked for Varian in the 1960s, but moved back to Oxford University in 1973. With his students he improved on Ernst's methods, in particular developing two- and three-dimensional NMR.

In the same period, mass spectrometry was opened up by new methods of sample insertion, which have permitted the use of solutions and large organic compounds such as proteins (Morris 2015). The earliest method was the "thermospray" technique introduced by Marvin Vestal and Calvin Blakeley at the University of Houston in the early 1980s. Eventually this technique was displaced by the "electrospray" method introduced by John Fenn of Yale, who won the Nobel Prize in 2002 for his process (Grayson 2011). As with NMR, there have been several further improvements in techniques, including tandem mass spectrometry, in which compounds that have been separated by their mass in the first mass spectrometer are then broken up in the second one. Traditionally the mass of the ion was measured by the curvature of its flight in a magnetic field. That method has now been partly replaced by the "time of flight" method, which measures the time that the ion takes to fly through the magnetic field. The time of flight method, like the latest versions of NMR, is particularly useful for large molecules such as proteins and other biomolecules. This is part of a general shift in organic chemistry away from complex but relatively small molecules to the study of extremely large molecules, such as proteins. Many modern mass spectrometers used for this purpose (such as the matrix-associated laser desorption/ionization time-of-flight, or MALDI-TOF, mass spectrometer) are still as huge as their predecessors in the 1940s. By contrast, the miniaturization of the quadrupole mass spectrometer, which uses an oscillating electrical field rather than a magnetic field to separate the ions, has enabled mass spectrometers to be combined with HPLC to produce desktop instruments that today can be found in almost any laboratory despite their relatively high cost. Like all modern instruments, these "hyphenated" HPLC-MS machines are computerized, and can access an enormous library of characteristic mass spectra patterns to produce instant identification of the compounds in the mixture.

The classical laboratory that first appeared in Germany in the 1860s survived almost unchanged until the 1990s. Health and safety along with the desire for flexibility were the major driving forces for fundamental changes in laboratory design. The change was driven by companies in the pharmaceutical industries that had larger construction budgets than most universities and were more safety-conscious than academic researchers, at least until to the 1980s. Two British pharmaceutical firms, AstraZeneca (then Fisons) and GlaxoSmithKline, built laboratories in the 1990s that became a model for universities seeking to construct new laboratory buildings. One such university was Oxford University, which completed its new Chemistry Research Laboratory (CRL) in 2004 (Morris 2015). From the outside, the plate-glass building maintains the trend ever since the Stauffer laboratories for laboratory buildings to look like office blocks rather than temples of science. Internally, however, there are major changes from the 1960s. The key difference is the total separation of the actual laboratories from the ancillary services (such as the drying of solvents and mass spectrometry) on one side and the open-plan non-experimental working areas on the other. Rather than everything taking place in the laboratory – including writing reports, using a spectrometer and drinking tea – only the actual experimental work can now take place there. The laboratories are separated from the office space by a glass wall, which is transparent so that someone in the office can see if something has gone wrong in the laboratory. The different areas even have different floor coverings to show if they are "clean" or "dirty." The laboratories are much smaller, but contain no fewer than eight fume cupboards together with a central bench. Practically all experiments are carried out in these fume cupboards. The aim was to have one fume cupboard per worker, but this has not in fact been achieved.

On one level, this may seem a relatively small change, but it has completely changed the culture of the laboratory. Rather than the laboratory being a complete way of life, with all the work being carried out there, it is now a confined space into which one goes just to carry out an experiment. By the same token, the office space is only used for certain activities. This means that the team spirit that was once prevalent in the laboratory has now largely been lost. In the CRL, they have tried to create a community spirit by building an indoor atrium with a cafe between the laboratory block and the office block that houses the department's administration and the professors' offices.

The other important but largely hidden feature of the CRL is the flexibility of the design. While the laboratories may seem immutable, they can in fact be quickly dismantled and converted into office space. In order to be able to do this, the laboratories incorporate two design features that were becoming commonplace from the 1970s onwards. One was the transfer of the utility lines for electricity, gas, and water (and nowadays computer cables) from under the floor to above-ground ducts which can be easily accessed and moved if necessary. This change was also the result of the shift from a wooden floor to a concrete floor, which had already taken place by the 1960s. The other was the introduction of the C-frame bench (Watch 2008: 155). This is a C-shaped steel frame from which cabinets can be hung under the benches. The frame can be fitted with castors that enable the benches to be easily moved. Furthermore, it can be combined with a cantilevered casework system that creates wall-hung storage space, in contrast to the traditional cupboards under the bench. Of course, this type of arrangement is very similar to modern offices or kitchens, leaving the fume cupboard as the only distinctive feature of the modern laboratory. The fume cupboards themselves are easily removed and can be replaced by "glove boxes," which are used for handling air-sensitive or radioactive chemicals. Instead of the chemical laboratory building being both externally and internally a permanent monument to the practice of chemistry, it has become an office block which is partly given over – for the time being at least – to rather recondite purposes.

There is one other seemingly minor change which may have implications for the future. Until recently, the ordinary chemist had to hand over samples to a specially trained technician before they were inserted into the mass spectrometer or NMR machine. However, in a modern laboratory even students can approach these delicate instruments and insert their samples, even with the very largest and most expensive mass spectrometers. Prosaically, this is not the consequence of a great improvement in the handling abilities of chemists, far less an outbreak of social equality in laboratories, but a result of the introduction of automated sample insertion devices. The chemists and students simply put their tubes in a stainless steel chain which then chugs its way towards the sample entry port of the machine and pops the sample in correctly. Meanwhile the chemists will have made their way back to the working area, where the results will be sent to their computers (or perhaps in the near future, their smartphones).

For the most part, automation in the shape of robot synthesis or automatic analysis has been largely confined to the pharmaceutical industry and clinical services (in the form of automatic blood analyzers). However, given the continual drive for improved health and safety, it cannot be beyond the bounds of imagination that at some point in the not so distant future, chemists will be deemed both too unreliable and too fragile to carry out chemical experiments. While experiments in universities may change too often and be too complicated to be completely automated, one can easily envisage experiments being carried out by robots controlled remotely from a computer in the chemist's home or office using webcams, in much the same way as Mars rovers are controlled from an Earth-based office. If this were ever to happen, the culture of experimental chemistry would of course be utterly changed. Yet in many respects it would only be the culmination of a process of smaller changes that have taken place over the last century.

CHAPTER FOUR

Culture and Science: *Materials and Methods in Society*

CARSTEN REINHARDT

INTRODUCTION

Berlin, 1900: Emil Fischer opens the new chemistry laboratory of the Friedrich Wilhelm University; a colleague congratulates him on this achievement, calling it the "Premier laboratory on Earth." This phrase emphasized Berlin's then-central role in the academic community of German chemistry, and alluded to the dominant position that Germany's chemical industries and sciences were enjoying worldwide (Reinhardt 2010). Fischer's laboratory was a temple of science, equipped with the most modern equipment for its day, and boasting four grand halls for laboratory work. There, the daily manual work of almost 200 doctoral students and postdoctoral fellows was performed, supervised by watchful instructors, and geared toward analysis and synthesis of novel compounds. Fischer's dream was to solidify the centrality of chemistry's position in the range of scientific disciplines through the impact of its laboratory methods, especially on the future of the biological sciences. Synthetic life was the ultimate goal, to be reached by chemistry's efforts. Connected to this aim was his plan to found a large central research institute that would connect all subdisciplines of chemistry under one roof. He wished to establish a single institute as the headquarters for his science, and in this way to keep chemistry in its position at the center of scientific, technological, and social development.

United States, late 1980s: The term *collaboratory* is born. This artificial noun, constructed from collaboration and laboratory, denotes the virtual cooperation of groups of scientists sharing laboratory technologies while physically distant from each other. Through overcoming the restrictive walls of the laboratories, disciplinary distances between the sciences are to disappear as well (Reinhardt 2006a). Of course, collaboratories are possible only on the basis of the Internet, or its predecessors, as huge amounts of data have to be shared. Moreover, the data to be shared have to be produced first, and this points to the second precondition of the collaboratory style: the generation of instrumental data, mostly spectra of various varieties. It would seem that the premier, central laboratory on earth has thus becomes dispensable towards the end of the twentieth century. But what about chemistry's assumed central role in the range of scientific disciplines, technical fields, and society and culture?

In this chapter, I will argue that chemistry's role in influencing the sciences, technologies, and society at large dramatically changed over the course of the twentieth century. For this purpose I will use Germany as a case in point for the first half of the century, and the United States for the second. As it did throughout its history, chemistry exhibits a characteristic duality in the twentieth century as well: it is both a science and an industry. Both in its professional practice and in public understanding, chemistry has been heavily influenced by this duality. During the late twentieth century, a third, less clear-cut dimension appeared on the scene: the English noun "chemical" is commonly used to refer to a hazardous substance. The impact of chemistry on public health and the environment now amounts in effect to a third dimension in the semantics of chemistry. In these three dimensions, chemistry's roles and functions are governed by two features: materials and methods. The promise of centrality refers to the provision of novel substances for the neighboring sciences and for society at large, and the offering of solutions to scientific and societal problems and challenges. Seen in this way, chemistry is a fundamental science, in supplying materials. It is also a networked science, in supplying methods for all the sciences.

In the first part of this chapter, I will show how chemistry gained a central role in society through its material productivity. Chemistry affected other disciplines and industries in providing necessary materials for almost any sector. As the first half of the century was dominated by the two world wars, this was done under extreme circumstances. Materials provided by chemistry were crucial for warfare, and thus were perceived as critical for national security, but also for social well-being. In the second part, I will attempt to show by close reading of two US reports on the state of chemistry from the mid-1960s and the mid-1980s, respectively, that chemistry lost its status of provider around the 1970s, even though this was when chemists began explicitly to claim a central role. This happened for two reasons: the pace of chemistry's material innovations slowed, and it was taken for granted. The environmental problems appearing clearly in this period caused chemists to adopt a defensive mode. In the 1960s it became clear that chemistry affected society not only in beneficial ways, but also through the environmental and health-related risks that it created. At the same time, however, chemical thought and practice offered solutions for many of these problems. Thus, in the third part of this chapter, I will analyze how chemical methods and chemical thinking expanded into many neighboring disciplines and technical fields. However, in the end, the adaptation of novel methods out of chemistry to fields ranging from physics to engineering and from medicine to environmental health was not perceived by the recipient fields and the public as originating in chemistry. Hence, chemistry did not reap the benefits from its expansion.

CREATING A BASELINE FOR SOCIETY'S NEEDS: MATERIALS BETWEEN ERSATZ AND MIRACLE

The strong relationship between chemical science and chemical industry rests on the isolation, transformation, and synthesis of substances – and dates back to the times even before the "second industrial revolution" of the late nineteenth century (Homburg 2018). Chemistry is deeply intertwined with industry, closer and even more extensive than physics is with electrical engineering, or biology with biotechnology. Artisanal, manufacturing, mining, agricultural, pharmaceutical, and medical cultures were impacted by chemistry mainly through the materials leaving the chemists' laboratories (Roberts and Werrett 2018). On this basis, chemistry became the largest scientific discipline, measured in terms of the number of chemists employed by universities, companies, and the government (Johnson 2008; Maier 2015: 11, 19). Throughout the twentieth century, chemistry's prominence relied on the popularity of its dyestuffs, pharmaceuticals, plastics, fibers, fertilizers, and explosives. The steady stream of new compounds linked chemistry's image to industrial progress, and the achievement of societal development.

The first half of the twentieth century effectively constituted one long war, at least when seen from a European perspective (Hobsbawm 1994). In terms of industry, the years from 1914 to 1945 constitute a period from brutal military action in 1915 and an all-out effort on all sides to win the "Great War," through an interwar period marked by a global regrouping of the industry, to an armament program that began in Germany in 1936 with the Four Year Plan, to another global catastrophe. In terms of science, the period marked the end of German dominance and the shift from German to (American) English as the lingua franca. Still, even after World War II, American chemists had to learn "chemical German" in order to read the older literature, and Soviet-Russian articles were increasingly translated into English from the 1950s (Gordin 2015).

How secure (and arrogant) German chemists felt of their chemical hegemony before the Great War is demonstrated by an event at mauve's fiftieth anniversary in 1906. Mauve was discovered in 1856 by William Henry Perkin, a young English chemist and entrepreneur; his marketing of the pale purple dye marked the beginning of the synthetic dyestuffs industry, and hence of the modern organic chemical industry (Travis 1993). Despite its origins in Britain, the heart of the synthetic dyestuffs industry soon moved to the Continent, first to France, then increasingly to Germany, and, to a lesser degree, Switzerland. At the turn of the century Germany held approximately 85 percent of the world market in synthetic dyestuffs. The market turnover of German synthetic indigo in 1906 was on the brink of surpassing the value of natural indigo from British India. In that fiftieth anniversary year, English chemists invited their Continental colleagues to a banquet in honor of the founding of their common enterprise a half-century earlier. Among those present from Germany were Emil Fischer, and the Bayer company's managing director, Carl Duisberg, the latter representing the second-largest and fastest-growing of the giant German companies, with a new manufacturing complex at Leverkusen on the Rhine (Reinhardt and Travis 2004).

Duisberg, who later became Chairman of the Supervisory Board of the huge dye syndicate IG Farben, presented at the dinner a straightforward cause for German supremacy: the success of the Germans was to be found in the fruitful combination of science and practice, as well as in their distinctive capacity for hard work, and thus it was a result of their inherent qualities as a race. It followed from Duisberg's argument that the shift of the center of industrial activity to Germany had been inevitable; the British had to come to terms with this situation rather than hope for an improvement in the British dye industry, which would never take place. Duisberg's rhetoric attributed success to the character and mentality of a nation, expressed in terms of the struggle for existence, rooted in social Darwinism and cloaked with nationalist language. This struggle was soon to become a reality, with the beginning of war in 1914.

Germans did not expect the war to last long. The German military command had planned rapid offensives, first in the west and then on the eastern front, on the assumption that the Russian army was weaker than the French. But the French army stopped the German advance at the Marne in September 1914, and the four-year horror of trench warfare began. The sea blockade by the British Navy was crippling for the Central Powers, as it cut them off from crucial supplies, especially Chilean nitrate as the best available source for nitrogen, desperately needed for both fertilizers and explosives. Equally critical was the deadlock on the front lines: to advance across open terrain against hostile machine gun fire proved to be an almost impossible task.

Both problems were largely solved by German chemists and engineers. Although they were not well stocked with supplies, the Germans could count on their existing chemical plants. Those working in the traditional style of organic synthesis, largely used for dyestuffs manufacture before the war, could be relatively easily converted into production lines for war gases and explosives. Fortunately for the production of explosives, BASF had introduced a new synthesis of ammonia from nitrogen (from the air) and hydrogen using high pressures and a catalyst. Ammonia could be used directly as a fertilizer in the form of ammonium sulfate, and the Germans found a way of oxidizing it to produce nitric acid, the basis of explosives manufacture. Chemical warfare began at Ypres in April 1915 when German troops used chlorine gas to attack the British lines. Responsible for the chlorine and for the idea of using it in battle was Fritz Haber, the same chemist who had designed the ammonia synthesis from atmospheric nitrogen. The Allied forces were swift to respond to the clouds of German war gases with their own gas attacks, and neither side was able to capitalize on the introduction of this new tactic (Johnson and MacLeod 2006; Friedrich et al. 2017).

Bound together by the symbolic figure of Haber, chemical warfare transformed chemistry in two major ways. It brought about a new coordination between scientists, engineers, and military officers. Haber put it metaphorically after the war when he described the prewar German Reich as a house in which the scientist, the engineer, and the officer lived together, but did not work together, just greeting each other on the staircase (Haber 1924). During the war, this changed rapidly with the establishment of key commissions for the provision of war-related chemicals. In addition to Haber, Fischer, the physical chemist Walther Nernst and many others joined this endeavor, which lasted until November 1918.

The impact of chemistry during the war changed its neighboring scientific disciplines, the military, and society. Both high-pressure chemistry and gas warfare introduced new interdisciplinary features into chemistry: the ammonia synthesis introduced continuous production, relied on advanced materials engineering and large-scale testing for catalysts, and was linked to new measuring devices. The development of gas warfare called for the interplay of physics, physiology, technology, inorganic chemistry, and organic synthesis. Haber's Kaiser Wilhelm Institute for Physical Chemistry and Electrochemistry, founded in Berlin in 1912, grew from five employees at the beginning of the war to 2,000 by 1917, 150 of them being scientists engaged in all stages of research and development (R&D), including deployment. After the war, having been accused of war crimes on one hand, and being awarded the Nobel Prize for 1918 on the other, Haber defended his actions with technocratic jargon, devoid of any ethics.

Thus, it became clear to all the countries involved in World War I that chemistry was crucial for each nation state's military capabilities in wartime. And it became clear that it was social, technical, and educational capabilities that matter, not a supposed national character. In addition to explosives, fertilizers (also based on ammonia) were much needed in agriculture to feed the starving population; ersatz materials of all kinds replaced or or at least extended normal foodstuffs; and construction materials based on German sources supplemented imported metals. During World War I, synthetic materials, formerly regarded as a variation and improvement on nature, became to be regarded as cheap and less desirable than their natural counterparts. The war brought about new alliances between science and the state, between science and industry, and between disciplines. Although chemistry made a major contribution to the war effort, the lack of food at home, the terrible consequences of chemical warfare, and the inferior quality of ersatz products meant that chemistry had a poorer reputation after the war than most chemists would have expected after all their wartime efforts.

Nevertheless, the interwar period saw rapid growth of chemical know-how among the nations involved in World War I. In particular, France, Britain, and the United States, stimulated by the unavailability of products of the German chemical industry, quickly developed their own organic chemical industries, partially on the basis of German know-how and partially on their own research (Steen 2014). Being traditionally strong in inorganic chemicals and electrochemistry, these nations thereby balanced their industrial portfolio. In Germany, the technological paradigm of high-pressure chemistry, which had been so successful in the case of the Haber–Bosch process, created technological momentum inside BASF and later IG Farben that led to a number of innovations, notably the high-pressure synthesis of methanol, and synthetic gasoline from coal. First in the US, and later in Germany, these developments contributed to the emergence of chemical engineering as a separate field, with its own education, context, and culture, which complemented the field of pure chemistry (Furter 1982; Cohen 1996). Chemical engineering, one might argue, was one of many spin-offs from academic chemistry, leading to independent neighboring subdisciplines, in both science and technology. Biochemistry, with its historical roots in both medicine and chemistry, is a good example (Kohler 1982). Thus, it became ever more difficult to maintain the unity of chemistry. These new specialties were the result of both intellectual and technological success.

Although the giant complexes of ICI at Billingham, of Du Pont at Belle (West Virginia), and of IG Farben at Leuna and Ludwigshafen were hard to overlook, these factories could be regarded as developments that were neither visible to the general public nor of much relevance to everyday life. But nothing could be further from the truth. The agricultural experiment station of BASF at Oppau (neighboring Ludwigshafen) created new fertilizers and pesticides, which reached the farmers' fields and the consumers' tables rapidly and consequentially. The message from the chemical industry was that chemistry

FIGURE 4.1 US Department of Agriculture poster showing a woman killing an insect with a pesticide, 1943. Hulton Archive/Getty Images.

created food, and this message was heeded after the famines of World War I. As this new abundance was created from fertilizers produced literally out of "thin air" and made possible by chemicals (i.e. catalysts) that were not consumed in the process, it was presented by the industry as a scientific miracle. Chemistry,

FIGURE 4.2 1945 illustration of crop dusting by helicopter. Hulton Archive/Getty Images.

it seemed, was the origin of a new cornucopia, produced by science, and at little cost (IG Farben 1938). Of course it was recognized that these new fertilizers were based on the use of vast amounts of coal, but this was regarded as perfectly natural. That the massive use of both fertilizers and pesticides would have serious detrimental impacts on both the environment and human health was not yet widely appreciated.

Soon, in addition to fertilizers and pesticides, entirely new substances entered the marketplace: plastics and (semi-)synthetic fibers comprised the second great wave of innovation in the early twentieth century. This was partially based on breakthroughs in the understanding of the composition and structure of these materials, which produced new theories regarding such polymers, or macromolecules as they began to be called. But many of the industrial breakthroughs were based on empirical work. By the late 1930s, polymer chemistry became another new subdiscipline, which linked physics, engineering, and chemistry in new ways (Furukawa 1998).

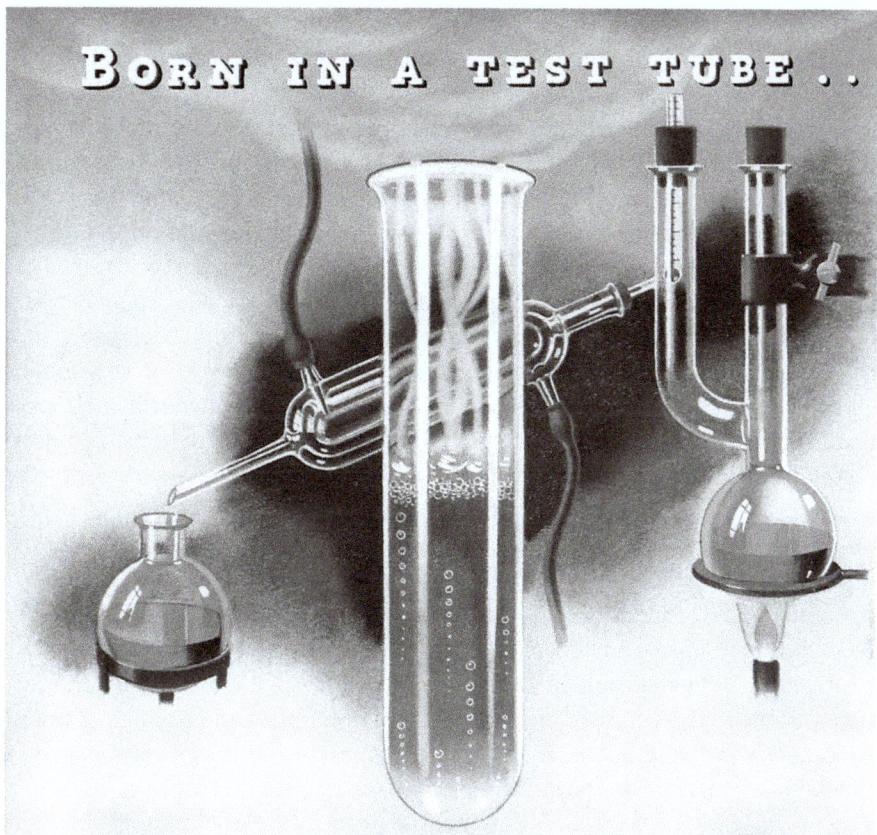

FIGURE 4.3 Advertisement celebrating the "age of synthetics," by British Celanese Ltd., 1946. Hulton Archive/Getty Images.

In medicine as well, chemistry helped to produce new advances. In the late nineteenth and early twentieth century, synthetic pharmaceuticals were mainly painkillers or sleeping aids, most notably Bayer's aspirin and the barbiturates invented by Emil Fischer, but now new, more powerful drugs began to appear (Lenoir 1988; Travis 1989; Hüntelmann 2011). The sulfonamides, the first truly effective antibiotics, were developed at IG Farben in Elberfeld, and this line of research was quickly taken up by firms in France, Britain, and the USA. A similar scenario happened with antimalarial drugs, and remedies for other tropical diseases. Now, with the crucial help of chemistry, medicine could conquer life-threatening diseases at home and in the tropics, contributing decisively to a colonizing nation's prestige and power. With the synthesis of some vitamins and the elucidation of their role in human metabolism, chemical substances became household items even for the healthy. Chemicals came to be seen to be essential for human nutrition – especially Vitamins C and D – even when the necessary dose as a dietary supplement was, as in the case of Vitamin C, largely overstated (Bächi 2009). Similarly, research on steroids led to advances in the understanding of reproduction and metabolism in general, creating the new scientific field of endocrinology, and leading to medical remedies for hormone-related diseases (Stoff 2012).

When in 1933 the giant Bayer trademark cross was illuminated for the first time on top of the headquarters of the Leverkusen plant of the huge chemical concern IG Farbenindustrie AG, it was considered to be a key event not only for the marketing of the most influential brand in pharmaceuticals, but also a sign of the pride of German industry in its products. The hope that this light would shine peacefully was betrayed almost immediately. It was in that year that German chemical industry changed course, and, even more forcefully than during World War I, it assisted the nation in its second war in a generation. German self-sufficiency (autarky) was a crucial precondition for a successful war, and the German chemical industry was essential for its development. Synthetic rubber (called Buna) and synthetic fuel based on coal were arguably the most important, but certainly not the only contributions (Birkenfeld, 1964; Morris 1994). In metallurgy, new alloys, based on German raw materials and known as German metals, were the basis of research in both science and industry (Maier 2007). In medicine, performance enhancers, such as amphetamine, were used by the military on both sides (Rasmussen 2009). During the war, on the basis of a famous accidental discovery by Alexander Fleming in the late 1920s, penicillin was manufactured on an industrial scale through an early use of biotechnology. All of this could be regarded as standard involvement in armaments research. In Germany, however, scientists, engineers, and physicians went much further. Sulfonamides and other medicines were tested on inmates of concentration camps, and many of them were killed during these horrific procedures. IG Farben built its own concentration camp in Auschwitz–Monowitz for the construction of a plant for synthetic rubber and synthetic fuel, killing tens of

FIGURE 4.4 The illuminated Bayer cross, Leverkusen, *ca.*1935. Imagno/Getty Images.

thousands of forced laborers in the process. Zyklon B, a product of Degesch, a company in which IG Farben held the majority of shares, was used to kill millions of Jews in the Holocaust (Rosenbloom and Althaus 2010). Chemistry, and chemical products, had a decisive role in all of this.

It is ironic that World War II came to be called the war of the physicists, due principally to an invention that arguably owed at least as much to chemistry as physics. In December of 1938, the chemists Otto Hahn and Fritz Strassmann, working in the Kaiser-Wilhelm-Institute for Chemistry in Berlin, decided on the basis of chemical analysis, that the products of their bombardment of uranium with neutrons were elements lighter than uranium, and not the "transuranium" elements they had intended to create. Their former colleague, the Jewish physicist Lise Meitner who escaped to Stockholm in the summer of 1938, together with her nephew Otto Frisch, then realized that the uranium nucleus had been split. In this way atomic fission was discovered; there was now a clear pathway to the atomic bomb, and atomic energy. Subsequently, physicists both in Germany and in the USA took over the projects to create an atomic bomb. In contrast to the US Manhattan Project, the German program, immeasurably smaller, did not succeed in making a bomb, and even fell short of building a viable nuclear reactor for electric power (Walker 1989). However, in both nations' projects the contributions of chemists were crucial, and led to the synthesis of the transuranic elements, neptunium and plutonium in the USA.

Despite these crucial contributions, up until World War II synthetic materials were the exception and not the rule on the mass market. Society was still dominated by traditional materials: wood, iron, ceramics, and natural fibers. When nylon stockings came to the US market in 1939, they created a stir. The dangers and risks of some synthetic materials were not well known, at least not publicly. The carcinogenic effects of aromatic amines had been known since the nineteenth century, but the general public did not become aware of that until much later (Stoff and Travis 2019). Similarly, the health risks associated with asbestos fibers became known in the 1940s, but they were not widely perceived until the 1950s and 1960s, likewise the risk of tobacco smoke (Oreskes and Conway 2010). The first lists of threshold limit values in occupational health appeared in the 1930s and 1940s, but received little attention. The situation changed, however, as the quantities and the numbers of chemicals ballooned in the 1950s and 1960s (Homburg and Vaupel 2019). The timespan from shoddy ersatz materials in 1917, to the miracle fiber in 1939, to the deadly poisons of *Silent Spring* in 1962, was only forty-five years.

Cold War societies relied on chemistry even more than during World War II. Arguably, the superpower confrontation was led by consumerism as much as by armed conflict and ideological debate. With a shift in raw materials from coal to oil and natural gas, the sales of petrochemicals experienced exponential growth in the 1950s and 1960s. New plastics appeared on the mass market, such as polyethylene and polypropylene, and then the polyurethanes. Specialty materials, such as Kevlar (among other uses, employed in bulletproof vests), Teflon, and a Teflon-based fiber, Gore-Tex, brought chemistry into most households. As a result of these developments, a report on the future of

FIGURE 4.5 Depiction of a couple walking inside chemistry beakers, 1952. GraphicaArtis/Getty Images.

chemistry by the US National Academy of Science in 1987 claimed that the Iron Age had in fact succeeded to the polymer age:

> You may think we're already there, with your polyester shirt, your polyethylene milk bottle, and polyvinylchloride suitcase. We walk on polypropylene carpets, sit on polystyrene furniture, ride on polyisoprene tires, and feed our personal computers a steady diet of polyvinyl-acetate floppy disks. In just the last 40 years, the volume of polymers produced in the United States has grown 100-fold and, since 1980 exceeds the volume of iron we produce. But the best is yet to come.
>
> (National Research Council 1987: 62)

This "best" future was materials science, which would create high-performance materials. Indeed, by the late 1980s, chemical products had entered the households and the lives of people all over the world. In the industrialized nations, there was a steady supply of new medicines, among them semisynthetic penicillin-class antibiotics, which saved many lives. The contraceptive pill changed sexual mores in many societies (Djerassi 1981). In the second half of the twentieth century there were few parts of private life that were left untouched by chemical materials. Culturally, this was often perceived as a loss, as in Jacques Tati's movie *Mon Oncle*, or seen satirically, as in Mike Nichols' *The Graduate* ("There's a great future in plastics"). But chemists could claim centrality, as chemistry in war and peace had arguably sustained the development of Western nations and secured their prosperity through the plethora of new materials. While this may have appeared to bode well for the future, it was ultimately not a winning strategy, as we will see.

MIDWAY: WINNING WARS AND PEACE, BUT THE CENTER IS LOST

The atomic bomb and the DNA double helix are the scientific icons of the twentieth century. Both depended heavily on chemical knowledge. However, it was physics in the first case, and biology in the second, that made the headlines, and in the 1950s and 1960s these were the disciplines that predominantly attracted the upcoming generations of students and substantial financial resources, while chemistry lost ground. Even though the materials created by chemistry proved to be crucial in Cold War countries, societal recognition and the highest scientific reputation was no longer to be gained in this way.

The reasons for this decline in the reputation of chemistry are difficult to determine. Was material productivity taken for granted, and relegated to the status of routine engineering rather than brilliant science? The brilliance that was in engineering, and the fact that much of new materials development depended on breakthroughs in basic science, was hidden from the view of the general public. Chemists saw themselves as providing a service, tellingly illustrated by the personification of academic chemistry supporting engineering in its role of promoting the world's affairs (Mauskopf 1993; Ede 2007a). Hence it was precisely the close relationship with industry that contributed to the reduction of the reputation of the science of chemistry, even before pollution became an important topic in public discourse. In addition to having a service mentality, a fractured academic landscape prevented chemistry from taking a full part in the exponential growth of science funding in the United States, and in other Western nations following the Korean War and the Sputnik crisis. Much of this growth went into "Big Science," centralized scientific facilities containing large instruments, which needed extensive funding and manpower, created much media attention, and fulfilled even greater military ambitions. Particle accelerators became the cathedrals of the Cold War scientific age, but laboratory ships in oceanography and giant telescopes in astronomy are also cases in point (Galison 1992; Hughes 2003). Compared to this, in the context of 1960s science funding chemistry was seen as a "cottage industry," based on small decentralized teams or even individuals, going about their jobs individualistically, which is portrayed as positive (National Research Council 1987: 223). This attitude led to severe disadvantages in opening the taps of governmental funding programs, which were increasingly centered on large-scale operations.

Chemists attempted to stem the tide. In a famous report on the state of their discipline published in 1965, a group of chemists in the National Academy of Science under the aegis of lead author Frank H. Westheimer portrayed chemistry as the "central science," attempting "to account for the properties and behavior of every material thing in the universe ... Chemistry is so pervasive that we could not hope to define its boundaries, much less to survey them" (National Research Council 1965: foreword). Nonetheless, the authors attempted to characterize the attitude that strongly characterizes chemistry: "a large part of chemistry deals with change. The essential work of the chemist is not just to observe change as it occurs in the world, but to control and change the material world according to man's needs" (1). In this task of control, the role of basic research was highlighted, for in the view of the authors, most "practical discoveries in chemistry" directly emerge from basic science published in peer-reviewed journals (3, 39–41). On this basis the pace of new discoveries was staggeringly high: "On the average, a new chemical compound is synthesized every five minutes" (4). The authors declared that decentralization was a boon and not a disadvantage for chemistry's productivity:

> Especially because chemistry is so flexible and so varied, because it is practiced effectively in so many laboratories, and is not tied to a few relatively expensive installations, we can look forward to a continuation of unpredicted innovations from creative scientists around the world.
> (5)

By the same token, one could also argue that chemistry was overstretched.

According to the authors of the Westheimer report, their science stood centrally "among the fundamental sciences, lying as it does in the continuum between physics, on the one hand, and biology on the other." Physics and biology, and the many new subfields, such as "biochemistry, geochemistry, chemical physics, atmospheric chemistry, [and] chemical engineering," are centered around a chemical concept: "The widespread and central use of the idea of molecules attests that this is the chemical age" (8–9). This was the case not just with the neighboring disciplines, but also with regard to industry, agriculture, medicine, and national defense. Chemistry stood in "the service of man" (5).

In this sense, chemists had a fundamental task: "supplying key materials and principles that are interwoven throughout today's technology, natural sciences, and culture. Through the development and refinement of concepts concerning molecules and their functions, chemistry provides a common resource for experimental science, comparable with the language of quantitative scientific thought provided by mathematics" (1). Thus, two claims were made in the report: chemistry was a resource, and chemistry was fundamental.

However, this integrality and centrality was not without strains, and it was not equally distributed between physics and biology. "Chemistry is the mother-science of biology" was a straightforward statement made in the report (105). However, biologists were less certain of this; Christophe H.W. Hirs from the Brookhaven National Laboratory commented:

> I am concerned by what I feel is a lack of awareness, on the part of chemists, particularly those on the faculties of our great universities, of the role to be played by chemistry in what will obviously be the golden age of biology. ... The biologist poised on the threshold of the great adventure is keenly aware of this. He would like the chemists to join in. But make no mistake about it; if they do not, he is perfectly willing to go it alone.
>
> (105)

The response of the chemists was swift: "Some members feel that the biologists have too narrow – too parochial – a view of science, and that they are urging chemists to do research immediately applicable in biology" (105).

After all, as the history of molecular biology, genetic engineering, and biotechnology shows, these fields originated largely outside of chemistry, sometimes even created in opposition to chemistry (Doogab 2015). The culture of "Big Pharma," shunned by many early self-made entrepreneurs in biotechnology, had a role in this, as probably do the attitudes of both academic chemists and biologists alluded to above. Whereas in the 1960s – with molecular biology already emerging and genetic engineering just about to – the relationship to biology was depicted by chemists as one of superiority, their attitude towards physics, the other major scientific neighbor, was much more balanced. Citing famous scientists from Marie Curie to Einstein, the Westheimer report averred that their relation was reciprocal: "gift-giving." For example, "quantum chemistry is a fundamental modern contribution to chemistry from physics, while semiconductors and transuranium elements are among the many gifts to physics from chemistry" (106). The relationship of physics and chemistry was thus one of equal partnership: "The two sciences are interdependent and support each other by continuous exchange of ideas and experimental techniques. There can be no doubt that their future courses will be closely interrelated" (107). In sum, chemists strove to influence biology in a top-down approach, while they saw their relationship to physics as an equal partnership.

However, the new fields of materials science, solid state physics, and nanotechnology, all with very substantial chemical input, developed their own directions and traditions, as did the new life sciences. The summary statement of the 1965 Westheimer Report presciently predicted:

> Since chemistry is the science of materials and changes in them, it necessarily interacts with other sciences. The general theory of chemistry – atoms, bound into molecules by electrical forces subject to the laws of quantum mechanics – underlies much of science outside of chemistry. The future of both the physical and the biological sciences will depend on chemists for materials and basic theory.
>
> (109)

Thus, the future was seen "outside of chemistry," and this is exactly what happened. The challenge for chemists was how to keep being in the loop with these new developments.

Soon, however, chemists became engrossed with self-inflicted problems. The so-called Pimentel Report (after its lead author George Pimentel) in its published version of 1987 opens with a similar line to its predecessor, the Westheimer Report, twenty years earlier. Its authors made the same point about "how central chemistry is among the sciences as they are applied to human needs. It shows how important chemicals are to our survival" (National Research Council 1987: 1). But what immediately followed was new. "And just what is a chemical? Perhaps you have your answer ready – DDT, Agent Orange, and dioxin are chemicals. Yes, indeed they are, just as much as sugar and salt, air and aspirin, milk and magnesium, protein and penicillin are chemicals. We ourselves are made up entirely of chemicals" (1). Thus, chemists had to counter the bad image they met in the media, after the publication of *Silent Spring*, the use of Agent Orange in the Vietnam War, and the Seveso (Italy) industrial accident. In order to do this, they highlighted the economic importance of their science, with its great "responsiveness to human needs," and – this too was new – in its "universal philosophical significance" contributing to the "ultimate understanding of life" (2).

With this latter comment, the chemists were resurrecting a century-old debate between vitalism and materialism. For some scientists, the quest of artificial synthesis of life and the explanation of life's hereditary mechanism by chemical means also suggested the lack of any need for organized religion. As a consequence, they were attacked by intellectuals rooted in the Christian tradition. The unraveling of the structure of DNA by James Watson and Francis Crick and the Stanley Miller–Harold Urey experiment showing how the synthesis of amino acids from simple molecules was possible under conditions present on Earth billions of years ago were often interpreted in this way, especially in the cultural atmosphere of 1950s England. In this way, chemistry was involved in the great debates between science and religion, within the framework of intellectual discourse, ideological pursuit, and philosophical scholarship (Bud 2013).

For chemists, however, "environmental quality" was foremost among the recognized human and social needs in the mid-1980s. Although it did not define the term, the Pimentel report emphasized the issue with a simple headline: "No Deposit, No Return, No Problem," predicting the positive impact of biodegradable plastics, which would make the "plastic waste problem ... going, going, gone ..." (National Research Council 1987: 4). For the authors of the report, it was important to counteract the public perception that detection of chemicals meant hazard. For the chemist, *"detection can be equated to protection"* (7). Most of the Pimentel Report stressed – as did the Westheimer Report twenty years earlier – the beneficial impact of chemistry on human needs, as defined by US postwar industrial society. Health, energy, materials, food, and national defense are key, and modern instruments take center stage in achieving all this.

But the report did not clarify where the center now was, where chemistry resides in the spectrum of disciplines. The continuum between physics and biology, with chemistry occupying the essential and central position, seems to have been lost. The penultimate sentence of the 1965 Westheimer report reads: "We, too, believe that the advance of science provides one of the mainsprings of American prosperity, health, and safety, and repeat that 'chemistry is a major tool for making the most of what we have to work with and to live on'" (National Research Council 1965: 194). The penultimate chapter of the Pimentel Report is titled "The Risk/Benefit Equation in Chemistry." Thus, increasingly the new frontiers for chemistry were the front lines of a public debate on its risk. However, there did exist, through the chemists' achievements in the middle of the twentieth century, additional and novel intellectual frontiers, greatly expanding chemistry's reach, but they were mostly not considered to be chemical.

PATHWAYS THROUGH THE DISCIPLINARY MAZE: METHODS AND HYBRID FIELDS

When during World War I the first blow to chemistry's positive image was struck by the deployment of chemical warfare, x-rays and electrons also appeared in chemistry. These important entities, emerging around 1900 in physics, were soon appropriated by the chemists, both as concepts for their ideas on molecules and reactions, and, equally importantly, as research techniques and methods. These applications, brought into analytical practice throughout the century, led to a dramatic expansion of chemistry's reach. However, this expansion came at a price, as chemistry in effect became a tool-kit for its neighboring disciplines and technologies. Even worse, these new techniques have often been used without naming them – or even regarding them – as chemical at all. Connected to the arrival of x-rays and electrons in both physics and chemistry was an influx of mathematical concepts and techniques, linked to the birth of a whole spectrum of new hybrid fields. Among the examples studied by historian Mary Jo Nye are physical organic chemistry, quantum chemistry, chemical physics, and x-ray crystallography (Nye 2018).

For some, the appearance of these new entities and instruments offered the opportunity for a reduction of chemistry to physics. Most eloquently, in 1939 the theoretical physicist John Clarke Slater called for such a merger, with physics on top:

> It is probably unfortunate that physics and chemistry ever were separated. … A few years ago, though their ideas were close together, their experimental methods were still quite different: chemists dealt with things in test tubes, making solutions, precipitating and filtering and evaporating, while physicists measured everything with galvanometers and spectroscopes. But even this distinction has disappeared, with more and more physical apparatus finding its way into chemical laboratories …. The author hopes that this book may serve in a minor way to fill the gap that has grown between physics and chemistry. This gap is a result of tradition and training, not of subject matter.
> (Slater 1939: v, viii)

Indeed, new subdisciplines soon populated the area between physics and chemistry. Most directly, quantum chemistry was an outgrowth of the new, quantum-mechanical concepts of atoms, molecules, and bonds. However, instead of becoming the trailblazer for chemistry reduced to physics, quantum chemistry developed its own independent traditions, deeply affected by different approaches toward mathematics, and by the momentum of visual imagination so typical of chemical thinking (Simões and Gavroglu 2001). Furthermore, chemical physics, emerging out of the new physical concepts and spectroscopic techniques during the 1930s, was populated mainly by chemists moving into the field, and not by physicists (Reinhardt 2004). Nuclear chemistry, an offshoot from earlier research on radioactivity, was from the beginning both a physical and a chemical field, drawing on the evidence generated by new apparatus, which had originated in physics (Roqué 2001). In addition to nuclear chemistry, radio-, geo-, and cosmochemistry emerged as truly interdisciplinary fields, completely integrated hybrids based on the classical disciplines (Kragh 2001). Chemistry thus joined its neighboring disciplines in exploring both the microcosmos of the atomic nucleus and the macrocosmos of the universe.

Also inside the arguably most powerful "infield" of chemistry – organic chemistry – deep inroads by both physics and biology were visible. Physical organic chemistry, first in the UK, and then in the US, emerged with its characteristic mechanistic thinking as a powerful tool complementing, and competing with, the traditional approach of organic chemists. New methods, aptly called "physical" to denote their origin in the alien field of physics, joined in revolutionizing the practice of organic chemistry, in both its analytical and synthetic dimensions. Moreover, bio-organic chemistry – a new field centered around large, biologically active molecules – completed this transformation of organic chemistry in the 1960s. Organic chemistry kept its name, but completely changed its conceptual base, its methodical arsenal, and its major research orientation (Morris et al. 2001). Major changes also occurred somewhat earlier – before World War II – in the field of biochemistry, with the development of a classical biotechnology and a biochemistry centered on hormones and vitamins as active metabolic substances (Marschall 2000; Rasmussen 2001).

A great rationale for chemistry's unity, the link between synthesis and analysis that classical, "wet" methods, grounded in chemical reactions, created, was thus dissolved in the mid-twentieth century (Morris and Travis 1997; Morris 2002a). Thereafter, analysis was largely based on the new "physical methods," with roots in the theory of physics and electronic instrumentation. This was not an isolated case. Most usage of instrumentation, as I have argued, can be classified by the main functions of isolation, identification, interpretation, and intervention, with imaging as a special type of identification. For isolation, we can identify instruments serving for the separation and purification of compounds, notably chromatographic techniques, electrophoresis, and the ultracentrifuge. Detection affects mostly different kinds of spectroscopy. Isolation and identification are often intertwined. In the course of the century, imaging techniques gained independence and their own technological momentum, with diverse techniques of microscopy, including electron microscopy, and magnetic resonance imaging and computer tomography with its important uses in medicine. The field of interpretation is affected of course by computation, and increasingly artificial intelligence. The last group, namely intervention, includes applications such as scanning tunneling microscopy in emerging nanotechnology (Reinhardt 2020).

All these functions came to be crucially important in the physical and life sciences, and they all rely more or less on chemistry as the major developer of concepts, techniques, and instruments. This caused the "great expansion" of chemistry during the twentieth century into its neighboring disciplines and technologies, as well as medicine. The "mental model" of molecular structures – the characteristic feature of chemistry (Steinhauser 2014) – was applied in biology, physics, nanotechnology, and materials science mainly with the help of these new methods. In the process, chemistry was not reduced to physics. Instead, the new instrumental methods tremendously increased the range of chemistry. This happened largely between the 1930s and the 1980s, when a huge number of novel instrumental methods were used first to complement and later to replace "wet" chemical methods in the isolation, identification, and interpretation of compounds. Thus, chemistry escaped the test tube, and soon scientists could do chemistry almost anywhere, from the atomic nucleus to the universe. In the process, we might argue, the epistemic core of chemistry remained intact, while the new physical methods were successfully adapted. However, the disciplinary landscape around chemistry changed dramatically. Often under the guise of "molecular sciences," chemical concepts and methods traveled far beyond their traditional terrain. Chemists and allied scientists generated a sophisticated support system for solid-state physics, materials science, nanotechnology, biotechnology and genetic engineering, and medicine (Bensaude-Vincent 2018). The historian and philosopher Bernadette Bensaude-Vincent, applying the concept of the rhizome, sees chemistry's expansion as a diffusion into other sciences, often under the guise of technology. In her opinion, four fields appeared at the end of the twentieth century, all of whom, while "impregnated" with chemistry, did not carry its name. These "terrains" are nuclear technology, materials science and engineering, synthetic biology, and nanotechnology. In Bensaude-Vincent's view, chemistry does not command a terrain with sharply defined boundaries. On the contrary, these new fields of technoscience influence and shape the "epistemic profile" of chemistry.

Chemistry and chemical technology were formative forces in the emergence of nuclear technology, with Du Pont, the most important industrial company involved. In the 1960s and 1970s, materials science was dominated by physics, chemistry, and metallurgy, merging into a new entity under the pressure of governmental needs. The new field of synthetic biology almost literally fulfilled Emil Fischer's dream, attempting to construct life based on the molecular tool-kit. Nanotechnology epitomized another one of chemistry's old dreams of complete control, this time at the scale of individual atoms. The earth system sciences is a field where chemists, geologists, climate scientists, and even social scientists have together constructed a new geological epoch, the Anthropocene, which is defined as the era of lasting human impact on the Earth's geology (Bonneuil and Fressoz 2016; Trischler 2016). With this development, Bensaude-Vincent remarks: "chemistry may well have turned the entire Earth into a laboratory, a world laboratory" (2018: 607). This therefore appears to be the most recent stage of chemistry's expansion.

CONCLUSION

This chapter has shown how chemists greatly influenced their neighboring scientific disciplines, technology, and society. In the first part of this chapter I focused on the role of materials created by chemical science and industry. Through its prime characteristic of material productivity, chemistry's impact has been huge, and until the 1960s, it was largely seen as beneficial. In the second part of the chapter I analyzed the important interlude between the 1960s and the 1980s, when chemistry's power was diminished, partly owing to environmental problems, and examined how chemists regarded their changing relationship especially with physics and biology, but also with the public at large. In the third and final part of the chapter, I examined the impact of chemical methods, both in the context of providing solutions to other disciplines as well as posing social and environmental challenges. Arguably, chemistry's impact was greatest when its methods and techniques were deployed in other fields. In addition, new scientific and technical subdisciplines were developed with the help of chemistry, ensuring that chemical thinking and practice was present in large areas of science and technology. However, much of this has not been considered as part of chemistry, but was called molecular biology, materials science, genetic engineering, or nanotechnology, to mention just a few of these fields. Hence, the impact of chemistry in the second half of the twentieth century was greatest when it was not considered to be chemical.

However, the chemists' claim to control the molecular world almost at will, and for the benefit of society, ensured that the world was indeed impacted. Chemistry has been able to provide materials and methods for meeting society's needs, but in the same way, it also created, or helped to create, severe health and environmental problems (Boudia and Jas 2014). The growth of the number of chemical substances, and the volume of their production, has been staggering. For a long time now, chemical pollution has been global. Meanwhile, the generic term "chemicals" has subsumed older categories, such as "toxic" or "hazardous" substances. Despite all this, chemists' efforts to attempt to shape their environment – through the chemical substances they synthesize and use – has not changed. One thing is certain: chemistry, in the form of its residues left behind by a century of science and industry, will stay with us and will continue to impact us (Boudia et al. 2018).

Thus, chemistry represents a combination of the scientific, the industrial, and the environmental mindset. Broadly construed, it can be described as "an enabling scientific approach that shapes – and sometimes creates – the disposition of the world's resources" (Reinhardt 2018: 564). From the very outset, making nature disposable refers to the fact that technological and scientific manipulations – which usually take place in laboratories – are recognized by chemists as their characteristic pathway to nature. Chemists reach knowledge of the natural world by controlling and transforming nature (Bensaude-Vincent 2018). In the public perspective, chemistry is seen as dominated by control, less a science and more a technology. Referring to physics and biology as the dominating scientific disciplines in public perception, the science writer Philip Ball argues:

> [C]hemistry, in contrast, seems to have little to offer in the way of grand themes. In fact, it often seems today not to be asking any questions about the world at all: it is primarily a synthetic science, a science bound up with many things. Even many scientists, if they have no real knowledge of chemistry, seem unable to find a way to fit this discipline into their vision of what science is about, namely the process of discovering how the world works. Most current distinctions that are drawn between science and technology will place today's chemistry squarely within the realm of technology, or at least applied science, concerned as it is much more with invention than with discovery.
>
> (2007: 98)

There is an irony, or a paradox, in the fact that chemically caused environmental problems are now upon us with a vengence, and with an impact that resembles natural disasters and plagues. The chemical world, whose growth was heralded in the 1980s Pimentel Report on opportunities of chemistry (National Research Council 1987), has become our naturalized environment. Some chemists considered this to be another opportunity:

> Moreover, it is now becoming apparent that many of the unanticipated environmental problems created by technology may be stated as chemical problems. Environmental pollution in particular may be understood in terms familiar to the chemist and chemical engineer, and a large amount of relevant chemical information already exists for use in seeking solutions to this complex social and economic problem.
>
> (Handler 1970: 22)

However, for many people, the environmental problems were seen not just as a new challenge for more (and better?) science. In discussing Don DeLillo's novel *White Noise*, Ball comments on "how the prosaic process of synthesis and artifice can generate something that resembles a *natural* hazard," and quotes DeLillo: "This was death made in the laboratory, defined and measurable, but we thought of it at the time in a simple and primitive way, as some seasonal perversity of the earth like a flood or tornado, something not subject to control. Our helplessness did not seem compatible with the idea of a man-made event" (DeLillo 1984: 127; Ball 2007: 106). Thus, at the end of the century, chemistry might have become (again) a "real" science, not a technology, trying to understand a world, although partially made by it, that is now out of its control.

CHAPTER FIVE

Society and Environment: *The Advance of Women and the International Regulation of Pollution*

PETER REED

INTRODUCTION

The period between 1914 and the present has seen remarkable demographic changes: the global population rose from 1.5 billion (1900) to 7.7 billion (2019), and since 2007 a greater proportion of this population lives in towns and cities than in rural areas. Over the same period, rapid advances in chemical science and industrial technology have led to the manufacture of a much wider range of chemical products that have brought unprecedented benefits for society's well-being, including medicines, food, clothing, furnishings, and household products. These consumer products are generally available in advanced industrialized countries, but not all regions of the world have yet benefitted, and day-to-day survival has remained a struggle especially in rural areas. In 2016 it was estimated that 10.7 percent of the world population lived below the international poverty line set by the World Bank (World Bank 2016).

Public concern about chemical products rose after the use of toxic chemical agents during World War I, and and these concerns are now focused on their impact on human health and the environment. Accidents involving highly toxic chemicals have regularly reinforced these concerns. From the 1970s governments and international organizations such as the United Nations and the World Health Organization have taken the lead in identifying harmful chemicals and having their use closely monitored, because health outcomes can often take many years to manifest themselves fully. Chemists (alongside others) have taken an active role regulating chemical products and mitigating environmental damage, while also investigating industrial accidents to understand why the event occurred and to revise protocols to reduce the risk of recurrence.

Educational opportunities in chemistry have expanded dramatically since 1914, both at the school and university levels. Women and minority groups in particular have taken advantage of these opportunities as industrialized societies in nations across the world have adjusted and allowed them to exercise their full capabilities and responsibilities. Many more have studied chemistry and pursued chemistry qualifications that have provided a pathway to leadership roles in academia, commerce, and industry that were previously the prerogative of men.

The following review of the impact of chemistry on society and the environment in the period between 1914 and 2018 embraces six overarching themes: chemical warfare agents and international conventions; chemistry in everyday life; chemistry and the environment; women and chemistry; and climate change and the "Anthropocene" epoch.

CHEMICAL WARFARE AGENTS AND INTERNATIONAL CONVENTIONS

In 1917, Richard Pilcher, registrar and secretary of the Royal Institute of Chemistry used the phrase "the chemists' war" to describe World War I because of the crucial role played by chemistry and chemists in the production of a wide range of chemicals as part of the war effort (Pilcher 1917: 411). These chemicals included explosives (trinitrotoluene (TNT), ammonium nitrate, and amatol), propellants (cordite), metal alloys for weapons, and military equipment and medicines. However, World War I was more notoriously remembered for the deployment of chemical warfare agents such as chlorine, mustard gas, and phosgene, and the horrors of their effects on military personnel. The association between chemistry and these agents raised concern in the public's mind, linking chemistry with damage and danger rather than with the major benefits its application could bring to the industrialized societies of nations.

At the conclusion of World War I there remained deep concern on the part of both the public and political leaders over the future proliferation and deployment of even deadlier agents. In 1922 representatives from the

USA, France, the United Kingdom, Italy, and Japan agreed to ban the use of suffocating, poisonous, and other gases, but the ban was never ratified (Szinicz 2005: 180). A further attempt in 1925 led to the drafting of the Geneva Protocol for the "Prohibition of the Use in War of Asphyxiating, Poisonous or Other Gases, and of Bacteriological Methods of Warfare," and it was signed by thirty nations. Although the major powers had stockpiles, there was no use of chemical agents on the European battlefields during World War II (Evert 2015). However, in the postwar period many countries built up stocks of toxic nerve agents that included sarin and VX. Representatives of the United Nations feared further proliferation, and drafted a new Convention on the Prohibition of the Development, Production and Stockpiling of Bacteriological (Biological) and Toxin Weapons and their Destruction; it came into force in 1975.

Some countries were not deterred. The 1988 attack by Saddam Hussein on the Kurdish people in Halabja, southern Kurdistan, earned worldwide condemnation. To further strengthen the international protocols, in 1992 the United Nations General Assembly approved the Convention on the Prohibition of the Development, Production, Stockpiling and Use of Chemical Weapons and on their Destruction, sometimes abbreviated to the Chemical Weapons Convention. The Convention is administered by the Organization for the Prohibition of Chemical Weapons, based in The Hague (Netherlands), which was awarded the 2013 Nobel Peace Prize. Swift action is mandated for infractions, and in 2013 Syria was pressed by the United Nations Security Council to destroy its stockpiles after deploying chemical agents that it had previously denied possessing. As at 2016, 192 countries had agreed to be bound by the 1992 Convention, and stockpiles around the world have steadily been reduced. However, in 2018 Syria was again accused at the United Nations of deploying chemical weapons against civilian populations. Major concern remains within the international community over the possibility of terrorist groups acquiring these agents and deploying them in a densely populated area.

Two incidents have heightened these concerns. In March 1995 members of a cult movement in Japan released sarin gas in the Tokyo subway, killing twelve people and injuring many more (Tu 1999). In Salisbury (England) in March 2018 two Russian-born British citizens were poisoned with a Novichok nerve agent known as A-234; one of these citizens had been a double agent for Russia and the United Kingdom and had settled in the UK in 2010. Both survived after spending several weeks in critical condition in hospital. Unfortunately, some three months after this incident two UK residents came into contact with Novichok; this was probably linked to the first incident, although the second took place in Amesbury, about 7 miles from Salisbury. One of these people died some three weeks later. The British government accused Russian agents of responsibility, which the Russian government denied.

CHEMISTRY IN EVERYDAY LIFE

In the period between World War I and the first decade of the twenty-first century consumer products changed daily life profoundly for most people living in industrialized societies (Emsley 2015). These consumer products have formed a wide spectrum: pharmaceuticals, synthetic fibers, plastics, synthetic rubber, cleaning products, fireproofing chemicals, agrochemical products, and food. These products have emerged in a haphazard manner as chemical research has advanced, chemical formulations have been assessed for human benefit and safety, and the chemical industry has refined its ability to produce such chemical products on an ever-larger scale to meet the demands of global markets. While most chemical products have brought benefits, many have proved sufficiently dangerous to warrant strict assessment before use and then tight regulation or even an outright ban. The public is largely unaware of the chemical contents of most products it uses regularly, relying on safety regulations and guidelines, especially where health and dietary factors are important.

In the early years of the twentieth century clothing was made of natural materials – cotton, wool, and to a much lesser extent, silk, but as the century progressed advances in chemistry and technology resulted in a series of synthetic fibers that included rayon, nylon, and polyester. Together with the availability of new dyestuffs, these fibers revolutionized clothing manufacture and fashion, and helped establish the increased global demand for clothing (Raitt 1966).

Rayon is a man-made fiber consisting of cellulose fibers (from wood or cotton) that have been modified in solution and then forced through a spinneret to create a number of fine filaments that can then be spun to increase their strength. Several specific processes have emerged. In the 1880s the cellulose fibers were dissolved in nitric acid, but the resulting nitrocellulose (guncotton) was flammable. Ten years later chemists Charles Cross and Edward Bevan discovered viscose rayon, when they used caustic soda and carbon disulfide to dissolve the cellulose fibers. Cuprammonium rayon was produced by dissolving cellulose in a copper solution with ammonia, after which the emerging filaments from the spinneret were forced into a bath of sulfuric acid (Blanc 2016: 24). This rayon did not use the dangerous chemical, carbon disulfide (see later section of this chapter), but it is more expensive to produce. By 1960 some 5 million tons of viscose rayon were produced globally (Raitt 1986). Because of its high tensile strength and its water absorption, rayon found wide application. Rayon staple was blended with cotton and wool for a range of different uses: mixtures of rayon and wool were used for carpets where being hardwearing and resistant to staining were important; and mixtures of rayon with cotton were used in sports clothing where moisture absorption was beneficial.

The first fully synthetic fiber to be discovered was nylon in 1935. Nylon is light and has high tensile strength; thanks to these properties it became widely used for stockings, clothing, ropes, tire-cords, and tarpaulins. Blends of natural fibers with nylon were used for clothing; trousers using a blend of wool and nylon could have a much sought-after permanent crease, and socks used the same blend to increase durability. Terylene (polyester) was discovered over the period 1939–1941 and has properties similar to nylon, but also with heat-setting properties that aided pleats and creases in clothing.

CHEMISTRY AND THE ENVIRONMENT

While the range of ever-more sophisticated chemicals has brought many benefits for consumers, the diversification of chemicals and their increased scale of production have all too frequently detrimentally affected the natural environment, the working environment, and the home environment. Chemists (working alongside others) have taken a leading role in the regulation and monitoring of chemicals in the different environmental settings to ensure healthy conditions are achieved. Too often the long-term impact of chemical commodities on the environment or on people's health was not studied or understood in sufficient detail before their sale and use. More advanced analytical instrumentation has played an increasingly important role in identifying and studying harmful chemical entities to support the work of regulators.

Air Pollution

Air quality after 1914 continued to be affected by harmful emissions from industrial processes and by the smoke from the burning of coal, but two major episodes resulted in important regulation, namely Donora (Pennsylvania, USA) in 1948 and London (UK) in 1952. Later, attention focused on the fuel additive tetraethyl lead, and then on how inefficient burning of fuel in internal combustion engines contributed to photochemical "smogs" in Los Angeles (California). While regulation is constantly under review, major concerns remain over air quality and human health, especially in emerging-economy countries that still rely heavily on coal.

The Donora event is often thought to have started the clean-air movement in the United States that led finally to the Clean Air Act in 1970. The episode started on October 27, 1948, when Donora (a town of about 14,000 residents some 20 miles from Pittsburgh) suffered intense smog, causing many to suffer respiratory difficulties. The town was accustomed to emissions of sulfur dioxide and hydrogen fluoride from the US Steel's Donora Zinc Works and its American Steel & Wire plant, but this event was exacerbated by a temperature inversion, in which warmer air trapped colder air containing the pollutants close to the land surface. The episode continued until October 31, after the plant closed and rain dissipated the pollutants, but not before twenty people died and many others suffered respiratory difficulties. The aftermath took a course that has

FIGURE 5.1 A US Public Health Service worker measures samples of air pollution from the American Steel and Wire Company's zinc works in Donora during a four-day test. Photograph by Bettmann/Getty Images.

played out on many other occasions. US Steel never acknowledged responsibility for the episode, although it settled many financial claims. Had the case gone to court the company was prepared to claim that the episode was "an act of God" (Hamill 2008).

In Britain, the Alkali Inspectorate (established in 1864) was responsible for the control of harmful emissions from industrial processes, but regulation of "black smoke" from the burning of coal was still outside its remit even though several major smog episodes had occurred between the two world wars. However, a major smog episode in London in December 1952 eventually brought about far-reaching parliamentary legislation. The impact of the episode was evident from newspaper reports recording the effects on London's daily life, in which it was noted that air quality monitoring equipment had failed due to the high level of pollution and the size of the particulate matter in the smog. Although mortality estimates were put at about 4,000, the government showed little concern. A later review of the episode concluded that the mortality rate might be closer to 12,000 (Bell et al. 2004). Its stance was that these smogs were natural events that had to be endured; the severe floods along the east coast of England in 1953 met with the same response. However, public pressure induced the government to set up a committee of enquiry in May 1953, but while the committee was still deliberating, the City of London Corporation secured a parliamentary bill to create smokeless zones across the City of London area (an area much smaller than the whole of London) that led the way for smokeless zones in other local authorities (Ashby and Anderson 1981: 110–11). While the government dithered, several members of Parliament resolved to take urgent action. A Private Member's bill received wide support across the political spectrum when it was debated by Parliament. Having lost the initiative, the government decided to draft its own legislation based on the Private Member's bill; this legislation became the Clean Air Act 1956 (Reed 2014: 173–5). The legislation's impact was subsequently enhanced by the switch to electricity and gas for heating, the adoption of domestic central heating systems, and the designation of smokeless zones across many parts of the United Kingdom. With major smog episodes eliminated, emissions from internal combustion vehicles would become the main cause of poor air quality in cities with continuously high traffic levels.

The smog associated with Los Angeles (California, USA) has a different cause. Motor vehicles' inefficient burning of petroleum products results in an aerosol smog (Reed 2014: 175). The situation was exacerbated by the air inversion associated with Los Angeles that kept the polluted air close to ground level. The first smog episode was in 1943. That cars (and other internal combustion engine vehicles) were responsible only emerged in the 1950s as a result of research by the Dutch-born Caltech biochemist, Arie Jan Haagen-Smit (Bonner 1989). By 1966 California had set stringent vehicle emission standards; Volkswagen cars were found to violate the standards and in 1969 their sale was prohibited until the standards were met. In 1967 the California Air Resources Board (CARB) was established, and the following year Haagen-Smit was appointed its chairman. The CARB has remained a leading regulator in setting stringent air quality controls; its work influenced the federal Clean Air Act of 1970 and has continued to inform the work of the federal Environment Protection Agency (EPA) established by President Nixon in 1970 (Lewis 1985). Several US states have followed California's regulatory approach. During the 2000s the CARB took the lead in controlling greenhouse gas emissions that scientists now believe are responsible for global warming of the planet (see below; California Air Resources Board 2017).

Acid Rain

Although acid rain (rain acidified by the oxides of sulfur from the burning of coal, and from the oxides of nitrogen) was identified as far back as 1859, it has continued to wreak havoc on the natural environment as well as on metal and limestone structures. Countries in Europe and in North America were regular offenders in the period up to the 1980s when legislation began to control acid rain. In more recent times, emerging industrialized countries such as China, India, and Brazil have struggled to reduce its impact because of their dependence on coal.

Transnational agreements have also been necessary, because acid rain (like other forms of air pollution) does not recognize national boundaries. Some of the worst examples of acid rain occurred in the 1970s in the Soviet Bloc countries, especially the German Democratic Republic where large quantities of low-grade coal (containing high levels of sulfur) were consumed in substandard industrial plants that lacked suitable filters. The acid rain released from tall chimneys traveled across the Scandinavian countries taking a toll on forests and woodlands. The UK's electricity generating plants were also accused of contributing to the damage. In response, the Geneva Convention on Long-Range Transboundary Air Pollution was established by the United Nations Economic Commission for Europe (UNECE). By 2017 the Convention had the participation of fifty-one parties, and addressed acid rain across the UNECE area, the Caucasus and Central Asia, and South-East Europe; information was also shared with other countries to achieve global regulation. Investigations have also shown that one of the main sources of acid rain in Europe is from the dimethyl sulfide produced in the extensive algae masses in the North Sea.

In North America both the United States and Canada were prompted to take action, having experienced the damaging effects of acid rain on freshwater and terrestrial ecosystems, buildings, and health. In 1980 the US Congress approved the Acid Precipitation Act that initiated an eighteen-year research and assessment program on acid rain management (the National Acid Precipitation Assessment Program), and then amended as the Clean Air Act in 1989. Later in 2005 the EPA issued the Clean Air Interstate Rule that set targets for reducing sulfur dioxide and nitrogen oxides across the eastern states and the District of Columbia by over 70 percent and over 60 percent, respectively.

CFCs and HCFCs

Chlorofluorocarbons (CFCs) were used widely from the 1920s as refrigerants and aerosol propellants, but in the 1970s they were identified as a possible threat to the ozone layer in the upper atmosphere, which protects the Earth from the sun's damaging ultraviolet radiation. CFCs are often referred to by their Du Pont trade name, Freon; Freon 12 (dichlorodifluoromethane) is one of the most common. With the manufacturing patent held by Du Pont due to expire in 1979, the United States banned the use of CFCs in aerosol cans in 1978, but the announcement in May 1985 that the British Antarctic Survey had detected a growing ozone hole over Antarctica the year before raised international concern. In 1987 the Montreal Protocol was drawn up to reduce the production of CFCs, but with adverse evidence mounting, in 1989 the twelve nations in the European Economic Community (EEC) agreed to ban production of CFCs by the end of the century. In 1990 the Montreal Protocol was strengthened to call for the elimination all CFCs by 2000, although CFCs were still permitted for certain medical applications. While alternatives to CFCs have been found, an illegal trade in CFCs has remained, especially in the Far East. Du Pont promoted their patented HCFCs (hydrochlorofluorocarbons), although later they, too, were shown to be damaging. In September 2007, 190 countries committed to the complete elimination of HCFCs by 2020.

Pesticides

Use of pesticides became widespread after World War II in an attempt to increase agricultural yields and aid food production. In Britain during the late 1940s farm workers were reported to have died from the use of the herbicide 4,6-dinitro-o-cresol, and following an investigation, Parliament approved the Agricultural (Poisonous Substances) Act of 1952 (Gay 2012: 92). In the early 1950s, two new insecticides (aldrin and dieldrin) were used in the UK, but evidence soon emerged of game birds dying, raising concerns about the ecological impact of organochlorine insecticides. Evidence also emerged from the Government Infestation Control Laboratory and the Laboratory of the Government Chemist of increased levels of chlorinated hydrocarbon resides in food such as lamb, mutton, beef, and milk. In 1965 the Natural Environment Research Council was established to focus research on the science of ecological systems. Britain was less affected by the use of DDT than the USA because of the government's promotion of the use of pyrethrins – extracts of the daisy-like *Pyrethrum* flower – that were found to be quicker in action than DDT and as a natural product was biodegradable, although it was expensive to produce (Morris 2019).

The chlorinated hydrocarbon DDT was used extensively during World War II by both Britain and the USA to protect their troops against insect-borne diseases, but also in the battle to control diseases carried by the many refugees in the aftermath of the war (Kinkela 2011). It was also found to be very effective against the malaria vector, which increased DDT's global use. In the immediate postwar period DDT was used widely in the USA for agricultural purposes, for instance to combat the Colorado potato beetle (potato industry) and the boll weevil (cotton industry), but because it was both effective and inexpensive, it was overused. Evidence gradually emerged that showed that DDT residues pass up food chains to harm top predators. Furthermore, insect species became resistant to the insecticide, reducing its efficacy, as the 1957 battle against midges at Clear Lake, California demonstrated. The dangers of DDT gradually mounted over time, public concern being expressed through a number of conservation organizations, such as the National Audubon Society. From 1957 federal controls were put in place on government land. Action groups against DDT were formed, including the Brookhaven Town group (with its base at the

Brookhaven National Laboratory) that in 1966 fought the use of DDT to kill mosquito larvae at Yaphank Lake (Long Island, New York; Dunlap 1981).

The publication of Rachel Carson's book *Silent Spring* in 1962 added to concerns over not only DDT, but also the indiscriminate and widespread use of all pesticides. Carson and her book were heavily criticized, not only by the chemical industry but also by some politicians and scientists. In advance of publication Carson had wisely asked several leading scientists in the field to read a draft and confirm the science. Although public interest was stirred by the pre-publication publicity and the book's serialization in the *New Yorker*, it was Carson's appearance on a TV program with high viewer ratings that really drew the public's attention to the issues (Lear 1997). This attention made it almost impossible for the regulatory authorities to dismiss Carson's arguments. In 1964 an amendment was made to the US Federal Insecticide, Fungicide, and Rodenticide Act regarding labeling of pesticides and provision of safety information. However, at the time the US Department of Agriculture had a conflict of interest, as it was in charge of regulating pesticides while also promoting the interests of the agricultural industry and farmers. This conflict was a factor in the establishment of the more independent EPA in 1970. In 1972 DDT was banned for agricultural use in the US (but not for the eradication of malaria). The Stockholm Convention on Persistent Organic Pollutants, signed in 2001 and implemented from 2004, followed the US approach on DDT (Downie and Templeton 2014). It was adopted into European Union law in 2008, and as of 2017 there were 181 signatories. The Convention regularly reviews potential pollutants and updates the list of chemical agents with restricted use.

Research on pesticides has been aided by the development of new analytical techniques able to measure minute quantities, not just a few parts per million (ppm) or even parts per billion, but by 2017, parts per trillion. Detection of DDT in foodstuffs in the 1950s employed the Schechter–Haller method developed in 1945 that made use of the colorimeter, a widely used laboratory instrument able to measure concentration by comparing the color associated with the sample against a standardized color chart (Morris 2009). Such a method could measure to about 2.5 ppm, but with the US Food and Drug Administration wanting to set maximum levels for pesticides on fruit and vegetables in line with levels tolerated by the human body, impetus was given to find more rapid and reliable methods. Ultraviolet spectrophotometers were found to be inaccurate and slow even though they had a detection limit of about 0.1 ppm. A "cranberry scare" in the US in 1959 prompted the search in both the USA and the UK for more reliable and consistent instruments able to detect minute concentrations of DDT (and other chlorinated hydrocarbons). Attention focused on the gas–liquid chromatograph invented by Archer Martin and Tony James at the National Institute for Medical Research in north London in 1951. The key to its application for pesticide measurement was its reliability and sensitivity (Morris 2009). This advance prompted the development of two different detector techniques: Dale Coulson's microcoulometry and James Lovelock's election capture detector. Instruments with these detectors were still in use in the 2010s in the USA and the UK, and levels of about 10 parts per quadrillion were detectable (Morris 2009).

Water, Soil, and Groundwater

The pollution of water, soil, and groundwater as a result of industry, agriculture, human sewage, major weather events, and consumer waste has continued. Better understanding of the health dangers and improved detection of chemical agents as described earlier have led to tighter regulation in more advanced countries of the world. In the twenty-first century, major incidents were more likely to result from accidental spillages, changes in sources of drinking water, or major weather events such as hurricanes. Even the oceans were experiencing increasing pollution due to the mounting quantities of plastic waste.

In Britain, legislation enacted during the nineteenth century remained in place (but with incremental tightening of regulations and an updated list of potential contaminants) until 1973, when Britain acceded to membership of the European Economic Community (EEC). Membership of the EEC required Britain to adhere to its environmental regulations, but it largely resisted applying them until the 1980s. Having signed the Single European Act in 1986, and upon ratifying the Maastricht Treaty in 1993, it became a member of the European Union (EU) and had to adhere to the EU's environmental policies and regulations. As far as Britain's water management was concerned, control of sewage treatment works remained at the local level with final effluents closely monitored on a continuous basis before their return to the main water courses. Most of the reservoirs created during the nineteenth century have continued in use, but with the steadily increasing demand for water, urgently needed management strategies have become a web of complex issues that include water pollution and treatment, water uses, and water quality (Tomkins 1996: 38–9).

The major water pollution issues faced by Britain in the nineteenth century came later in the United States because of its later start to urbanization and industrialization. Although incidents of industrial pollution were reported in local newspapers, no thorough survey was completed until 1922, when the American Water Works Association published a report by its Committee on Industrial Wastes in Relation to Water that documented that pollutants had damaged at least 248 water supplies in the United States and Canada (Tarr 1996: 360–1). Control of such pollutants was a state responsibility before World War II; the Oil Pollution Control Act of 1924 protected fishing and other commercial activities rather than public health. The New Deal of the 1930s provided federal funding for construction of urban sewage systems and water treatment plants: by 1939 the American population with access to sewage treatment rose to 39 million compared with 21.5 million in 1931 (Tarr 1996: 370). Following World War II the range of pollutants widened to include chlorinated hydrocarbons, synthetic detergents, and a range of carcinogenic chemicals. The regulations put in place through the various Clean Water Acts of the 1970s strengthened efforts to clean up many of the major rivers, including the Delaware and the Potomac.

Groundwater is found in aquifers deep within the earth in the cracks and pores of rocks, and forms an important part of the hydrologic cycle. In the United States groundwater has remained an increasingly important source of drinking water; nearly 50 percent of the US population relies on wells or springs for their drinking water (Pye and Patrick 1983: 713). Groundwater has always been susceptible not only to depletion, but also to contamination; increasing industrialization, urbanization, and agriculture have exacerbated the damage to this important source of drinking water. Pathogenic organisms and toxic chemicals have harmed ground water and damaged public health. Remedial action to remove or mitigate these agents proved expensive and often extremely difficult (if not impossible). In 2006 the EPA established the Ground Water Rule to address bacterial contamination associated with ground water, but other pollutants remain outside current regulation (EPA 2008).

A major episode affecting groundwater was contamination of the Love Canal landfill (near the city of Niagara Falls in New York State) by highly toxic industrial waste between 1942 and 1953 (New York State 1981). In the 1950s the site was used (without any cleanup) for a school, and later for housing. By the 1970s serious odors and obnoxious fluids were evident, and by April 1978 the area was declared a threat to human health. In 1980 the state authorities established the Love Canal Area Revitalization Agency to oversee remediation, with funding from the federal Superfund Act. The work took twenty years to complete at a cost of $400 million (Depalma 2004; Newman 2016).

The synthetic dyestuffs industry has also contributed to large-scale pollution of land, ground water, and waterways. The principal sites were those closely associated with the industry in Germany (Rhine valley), Switzerland (Basel), France (Lyon), and Britain (Manchester). It was not until World War I that the industry began to develop in the United States, and its subsequent rise was rapid. The dyestuffs industry was not the sole culprit; its intermediates were used by other industries, including pharmaceuticals and agrochemicals. Many residues and effluents from the industry proved to be highly toxic: carcinogenic dyestuff intermediates included beta-naphthylamine and benzidine, and were found to be resistant to microorganisms employed in waste treatment plants. Until the 1970s dye-making waste was rarely treated. This was during a period when major advances were made in the rapid detection of pollutants to levels of parts per million, using techniques based on gas chromatography (GC) as described above (Travis 2002a: 37).

By the 1960s increased public awareness and concern was heightened by the pictorial coverage of incidents showing colored river water or the large-scale death of fish in waterways. The establishment of the EPA in 1970, followed by further legislation including the Federal Water Pollution Control Act of 1972, began to address the outstanding issues. However, in the mid-1990s the pollution problems still remained acute; it was estimated in the United States that there are up to 400,000 sites threatening groundwater and soil contamination, with associated clean-up costs of $1 trillion (Travis 2002a: 49).

The contamination of water by lead has remained a major issue to the present day in the United States (Brown and Margolis 2012: 1). Even though lead levels were reduced from the 1970s due largely to the ban on tetraethyl lead additive in gasoline (petrol) and on the use of lead in paint, lead still continues to find its way into household water because of old lead piping and fixtures. Levels in children remain a major concern because accumulations of lead are known to affect their health, especially brain functioning. In 2014 the city of Flint (Michigan) changed its source of drinking water from Lake Huron and the Detroit River to the Flint River as a cost-saving exercise, but this change resulted in a sharply elevated level of lead, which forced a switch to bottled water and other sources. Action to reduce the lead in Flint's drinking water to the EPA-approved level was only completed in 2018. Meanwhile during 2017 in California the drinking water of many schools was found to have levels of lead above the 15 parts per billion stipulated by the EPA. The drinking water systems were immediately closed down until the old lead piping was replaced. According to many public health experts, the EPA-permitted level for lead is still too high and should be as close as possible to zero, given the likely long-term effects on children.

The oceans have largely been able to absorb pollutants without serious damage, but plastics waste has become a mounting issue. Plastics pollution of the oceans was first reported in the 1970s, but it took forty years before studies were undertaken to quantify the scale of the pollution prompted by concerns over the effects on the marine environment. Commercial plastics developed during the 1930s and 1940s emerged to occupy a very prominent place in the consumer marketplace by the early 2000s. For example, food hygiene regulations have contributed to the use of plastics as packaging materials. In the United States in 1960 plastics made up less than 1 percent of municipal solid waste by mass, yet by 2005 for sixty-one countries (58 percent of those with available data) plastics comprised at least 10 percent (Jambeck et al. 2015). But increasingly plastics are finding their way into the oceans through contributory waterways or through weather, and the scale is mounting to a dramatic level both on the sea surface and the sea floor. A research study that made use of data gathered by twenty-four expeditions across the global oceans between 2003 and 2013 concluded that more than five trillion plastic pieces weighing over 250,000 tons were floating at sea (Eriksen et al. 2014). In 2017 it was estimated that each year some nine million tons of plastics find their way into the oceans. By December 2017, 192 countries had signed the non-binding United Nations Clean Seas campaign, a first step in tackling the plastics pollution affecting the global oceans (Parker 2018: 40, 69). With the growth of national economies across the world the use of plastics will surely increase, requiring new waste management strategies not only in industrialized countries, but also in developing countries.

Working Environments

Most industrial processes have been conducted with little or no detrimental effects on workers within factories or on people living in the immediate surrounding areas, because of sound safety and working practices, often enforced by strict statutory regulation. However, workers in certain industries or in poorly operated factories have had to work under adverse conditions. Such conditions have endured because of poor management, lack of adequate regulation, or the sometimes decades-long delay in confirming a causal link between working practices and a detrimental health condition. The cases of asbestos and of carbon disulfide in the manufacture of viscose rayon illustrate how the tension between economic enterprise and regulation to protect workers often play out over long periods of time.

Asbestos is a naturally occurring silica mineral that exists in several different forms, each having a distinctive color, but all sharing a fibrous constitution. It has several remarkable properties that have led to its widespread use, including its tensile strength and its resistance to heat, fire, and corrosive agents. It was due to these properties that asbestos was extensively used in construction projects from the end of the nineteenth century, when asbestos began to be mined on a large scale. Asbestos's use accelerated during the 1940s in the USA and in the 1950s and 1960s in the USSR. It was estimated in 1982 that about 30 million metric tons of asbestos had been used in the United States since 1900 (Craighead and Mossman 1982: 1446).

When asbestos is installed or removed from buildings, fibers released into the air get into the lungs of workers, causing asbestosis, mesothelioma, and lung cancer. The first case of asbestosis was diagnosed in England in 1924 and led to Parliament approving the first asbestos industry regulations in 1931. The use of asbestos was banned in 1999 by the Asbestos (Prohibition) (Amendment) Regulation, while the Control of Asbestos Regulations of 2006 governed the removal and disposal of asbestos. Although the US government had known of the health concerns, industry resisted, so American asbestos regulation experienced delays. The EPA drew up regulations for the asbestos industry in 1991, but a court order stopped enforcement. Worldwide consumption declined from about 4.73 million tons in 1980 to about 2.11 million tons in 2003 (Virta 2006: 17). The use of asbestos in new construction is now banned in many developed countries, including the European Union, Australia, New Zealand, and Japan, but it still remains widely used in China, India, Russia, and Brazil (Virta 2006: 17).

Carbon disulfide is a very efficient solvent, but its toxic effect on humans has been known since it was first produced in the 1840s. Identified health symptoms included headaches, muscle weakness, and bodily numbness (Blanc 2007). In 1892 a new method of producing artificial silk was patented in Britain that used carbon disulfide to convert cellulose pulp (from cotton linters and waste) into viscose rayon. This industry quickly took off on a very large scale in the United States (and elsewhere), even though concerns over workers' health were raised as early as 1904 (Jump and Cruise 1904). Production continued in the US and Britain because of the high profits: in the US, Du Pont's net profits from rayon production between 1922 and 1925 averaged 32 percent. Health concerns were downplayed because of the industry's contribution to national and international trade and the unemployment that would result should the factories be closed down. This was another case of economic gain overriding workers' health, alongside asbestos. However, the industry expanded to many other countries including Belgium, France, and Italy well into the 1940s (Blanc 2016: 164). Even investigations in Italy in 1930 linking carbon disulfide with Parkinson's disease did not persuade manufacturers to improve working conditions.

Other applications of carbon disulfide, such as grain fumigation, attracted the attention of the regulators in the United States, and in 1985 the EPA banned its use. Meanwhile, carbon disulfide was used in the manufacture of a new soil-treatment fumigant, metam sodium. It was found that the fumigant broke down in air with the release of carbon disulfide and methyl isothiocyanate (also released in the Bhopal incident in India in 1984 – see below). Over many years, agencies in the US have been at variance in enforcing an occupational limit for carbon disulfide: in 1977 the National Institute for Occupational Safety and Health, a research agency established by Congress, recommended a limit of 1 part per million, but the Occupational Safety and Health Administration (within the Department of Labor) still retains a limit of 20 parts per million.

Control of Toxic Substances

Governments have been obliged (sometimes under public pressure) to adopt regulations to reduce risk to human health and the environment by toxic substances. In the United States and Europe, important regulatory frameworks were adopted after World War II. In the United States, the Toxic Substances Control Act (TSCA), passed in 1976 and administered by the EPA, was approved in response to a 1971 report by the Council on Environmental Quality, which drew urgent attention to the prevailing lax regulation of chemicals. Under the TSCA, manufacturers are required to submit notification to the EPA before manufacturing or importing new chemicals for use in commodities. However, the TSCA has been criticized on a number of counts: the large number of chemicals in the US market; the high cost of assessing so many chemicals; the initial inventory of some 62,000 chemicals that are "grandfathered" and thus do not require EPA testing; and much of the information submitted by manufacturers is proprietary so it is not released to those wanting to challenge the EPA's conclusions. The EPA has been accused of not providing sufficient protection for children, pregnant women, or occupational workers. Several states have adopted their own more stringent regulations, exacerbating an already complex situation. In 2016 Congress approved the Frank R. Lautenberg Chemical Safety for the 21st Century Act to amend the TSCA's protocols for chemical assessments, begin prioritized review of existing chemicals, and improve public information transparency (EPA 2016).

The EU introduced the Registration, Evaluation and Authorization of Chemicals (REACH) in 2007 to improve the regulation of chemicals. The twenty-eight member countries of the EU – with a combined population of around 510 million (in 2017) – have agreed to REACH, which required all companies manufacturing or importing chemical substances into the European Union in quantities of 1 metric ton or more per year to register these substances with the European Chemicals Agency (ECA) based in Helsinki. Some 143,000 chemical substances were registered in advance of a December 2008 deadline. The ECA also compiled a list of chemical substances of very high concern (SVHC); as of 2017, there were 173 chemical substances awaiting thorough evaluation.

Public concern over the impact of chemicals on health have remained high, and regulatory bodies have had to review their procedures regularly to ensure that control limits for chemicals are stringently assessed. The long-term effects of chemicals on children and other vulnerable groups have also brought renewed determination and revised procedures on the part of regulatory bodies.

Major Industrial Incidents

Poor operation of complex chemical factories has resulted in major environmental disasters, with the loss of many lives, including incidents at Flixborough, England, in 1974; Seveso, Italy, in 1976; and Bhopal, India, in 1984. On June 1, 1974 an explosion occurred at the Nypro plant at Flixborough owing to the escape of cyclohexane and the formation of a cloud of inflammable vapor. The event was caused by a faulty by-pass system installed some three months before to replace a faulty reactor vessel. Twenty-eight plant workers were killed and thirty-six suffered serious injuries. Many in the local community were injured as well, and the surrounding area suffered extensive damage. The report by the UK Health and Safety Executive (HSE) drew attention to a number of defects in plant layout, maintenance, and operating procedures, and put in place revised procedures for the design and operation of such plants to improve safety and reduce risk (HSE 1975).

Two years after Flixborough, a major incident occurred at Seveso, 12 miles north of Milan, during the production of 2,4,5-trichlorophenol at a plant owned by a subsidiary of the Roche Group. The malfunction of a reactor relief valve resulted in the release of about 6 tonnes of a chemical mixture, including 2,3,7,8-tetrachlorobenzo-*p*-dioxin (TCDD) and related isomers, that was dispersed over an area of 18 square kilometers. TCDD – today usually referred to simply as "dioxin" – is extraordinarily toxic; within days of the release 3,300 animals were found dead and fifteen children were treated for chloracne (skin inflammation). A study in 1991 drew attention to the long-term health effects of those exposed to the dioxins; these included chloracne and birth defects, although later studies also made reference to various forms of cancer (Bertazzi 1991). Extensive soil decontamination was carried out to reduce the dioxins to safe levels and in 1980 a compensation agreement for 20 billion lire was signed, although the Italian government had earlier approved loans up to 112 billion lire to meet claims. The incident also brought improved industrial safety regulations, approved by the EEC in 1982 and called the Seveso Directive.

On December 3, 1984, Union Carbide India Limited's pesticide plant in Bhopal accidently released about 40 tons of deadly methyl isocyanate, killing more than 3,800 people (Broughton 2005: 6). Methyl isocyanate is a colorless, poisonous, and lachrymatory agent and is extremely harmful to human health with long-term morbidity and debilitation. Union Carbide Corporation eventually paid compensation of $470 million (USD), but this is surely inadequate compensation given the long-term impact on health. India at this time was in the early stages of industrialization, and successive governments have since revised regulation to avoid a similar accident in the future.

Indoor Environments

Links between health concerns and outdoor air quality have been studied since the nineteenth century, but it was only in the 1970s that attention became focused on indoor environments, namely homes, offices, factories, and schools. The Regional Office of Europe of the World Health Organization (WHO) has been very active investigating the links between health and the various agents found in indoor environments. Its first report, *Air Quality Guidelines for Europe*, was published in 1987, with a second edition in 2000. A more recent report highlighted dangers to health due to benzene, carbon monoxide, formaldehyde, naphthalene, nitrogen dioxide, polycyclic aromatic hydrocarbons, radon, trichloroethylene, and tetrachloroethylene (WHO 2010).

Asthma has been known since Egyptian times as a disease affecting the airways of the lungs. In 2017 the World Health Organization estimated that asthma affected 235 million people globally, and had caused the death of 383,000 in 2015; it is especially debilitating for children (WHO 2017). Although both genetic and environmental factors are thought to be causative factors, research has focused on the role of air pollution. Although some research has shown a correlation between asthma and outdoor air quality, a clear causal link has eluded researchers. More recent investigations have focused on the home environment where children spend a disproportionately large portion of their lives. The home environment is complex, and is influenced by a wide range of contaminants, including volatile organic solvents, radon gas, particulates, as well as biological agents (including mites, pollen, bacteria, and viruses). The experimental control of such environments for researchers has proved challenging, but chemists have continued to work alongside public health specialists and epidemiologists to better understand the cause of this debilitating and often life-threatening disease.

WOMEN AND CHEMISTRY

Before World War I few women were employed in the chemical industry because such work was "regarded as incompatible with both their femininity and gentility" (Horrocks 2000: 354). However, the war brought major changes for women: those qualified as chemists began working in industrial laboratories, and large numbers were employed in chemical manufacture; in particular, the increased demand for cordite and explosives such as TNT employed thousands (Rayner-Canham 1998: 165). In Britain, a number of government factories were built to meet the demand of the battlefield, and because most men were at war, large numbers of women were recruited to work in these factories; the proportion of women was as high as 88 percent, although their wages were only two-thirds of those of men doing the same work. The work undertaken by women could be debilitating and dangerous: particularly hazardous were the preparation of "devil's porridge" (the mixing of nitroglycerine and collodion during the production of cordite) and TNT, where the women were called "canary girls" because their skin turned yellow from exposure to nitric acid.

There were few qualified women chemists in this period. One who gave distinguished service was May Sybil Leslie (1887–1937). At the start of the war she was an assistant lecturer and demonstrator in the chemistry department at University College Bangor (north Wales), but with the pressing demand in government factories Leslie was appointed Research Chemist at the Litherland Factory in Liverpool, and was later promoted to the post of Chemist in Charge of Laboratory. At the end of the war, the government munitions factories closed almost immediately; the women workers returned to their traditional role, while Leslie was appointed Demonstrator in the Department of Chemistry at the University of Leeds (Rayner-Canham 1998: 166–9).

The interwar years were a difficult time for women chemists in both Britain and the United States. Discrimination against women increased in industry, as men realized that jobs previously assigned to men were within the capability of women (Roberts 1988: 67). In academia during 1919–1920, American women formed 47 percent of the undergraduate chemistry enrolments, and by 1929, 10 percent of chemistry doctorates were awarded to women. The Depression lowered the overall recruitment of chemists by industry, so the supply of male graduates met the demand. Common attitudes emerged against recruiting women in industry: training investment was lost when women left to get married; women were deemed not suitable for research (a view often expressed by male researchers); and women were unlikely to have the "flash of inspiration" associated with male researchers. While this hostile view of women prevailed, many qualified women chemists found opportunities in routine chemical analysis, and as chemical librarians or technical editors, roles then considered appropriate for women (Rayner-Canham 1998: 179, 199).

World War II brought additional opportunities compared with World War I. In the United States, the chemical industry employed women to replace men who were fighting on the many war fronts, and academia also recruited women in large numbers to faculty positions (Rayner-Canham 1998: 201). Unlike during World War I, however, industry was largely cautious about employing women. Walter J. Murphy, editor of *Industrial and Engineering Chemistry*, spoke for many in industry when he addressed prospective female applicants, saying (in Margaret Rossiter's paraphrase): "They should realize that women, in general, had so many weaknesses that an employer would find it hard to take them seriously. Almost any male chemist was preferable to any woman ... " (Rossiter 1995: 17–18).

Similarly, the postwar years to the 1950s were not promising for women chemists. In the USA, fewer women were employed in academia, government, and industry, and there were few protests (Rossiter 1995: 49). However, there were many outstanding women chemists in the UK and the US during this period; two of these were Dorothy Crowfoot Hodgkin (1910–1994) with her work in crystallography, and Emma Perry Carr (1880–1972) with her contribution to chemical education.

Dorothy Crowfoot had an outstanding school record and gained a place at Somerville College, Oxford, to study chemistry. She was one of a small group of four scientists in the year group at Somerville (Ferry 1998). For part two of her honors degree, Crowfoot studied crystal structures using recently acquired x-ray equipment. Graduating with first-class honors, she worked in John Desmond Bernal's laboratory at Cambridge and used x-ray diffraction techniques to show that proteins had a regular structure. She completed her Ph.D. in 1936. Hodgkin then continued her x-ray crystallography research at a number of different laboratories leading to elucidation of the structure of penicillin, Vitamin B_{12}, and insulin. She received many honors: being elected a fellow of the Royal Society (1947); and awarded the Nobel Prize for Chemistry (1964) (Ferry 1998).

Emma Carr had studied chemistry at Ohio State University and Mount Holyoke College before completing a chemistry degree at the University of

FIGURE 5.2 The British chemist Dorothy Hodgkin showing the model of protein molecular structure. Anni Settanta. Photograph by Mondadori Portfolio via Getty Images.

Chicago in 1905. After a period back at Mount Holyoke she returned to Chicago to complete her Ph.D. in 1910. On her return to Mount Holyoke she was appointed associate professor and then in 1913 full professor and head of department (Rayner-Canham 1998: 188–9). Carr was a charismatic teacher who valued collaborative work, and was also an outstanding head of department, establishing a chemistry course on a par with that at Yale University and with an emphasis on research (Rayner-Canham 1998: 189). Her collaborative approach to research on the structure of organic molecules using spectroscopic analysis established Mount Holyoke as a research institution and laid the foundation for the role of research in the study of chemistry. In 1935, Carr was awarded the Chemical Foundation's medal in recognition of her distinguished contribution to chemistry.

It was only during the second half of the twentieth century that women regained the ground and were able to become "chemists as well as women" (Rayner-Canham 1998: 202). The American Chemical Society workforce data survey in 2000 showed a shift in demographics over the previous fifty years, with many more women in the chemical workforce.

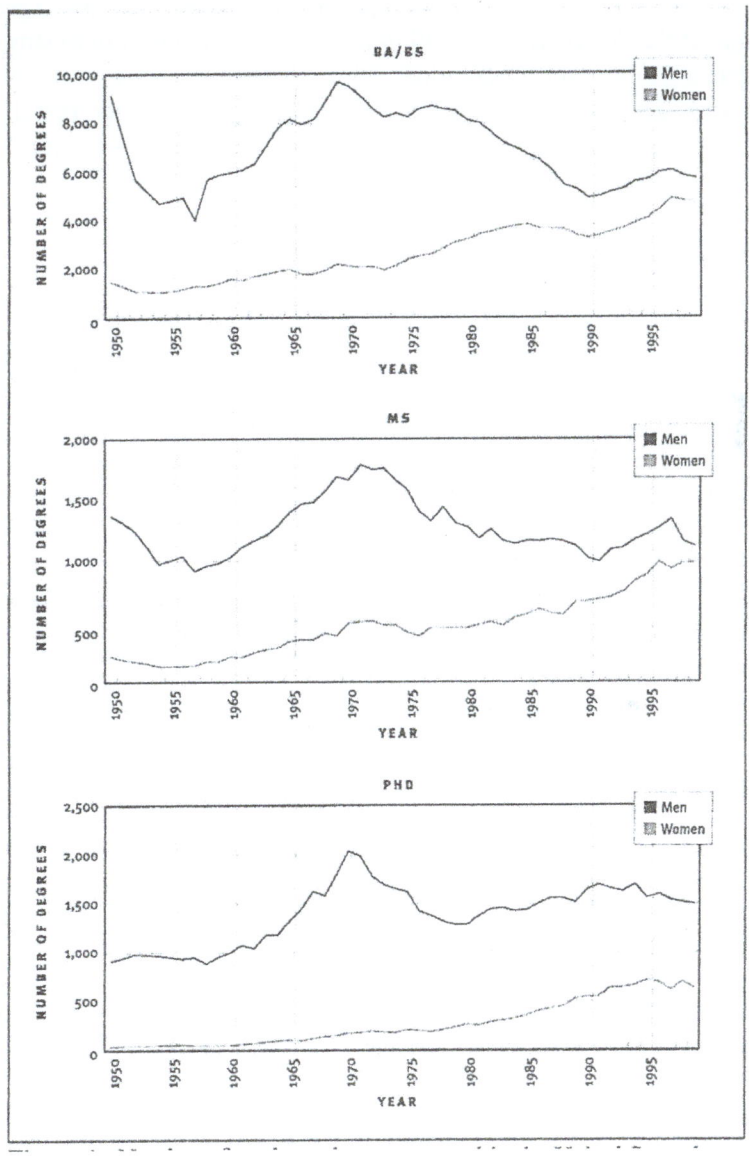

FIGURE 5.3 Number of graduate degrees granted in the United States by year and gender, 1950–1999. © American Chemical Society.

By the year 2000 women were earning more bachelor and master's degrees than men (55 percent compared with 34 percent) while men were attaining more doctorate degrees (64.4 percent compared with 44.7 percent) and women were less likely to be in full employment than men (see Table 5.1; Hinkle and Kocsis 2005: 4). The distribution of women and men across the main sectors of industry, government, and academia was equal to those at the bachelor level although more women were in academia. However, there was a marked difference at the doctorate level, with fewer women in industry but many more in academia (Hinkle and Kocsis 2005: 4–7).

The demographics of all chemists by gender and ethnicity reveal some changes over the period from 1990 to 2000 (see Table 5.2). Women chemists were younger than male chemists, and the data for ethnicity show a decline in white women chemists (from 89.0 percent to 80.8 percent), but a marked increase in Asian women chemists (from 6.2 percent to 11.9 percent; Hinkle and Kocsis 2005: 8–9).

The European Commission reported in 2013 that "while more and more women are reaching senior levels in science and engineering, the aspiration of gender equality is not yet fulfilled ... and there are nowhere near enough women at the top level of science and research" (European Commission 2013: 1). Data from most of the EU countries in 2010 (with the exception of Greece, Ireland, Malta, and Poland) showed that the proportion of senior academic positions (grade A) filled by women ranged from Romania (36 percent) and Latvia (32 percent) through Italy and Sweden (20 percent), Britain and Spain (17 percent), Germany (15 percent) to Luxembourg (9 percent; European Commission 2013: 2).

CLIMATE CHANGE AND THE ANTHROPOCENE EPOCH

Since World War I there has been increasing concern among scientists that the Earth's climate is changing. Such changes are not short-term, but are statistical variations of weather measured over decades or longer with many contributory factors. Evidence for such variation emerged from research undertaken by John Tyndall in Britain and Svante Arrhenius in Sweden during the nineteenth century (Fleming 1998: 70, 76–80). During the first five decades of the twentieth century there was much speculation about the theories of Tyndall and Arrhenius alongside other emerging theories (Fleming 1998: 108–9). By the 1950s concerns over rising sea levels, loss of habitat, and changing agricultural zones were regular features in scientific journals and the popular press. In 1961 the British engineer Guy Callendar, who as early as 1938 had drawn attention to the possible role of anthropogenic carbon dioxide from the burning of fossil fuels, concluded a longitudinal study that quantitatively confirmed the link.

TABLE 5.1 All chemists by employer, highest degree, and gender, 1999–2000 (%). © American Chemical Society

			All chemists				All men				All women		
			Year				Year				Year		
			1990	1995	2000	1990	1995	2000	1990	1995	2000
Highest degree	Bachelor chemists	Industry	73.2	77.9	83.1	74.5	79.8	83.7	69.9	73.8	81.9
		Government	10.0	8.0	7.6	9.7	7.7	7.7	10.9	8.7	7.3
		Other non-academic	13.0	5.0	4.2	12.7	4.7	4.1	13.7	5.8	4.4
		Academic	3.8	9.1	5.1	3.1	7.8	4.5	5.5	11.7	6.3
		Total	100.0	100.0	100.0	100.0	100.0	100.0	100.0	100.0	100.0
	Masters	Industry	64.9	66.3	69.9	68.1	70.6	73.7	55.7	55.9	62.1
		Government	10.3	8.9	8.2	10.1	8.6	8.4	11.0	9.5	7.8
		Other non-academic	10.2	5.7	4.9	10.1	5.5	4.2	10.7	6.3	6.1
		Academic	14.5	19.1	17.0	11.8	15.3	13.6	22.5	28.3	24.0
		Total	100.0	100.0	100.0	100.0	100.0	100.0	100.0	100.0	100.0
	Doctorate	Industry	49.1	49.8	52.4	50.3	51.9	54.3	40.2	38.9	43.6
		Government	8.0	7.9	7.1	7.9	7.6	7.0	8.9	9.6	7.6
		Other nonacademic	7.1	4.6	4.7	7.1	4.4	4.7	7.4	5.6	4.9
		Academic	35.8	37.6	35.8	34.7	36.1	34.0	43.5	45.8	43.9
		Total	100.0	100.0	100.0	100.0	100.0	100.0	100.0	100.0	100.0

TABLE 5.2 Demographics of all chemists by gender (%). © American Chemical Society

		All chemists			All men			All women		
		Year			Year			Year		
		1990	1995	2000	1990	1995	2000	1990	1995	2000
All chemists										
Age	20–29	11.0	9.4	7.0	8.1	6.9	4.7	24.0	18.5	13.7
	30–39	32.4	30.8	27.4	30.8	28.2	24.3	39.6	40.2	36.9
	40–49	28.3	27.9	29.8	29.7	29.1	30.3	22.3	23.9	28.4
	50–59	19.6	23.1	26.1	21.7	25.7	29.3	10.4	13.8	16.5
	60–69	8.6	8.4	9.5	9.7	9.8	11.2	3.7	3.4	4.4
	70 or older	0.0	0.3	0.2	0.0	0.3	0.2	0.0	0.2	0.1
Race/ethnicity	Hispanic	1.4	2.2	2.7	1.3	2.0	2.3	2.0	3.0	3.7
	Non-Hispanic:									
	White	90.5	84.7	83.9	90.9	85.4	84.9	89.0	82.3	80.8
	Black	1.2	1.4	1.8	1.0	1.1	1.5	1.9	2.2	2.6
	American Indian	0.3	0.2	0.2	0.3	0.2	0.1	0.5	0.2	0.3
	Asian	6.0	10.2	10.5	6.0	9.9	10.0	6.2	11.2	11.9
	Other race	0.6	1.3	1.0	0.6	1.4	1.1	0.5	1.2	0.7

In the United States, Roger Revelle of the Scripps Institution of Oceanography advanced the link between climate change and anthropogenic carbon dioxide from fossil fuels during the 1950s; he became a leading figure in the National Academy of Sciences Climate Research Board and the Committee on Climate of the American Association for the Advancement of Science (Fleming 1998: 107, 122–4). In 1988 James Hansen of NASA (National Aeronautics and Space Administration) announced to Congress that "global warming has begun" (*New York Times* 1988: 1, col. 1). By this time the vast majority of climate scientists across the world recognized the role of anthropogenic carbon dioxide, and sought the support of governments to reduce reliance on fossils fuels and set targets for climate temperature increases over the next fifty years or so.

On May 9, 1992 the United Nations Framework Convention on Climate Change (UNFCCC) was adopted. The UNFCCC set non-binding agreements on greenhouse gas emissions for signatory governments, but there was no enforcement mechanism. Since then, countries signing the UNFCCC have met regularly to review progress towards the agreed targets. Because progress was too slow, in 1997 the Kyoto Protocol was formulated and adopted, setting binding levels on developed countries that proved difficult to enforce. In 2015 representatives of 196 countries met in Paris and approved the Paris Agreement limiting global warming to less than 2 °C, while pursuing an aspirational limit of 1.5 °C. There are now 197 parties to the UNFCCC, and by November 2017 (with Syria finally agreeing to sign) all parties committed to the terms of the Paris Agreement. However, the United States administration in June 2017 announced its intention to withdraw, but this step was rescinded by the new administration in 2021. In response, many states in the US (California, New York, Washington, Oregon, Maine, New Hampshire, Massachusetts, and Maryland) have pledged support for the Paris Agreement and have adopted policies to meet the targets, in particular the temperature rise. The ongoing challenge for every signatory is whether they can reduce their reliance on fossil fuels within the allowed time frame and switch to other energy sources, such as renewables (solar, wind, and tidal) or nuclear. The shift requires long-term planning and commitment over many decades.

With humans' increasing influence on the environment and ecology, the word "Anthropocene" has been adopted to designate this new era in the Earth's history. This word is derived from *anthropo* (from the Greek word meaning human) and *cene* (from the Greek word meaning new). The biologist Eugene Stoermer and the atmospheric chemist Paul Crutzen are credited with using and popularizing the word from 2000 onwards. Since 2008 different groups of scientists have debated whether the Anthropocene epoch should be adopted to follow the current Holocene epoch, given the level of humans' impact on the planet, and use of the term Anthropocene has remained informal because final approval is required from the International Commission on Stratigraphy.

Gathering of relevant data led the Anthropocene Working Group meeting in 2016 to recommend adoption of the Anthropocene epoch to the International Commission of Stratigraphy, although as of 2020 a final decision had still to be taken.

There is still considerable debate among scientists as to when the Anthropocene epoch should be regarded as having begun. Paul Crutzen and the Intergovernmental Panel on Climate Change have proposed the start of the Industrial Revolution (around 1750), while James Lovelock has proposed the introduction of the Newcomen atmospheric engine in 1712. Final adoption of the Anthropocene by the International Commission of Stratigraphy will require geological markers such as radionuclides from nuclear bomb tests that are present in rock sediments and the "technofossils" (for example, bottles, cans, and mobile phones) found in such abundance (Zalasiewicz et al. 2017–2018: 18–19).

CONCLUSION

The world's population is projected to reach nine billion by 2050, and while standards of living are slowly rising, serious socioeconomic imbalances still prevail in many parts of the world. As chemical science and process technology continue to advance and production of beneficial commodities increases, the challenge will remain for those in the underdeveloped countries to also benefit.

The world as a whole also faces several fundamental issues. In 2015 The Royal Society of Chemistry published a report entitled *Public Attitudes to Chemistry* that listed some of the most urgent issues. These were: finding sustainable sources of energy to reduce dependency on oil; ensuring there is enough food for the world's population; ensuring access to safe, clean drinking water; combating the rise in bacterial resistance to antibiotics; and reducing pollution (Royal Society of Chemistry 2015: 24). The public response expressed confidence in chemistry and chemists playing a major part in resolving them during the twenty-first century. The list provides a useful checklist against which to assess progress, although other issues are surely to be added in the future.

CHAPTER SIX

Trade and Industry: *The Growth, Diversification, and Dissolution of a Global Industry*

PETER J.T. MORRIS AND ANTHONY S. TRAVIS

INTRODUCTION

The chemical industry grew enormously during the nineteenth century, but continued to expand and diversify in the twentieth century. As it is impossible to give a comprehensive survey of the complex history of the chemical industry in the space available, we have restricted ourselves to an explanation of how the culture of the chemical industry changed between 1914 and 2019. The industry shifted from a strongly nationalistic stance in the interwar period to an increasingly globalized and fragmented industry around the millennium. To understand these shifts we have to examine the impact of the two world wars, both of which shaped the development of the industry, as did the Oil Crises of the 1970s.

WORLD WAR I AND THE CHEMICAL INDUSTRY

August 1914 marked the start of an abrupt change in the affairs of the world. The fitful moves toward a global culture of internationalization and interdependence that had been fostered during the previous half century gave way to a separation of nations into opposing camps. This process had a major impact on the trade in chemicals, especially for those countries that had been economically dependent on what were now enemy states (MacLeod and Johnson 2006). The United States, although not a belligerent until April 1917, suffered greatly from the disruption in transatlantic trade (Steen 2014). Thus, for example, there were shortages of synthetic dyestuffs and organic intermediates that previously had been imported from Germany. On the outbreak of war Germany banned several export products, including processing equipment and Stassfurt potash, a vital fertilizer component.

Germany also had to respond to shortages, most particularly of Chilean nitrate, the nitrogen-rich mineral, following defeat of its Southeast Asia Squadron at the Battle of the Falkland Islands in December 1914 (Travis 2018). This commodity was not just the most important fertilizer, but was essential for the manufacture of nitric acid, indispensible for the production of both dyes and explosives. The response in Germany was expansion of two recently developed nitrogen-fixation processes, the Adolf Frank–Nikodem Caro calcium cyanamide synthesis and the Fritz Haber–Carl Bosch process that produced ammonia directly. The ammonia was used to make ammonium nitrate, a component of explosive mixtures as well as being the crucial ingredient for explosives such as trinitrotoluene (TNT) and picric acid. By contrast, by-products from the coke

FIGURE 6.1 There was a massive explosion at the Oppau Haber–Bosch factory, adjoining the BASF factory in Ludwigshafen, on September 21, 1921 when a fertilizer silo exploded. Over 560 people were killed and the damage to property was estimated to be $7 million. Photograph by ullstein bild Dtl. via Getty Images.

industry and gas works provided farmers with ammonium sulfate fertilizer. The Hindenburg program of July 1916 led to an increase in production of Haber–Bosch synthetic ammonia, partly through the construction of a large new ammonia factory at Leuna, near Leipzig.

However, all this was not enough to save the country from starvation and defeat. For their part, the Allies tried but failed to replicate the Haber–Bosch process during the war, because it was so technically complex. They could, however, rely on nitrate rock from Chile, despite German submarine attacks on merchant shipping. France also had the advantage of domestic manufacture of calcium cyanamide.

The overall reaction to such shortages and especially the ongoing needs of the Western Front, which from early 1915 stretched from the English Channel across France to the border with Switzerland, led to new initiatives in the organization of science and technology. States were heavily involved in the manufacture of chemicals, both directly and through sponsorship, in order to maintain strategic supplies. The German industrialist Walther Rathenau created a system of war material organization that was eventually imitated by all belligerents. There was a vast scale of production, administered, in Britain, for example, by the explosives department headed by Lord Moulton. State-funded organizations were set up to promote science and technology, in particular the forerunners of the British Department of Scientific and Industrial Research (1915) and the American National Research Council (1916). In Germany, chemical research for military purposes was conducted at the Kaiser Wilhelm Institute for Physical Chemistry and Electrochemistry headed by Fritz Haber, inventor of the ammonia process, who introduced large-scale gas warfare in April 1915.

Germany could rely on its large synthetic dye factories that were converted for manufacture of explosives. In 1914, Germany controlled most of the world production of synthetic dyes, which placed Britain, France, and the United States at a distinct disadvantage. Their embryonic dye industries were revived, and stepped up manufacture from organic intermediates of dyes, drugs and, from 1915, poison gases. The fate of these industries was an important factor in the international rivalry of the postwar period.

SEEKING DOMINANCE, 1920–1945

In the interwar period, particularly from the mid-1920s, there were three key characteristics of the chemical industry (Haber 1971; Travis et al. 1998). First, and above all, each of the major firms was very much tied to its own country, including IG Farben in Germany, ICI in Britain, and Du Pont in the USA. These firms arose as leaders in chemical manufacture in the 1920s. IG Farben (originally the Interessengemeinschaft Farbenindustrie, or in English the Dye Industry Syndicate) was a merger of the three major German synthetic dye companies – BASF, Bayer, and Hoechst – with some of their smaller competitors such as Cassella and Weiler-ter Meer (Hayes 1987; Lesch 2000; Abelhauser et al. 2004). It was first set up as a trust in 1916 during the war, partly to fund the construction of the Leuna works, then became a single corporation in 1925. The formation of Imperial Chemical Industries (ICI) in 1926 was a defensive reaction to this merger. It brought together Brunner, Mond & Co. (a manufacturer of sodium carbonate by the highly efficient Solvay process) and the explosives and metals manufacturer Nobel with the government-sponsored British Dyestuffs Corporation and the nearly moribund United Alkali Company that made sodium carbonate by the obsolescent Leblanc process (Reader 1970). Du Pont was the former American explosives trust that had been broken up by the US Supreme Court in 1912; its civilian explosives operations had been split between two new firms, Hercules and Atlas (Dyer and Sicilia 1990). While Du Pont retained the production of military explosives on the grounds of national security, the firm concentrated on the chemical industry following the end of the war (Hounshell and Smith Jr. 1998; Ndiaye 2007). Another important American company was the Allied Chemical and Dye Corporation, formed by the merger of five firms in 1920. For a while it looked as if this firm might represent the future of the American industry, but it soon declined in importance. Another major firm was Union Carbide, which pioneered the production of chemicals from American natural gas in the 1920s.

Elsewhere, CIBA was the largest company in Switzerland, and Solvay S.A. was the largest in Belgium, with important operations in other countries (Bertrams et al. 2013). By contrast, France was still a country of relatively small firms (Aftalion 2001). The two largest firms, St. Gobain and Kuhlmann, were both inorganic chemical manufacturers; St. Gobain also made glass. The industry in Italy had been traditionally based on superphosphate fertilizer. The former copper mining firm Montecatini was converted into a chemical company by the energetic Guido Donegani, who first took up superphosphate fertilizer production, then moved into new areas such as nitrogen fixation for fertilizers and dyestuffs (Petri 1998). In Japan the large Nichitsu corporation switched in the 1920s from the manufacture of cyanamide to synthetic ammonia (Molony 1990; Travis 2018). Nevertheless, the largest firms had ambitions outside their own countries, especially ICI in the British Empire and IG Farben in America. This was a major complication, as it prevented the firms creating mutually exclusive geographical spheres of influence.

A second key characteristic of this period is the circumstance that the largest firms sought to monopolize both their home countries and specific product areas in ways not seen before or since. This cultural trend was tied to a belief prevalent throughout the developed world that chemistry and the chemical industry was about to transform society, a view demonstrated by the slogan of the 1933 Chicago World's Fair: "Science Finds, Industry Applies, Man Adapts"

FIGURE 6.2 Soderberg electrode hall in Montecatini's aluminum factory at Bolzano, 1959. Montecatini diversified into a large number of sectors including light metals. Photograph by Mondadori via Getty Images.

(Ganz 2008: 57). Chemistry was very much the science to which this motto, and this movement, related. In keeping with the times, Du Pont introduced its famous slogan "Better Things for Better Living ... Through Chemistry" (Rhees 1993). The major firms eschewed any interest in immediate profits: market dominance was to come first at whatever cost, profits would then follow.

The third key aspect of this period was the rise of the automobile, especially in America, but also to a lesser extent in Germany and Britain (Greenaway et al. 1978). Automobiles relied heavily on products of chemical industry, and the major chemical firms were all linked in varying degrees to the expanding automobile industry. The introduction of quick-drying paints solved a major bottleneck in automobile manufacture by assembly line. The first artificial lacquer was based on nitrocellulose (or guncotton), a product already used in movie film and explosives manufacture. The lacquer was dissolved in an organic solvent such as ethyl acetate or butyl acetate. Spray painting of vehicle bodies was used from 1923, but slow drying of the finish tied up the entire production line. The problem was solved when "alkyd" quick-drying paints were introduced later in the 1920s. In countries with cold winters, such as the USA, Germany, and Britain, "antifreeze" was added to the radiator water, initially methanol, and then ethylene glycol, introduced as "Prestone" by Union Carbide in 1927. ICI, through its involvement in the metals industry, also manufactured car components.

Du Pont was a major shareholder of General Motors (GM), initially because of Pierre S. du Pont's role as a leading investor in the automobile firm (Chandler and Salsbury 1971). After World War I, with the encouragement of John J. Raskob who – like Pierre – had worked for both companies, Du Pont invested heavily in GM. In the interwar period half of Du Pont's profits came from its GM dividends, which the firm then used to move into new chemical fields. Both chlorofluorocarbon refrigerant liquids and the lead tetraethyl anti-knock compound were a result of this industrial alliance (McGrayne 2001). With the encouragement of Du Pont and the enthusiastic support of Harry McGowan, the former head of the Nobel explosives firm, now part of ICI, the latter also invested in General Motors stock. As a counterbalancing move, Ford bought IG Farben stock.

As automobile numbers soared in the 1920s, there was growing concern about whether there would be sufficient gasoline to keep all these vehicles on the road, because it was considered likely that the existing oil fields would dry up (Yergin 1991). This concern led during the late 1920s to the development of processes to convert coal to petroleum products, and alliances between IG Farben and two major petroleum companies, Standard Oil of New Jersey and Shell. As it turned out, the oil reserves never ran out, as the huge East Texas oilfields were opened up in 1930, followed by the discovery of even greater oil reserves in the Arabian Peninsula in 1938.

Before 1914, with the important exception of coal tar dyes, production of organic chemicals had been insignificant in terms of volume, with the partial exception of methanol and acetone, both obtained by the destructive distillation of wood. With the growing use of acetone in explosives and as a solvent and the use of methanol as antifreeze, wood distillation – notably by the large German firm HIAG – enjoyed a brief boom in the early 1920s (Chandler 1990: 482–3). The development of a process to make acetone from acetylene during World War I by the Hoechst company and the high-pressure synthesis of methanol from carbon monoxide and hydrogen by BASF in 1923 – a spin-off from the Haber–Bosch process – led to the collapse of the wood distillation industry by 1930. BASF followed up this success with an industrial process for making butanol from acetylene, in direct competition with the Weizmann process (fermentation of starches to butanol and acetone) used by the Commercial Solvents Corporation in the United States.

This brings into focus the issue of raw materials for organic chemicals. Wood was still a potential raw material, thanks to a process which used acids to convert wood cellulose into sugars developed by Friedrich Bergius, better known for his work on the coal-to-oil process. While the process was used in Britain briefly and Bergius set up a plant in Germany, it was not a commercial success (Jones and Semrau 1984). This left coal, abundant in Germany, Britain, and the USA, as the most readily available source. Coal was the raw material for the Haber–Bosch process, thanks to the introduction of the "water shift reaction," which converts the carbon monoxide/hydrogen mixture produced by the action of steam on red-hot coke into carbon dioxide and more hydrogen. A variant of this process yielded the carbon monoxide/hydrogen mixture needed for synthetic methanol.

However, coal was not the ideal feedstock for organic chemicals, except as coal tar for aromatic chemicals. Hence the introduction of calcium carbide, made from coke and lime, was a major breakthrough (Miller 1965; Morris 1983). It was first used from around 1900 to produce acetylene (by the action of water on the carbide) for household lighting. After this short-lived boom collapsed, calcium carbide found new uses, namely for oxyacetylene welding and to make the fertilizer calcium cyanamide. Just before and during World War I, three allied German firms – Hoechst, Griesheim Elektron, and the Consortium für elektrochemische Industrie (Wacker) – energetically developed acetylene as a chemical feedstock. By the 1930s, acetylene was the basis of many organic chemicals and polymers in Germany, although it was not widely used elsewhere. In the late 1920s the American firm Union Carbide, however, began the manufacture of organic chemicals, notably acetone and ethylene glycol, from another starting material, namely natural gas from West Virginia. In the 1930s, several major oil companies – Shell, Standard Oil of New Jersey (with the help of IG Farben), and the Anglo-Persian Oil Company (later renamed British Petroleum) – began to use off-gases from petroleum refineries to make organic chemicals. By contrast, in the late 1930s ICI first made polythene (discussed later in this chapter) using fermentation alcohol from the major British spirits firm Distillers, which had moved into the chemical industry with Discol motor fuel (a mixture of alcohol and petrol), and set up British Industrial Solvents to make and market alcohol-based chemicals (Weir 1995).

Fermentation alcohol is of course made from grains such as barley or maize. The whole question of making organic chemicals from similar plant sources became highly politicized in the 1930s (Pursell 1969; Finlay 2003). In the United States, the prominent Georgian chemist Charles Herty had long been concerned about the economic position of the southern states, and in particular sought to promote the use of the southern pine as a raw material. In the early 1930s, he made paper from southern pine trees. This breakthrough led to the setting up of the Farm Chemurgic Council in 1935 to promote the use of American agricultural products in industry and in particular the chemical industry. The promotion of "chemurgy" brought together Herty, his friend William Hale of Dow, industrialist Henry Ford, African-American chemist George Washington

Carver, and journalist Wheeler McMillen. Ford was already attempting to use soybean-based plastics for the body of his automobiles (Wik 1962). This movement gave rise to a polemical literature with colorful titles such as Hale's *The Farm Chemurgic: Farmward the Star of Destiny Lights our Way* (1934), and McMillen's *New Riches from the Soil* (1947), as well as to the setting up of four research stations within the US Department of Agriculture. Chemurgy found a new impetus when the issue arose of how American synthetic rubber would be made during World War II (discussed below).

Just as dyestuffs had given the chemical industry its major impulse in the late nineteenth century, high-pressure chemistry was the hallmark of the industry in the interwar period (Travis 2018). As BASF refused to license the Haber–Bosch process, inventors outside Germany sought to imitate it. Italian chemist Luigi Casale in 1921 developed a process which used higher pressures than the Haber–Bosch process. He was forced out of the Italian industry by Donegani, but his process was widely used elsewhere. Donegani favored the process developed by another Italian, engineer Giacomo Fauser. There was also the process of Georges Claude, who invented neon lighting; it was mostly used in France (Smil 2001). Japanese investigators developed a similar ammonia process in the late 1920s. Nitrogen fixation had obvious military value for explosives. The rising level of production, however, came up against the economic woes of the agricultural sector, especially after the Wall Street Crash of 1929. The major ammonia manufacturers set up a cartel, which rather ineffectively shored up prices and reduced production. Nevertheless, ammonia remained attractive to fascist Italy from 1925, communist Russia from 1928, and to Japan's colonization program in Korea from 1930.

One way of taking up this spare capacity in ammonia production was to use high-pressure technology to make synthetic gasoline from coal (Krammer 1978; Stranges 1984; Stranges 2000). IG Farben acquired the coal hydrogenation process of Bergius in the mid-1920s and developed it at the Leuna works. However, it turned out to be more troublesome than the Haber–Bosch process, and, as a result of the Depression, the price of gasoline fell. Carl Bosch, the Chairman of IG Farben, turned to the new Nazi regime in 1933 for financial support for the process. Keen to promote self-sufficiency and gain secure supplies of aviation fuel for the Luftwaffe, the Nazis signed a contract with IG Farben which was not particularly favorable to the firm, but made the huge Leuna plant economically viable.

Meanwhile, chemists at the Coal Research Institute in the Ruhr came up with a different strategy – the Fischer–Tropsch process – that made synthetic gasoline from carbon monoxide and hydrogen, which, as gases, were easier to process (Rasch 1989). ICI, always in awe of IG Farben's technological prowess, obtained a license to use the Bergius process at its Billingham factory near Stockton-on-Tees, which since 1924 made ammonia using a modified

Haber–Bosch process (Reader 1975; Stranges 1985). The driving force in Britain was to find a new way of consuming coal and thus keep coal miners in work (Bud 2018). The plant was technically successful, but it was not commercially viable despite government support. It was used to make aviation fuel during World War II. The major oil companies, in particularly Standard Oil of New Jersey and Shell, were also interested in the field, partly as a defensive measure in case petroleum did indeed run out, but also for its value in improving the yield of gasoline from crude petroleum. IG Farben and ICI joined forces with Standard and Shell in 1931 to form a holding company, International Hydrogenation Patents Company, to exploit the process (Reader 1975; Jonker and van Zanden 2007).

An increasing global demand for material goods in the interwar period placed pressure on the supplies of natural materials, ranging from cotton and wool to wood and rubber. To meet this challenge, chemical firms took up the rudimentary pre-1914 work on synthetic materials and strove to make them commercially viable (Mossman and Morris 1994; Russell 2000). Polyvinyl chloride (PVC) was particularly attractive, as chlorine was cheap and vinyl chloride could be readily made from acetylene (Kaufman 1969). The problem was working the material, because it decomposed when heated. There were two possible solutions. The manufacturer could either add organic chemicals (called plasticizers) to make the material pliable and use it as a substitute for rubber for certain applications such as wire coatings, a route taken by the American rubber firm BF Goodrich and in Germany by IG Farben. Or, like Union Carbide, they could add stabilizers and use it as a rigid material to replace wood and other construction materials. By the late 1930s, there was some success on both fronts, but the heyday of PVC lay in the future.

By the mid-1920s, it was clear that it was possible to make organic chemicals which were similar to isoprene, the building block of natural rubber, and convert them by polymerization into rubber-like materials (Morris 1994). There were two major drawbacks: these materials were not very good rubbers; and they were expensive in a period when the price of natural rubber had more or less collapsed following misguided attempts by the British government to manipulate the price. After three years' work, chemists at IG Farben discovered that these artificial rubbers could be improved by adding styrene, a compound which increased their wear resistance. At the same time, Du Pont accidentally made a chlorine-containing rubber that was both cheap and easy to make (Smith 1985). It also turned out to be resistant to oil and gasoline (unlike natural rubber) and hence very useful to the automobile and aircraft industries. Although Duprene (later called neoprene) was the first synthetic rubber to show commercial potential, it was several years before any profits materialized. IG Farben wished to concentrate on its rival to Duprene, called Buna N, but was directed by the Nazi regime to produce the cheaper Buna S, which could be used to make tires.

The related firms of Röhm in Germany and Rohm & Haas in the United States developed a substitute for glass called Plexiglas based on methyl methacrylate (Hochheiser 1986). It was discovered independently by chemists at ICI and given the name Perspex. This plastic substitute for glass became increasingly used for military aircraft canopies, because it was light in weight and strong, and did not produce dangerous fragments when broken. Under an exchange agreement with Du Pont, ICI licensed the technology to the American firm, which sold polymethacrylate under the name Lucite.

Another synthetic material whose importance would grow as a result of World War II was low-density polyethylene, or polythene as ICI called it (Wilson 1994). This waxy plastic was created by chance as a result of an attempt to carry out organic chemical reactions at high pressures. Full-scale production began in September 1939. As an excellent electrical insulator, it was ideal for the new radar equipment and as a covering for underwater cables. In the 1930s Walter Reppe of IG Farben carried out extremely dangerous research on the reactions of acetylene at high pressures and created polyvinylpyrrolidone (PVP) in 1939 that saw promise as a blood plasma substitute (Morris 2005). He also developed a new chemical route to butadiene, one of the building blocks of Buna S rubber (Morris 1998).

Although developed around the turn of the century, viscose rayon – produced from wood pulp and cotton linters ("cotton wool") – entered its glory days in the 1920s (Blanc 2016). Promoted as artificial silk, in many respects it was more like cotton, which it resembled chemically. Courtaulds, the English crepe silk firm and holder of the original patents, set up the American Viscose Corporation (AVC) in the United States, which rapidly became one of the most profitable subsidiaries in history, making far more money than the parent factory in Coventry (Coleman 1969). However, the US government insisted in 1941 that AVC be sold off before it would support Britain with the lend–lease program. A shiny form of rayon – acetate rayon – was introduced in the 1920s in both Britain and the United States by the Swiss brothers Henri and Camille Dreyfus. Du Pont decided to enter the field aggressively, obtaining French rights to a rayon process and for cellophane transparent material (Ndiaye 2007).

Having successfully established a foothold in the new industry, Du Pont now launched a program to develop a completely synthetic fiber. Rather than attempt to replicate the complex chemical structure of silk, Du Pont chemists made simpler analogues called polyamines. In 1935, the brilliant chemist Wallace Hume Carothers, with the help of colleagues including Julian Hill, succeeded in producing a filament fiber from an organic amine and an organic acid, each of which contained six carbon atoms; Du Pont accordingly named the new polymer nylon 66. A sufferer from depression, Carothers committed suicide in 1937, just a year before nylon was successfully brought to the market (Hermes 1996).

FIGURE 6.3 "Cheese at Its Tempting Best – in Cellophane: Cellophane Shows What It Protects! Protects What It Shows!" A 1949 advertisement for DuPont Cellophane from *The Saturday Evening Post*. Hagley Museum and Library with permission.

Convinced that it held a commanding patent position, Du Pont anticipated many years of uncontested supremacy in this field. In particular, it intended to drive IG Farben a hard bargain for a nylon 66 license, hoping in return to obtain a license for IG's polystyrene patents. Du Pont executives were therefore stunned in 1938 when their German counterparts offered them a license for the related nylon 6 (Achilladelis 1970). This version of nylon had been developed without official permission by Paul Schlack at IG Farben's Aceta cellulose acetate factory in Berlin. The two companies signed a cross-licensing agreement in May 1939. In the early twenty-first century about twice as much nylon 6 is produced than nylon 66, mainly because it is easily worked and cheaper. Just as ICI gave Du Pont Perspex, Du Pont licensed nylon 66 to ICI, which set up British Nylon Spinners jointly with Courtaulds in 1940.

The development of these new areas – as important as they were – should not obscure the fact that most of the profits of the chemical industry still came from the traditional areas of synthetic dyes and pharmaceuticals, especially during the Great Depression of the 1930s (Quirke 2005). While the rate of innovation of new dyes had slowed down, important new classes of dyestuffs were still being created (Morris and Travis 1992; Travis 2004). For example, in 1928, chemists working at the Grangemouth works of Scottish Dyes – which had just become part of ICI – discovered a blue coloration in a chemical reactor vessel. This was soon shown to be an iron-containing compound related to natural pigments like chlorophyll and hemoglobin. The copper salt of this chromophore was marketed as Monastral Fast Blue in 1934. IG Farben immediately launched its researchers on a mass program aimed at the production of this new colorant. In the same year, it marketed its own version as Heliogen Blue B, made by an improved method, much to the amazement of ICI's dyestuffs group (Reader 1975: 332). ICI later acquired a license to use the German process.

In the pharmaceuticals industry, IG Farben used bacterial staining by dyes to seek out a dye-related chemical to kill infections within the body. In 1935, following a long screening campaign, the firm brought out Prontosil, an *azo*-sulfonamide dye with physiological action (Lesch 2007). This antibiotic was based on the work of Gerhard Domagk and his colleagues. The Pasteur Institute in Paris showed that Prontosil was broken down in the body into a simple compound called sulfanilamide, which was the active substance. As this compound could not be patented, other firms quickly moved into the field. In particular, the English firm May & Baker – thanks to its connections with the Institut Pasteur in France – developed an improved sulfonamide, sulfapyridine, called M&B 693. These so-called sulfa drugs brought about sharp declines in death rates from puerperal fever and pneumonia. They were cheap to produce and could be administered orally, although they also had significant side effects. These successes brought additional chemical firms into the pharmaceuticals business, including ICI in Britain and the Swiss dye-makers. A second generation of sulfa drugs that was developed after World War II reduced these side effects and helped to conquer tuberculosis in the 1950s. Domagk was awarded the 1939 Nobel Prize in physiology or medicine, but was forbidden by the Nazis from accepting it; he belatedly received the award in 1947.

WAGING WAR, 1939–1945

The main impact of World War II on chemical industries and trades was an intensification of what had been achieved during the previous decades, rather than wholly new innovations. In Germany, the two key areas were the production of motor fuel from coal – by both the Bergius coal hydrogenation process and the Fischer–Tropsch process – and synthetic rubber. The output of the Fischer–Tropsch plants doubled in the first year of the war, but thereafter stalled (Stokes 1985). There were two basic reasons for this: the process relied on coke as a raw material rather than the more readily available lignite, and coke was needed by the steel industry. Furthermore, the output of the Fischer–Tropsch process was a low-octane gasoline rather than the high-octane aviation fuel that was essential for military aircraft. By contrast, from a much higher base, the production of synthetic oil from coal in Germany by the Bergius process nearly trebled between 1940 and mid-1944 to around 450 million tons a month. However, after the United States changed its bombing strategy to target the synthetic oil industry in May 1944, production soon slumped to almost a tenth of its peak, and at times even less. By this time, with its aircraft destroyed or grounded, Germany's demand for aviation fuel was declining. One of the wartime plants – Wesseling, near Cologne – became an important petroleum refinery for the chemical industry after 1945.

As for synthetic rubber, IG Farben had two factories up and running by 1940, and within a year was building two more. A plant within the chemical complex at Ludwigshafen (adjoining Mannheim on the Rhine) was completed by 1943, but it was so badly damaged by Allied bombing that it was closed down in October 1944. A factory in occupied Polish Upper Silesia, close to Auschwitz (Oświęcim in Polish), assumed to be outside the reach of Allied aircraft, was bombed by the Americans in late 1944. This and sabotage by its slave workers partly explains why it was still not up and running when it was overrun by the Soviet Army in late January 1945. Although the Oświęcim factory was completed by the Polish government and became an important supplier of synthetic rubber for the Soviet bloc, it is best known for IG Farben's brutal use of concentration camp labor in its construction (Wagner 2000). IG Farben also manufactured a completely new class of chemical weapons, nerve agents, which were discovered during the development of organophosphate insecticides at the Elberfeld factory (Harris and Paxman 2002). They were never deployed during the war, for fear of retaliation by the Allies, who were erroneously thought to have similar chemical weapons. New and even more powerful nerve agents were developed by Britain and the Soviet Union after 1945.

After the United States was cut off from the Far Eastern rubber plantations by the rapid advance of the Japanese army there was a drive to produce synthetic rubber on the basis of IG Farben's earlier research, which had been shared with Standard Oil of New Jersey (Herbert and Bisio 1985). American chemists soon developed a superior version of Buna S rubber, called GR-S, but vigorous debates emerged about the best raw material to use (Tuttle 1981; Finlay 2009). The agriculture lobby and its "chemurgist" allies (advocates of agriculturally derived chemicals) demanded the use of fermentation alcohol, a route used in the Soviet Union since the early 1930s, but the petroleum industry (and most of the experts) favored the use of butane off-gases from petroleum refineries, because this was far cheaper and easier to expand in the future. In the end, both routes were used, which was fortunate because the petrochemical process only came on stream after D-Day. Up to that time most of the Allies' meagre supply of synthetic rubber was made using the alcohol route.

The Japanese stranglehold on the Far East also cut the Allies off from quinine, as this antimalarial drug was largely a Dutch monopoly based in Java (Slater 2004). There was a pressing need for antimalarial drugs for the forces fighting the Japanese in eastern Asia and the Pacific; initially, casualties from the disease in the Pacific theater far outnumbered war-related casualties. The pharmaceutical industry developed novel antimalarial drugs based on prewar German research. The Americans concentrated on the production of the IG Farben drug Atabrine (mepacrine), although it suffered from side effects such as turning the skin yellow, which was alarming to the patient although it was harmless. Winthrop Chemical came up with a chloroquinoline called Sontochin, but it was largely overlooked until the same compound was found on German prisoners of war in Tunisia. Winthrop also developed a related drug, chloroquine (previously synthesized in Germany), which was found to be superior.

As some servicemen were still reluctant to take these drugs, it was fortunate that another solution to the malaria problem was now available, namely the insecticide DDT (Dunlap 1981; Simon 1999; Kinkela 2011). Paul Müller at the Swiss dye firm Geigy had developed DDT in 1939 with the aim of eradicating the Colorado beetle. DDT was inexpensive because it was made from two cheap chemicals: chloral and chlorobenzene. Not only could DDT be used against mosquitoes – especially in their breeding grounds of stagnant water – but it was also highly effective against the body louse, which caused the fatal disease typhus. A disease of overcrowding, typhus became rampant in Europe in the concentration camps and ghettos set up by the Nazis; even after liberation the disease spread among the rapidly growing population of refugees and "displaced persons." It was thanks largely to the widespread use of DDT that a major typhus epidemic was avoided. Bizarre as it may seem in the light of the more recent history of DDT, it is not surprising that Müller was awarded the Nobel Prize for physiology or medicine in 1948.

At Oxford University, the Australian pathologist Howard Florey, the German émigré Ernst Chain, and the English biochemist Norman Heatley succeeded in making penicillin from penicillium mold in 1940 (Bud 2007). Their first attempt to use it as an antibiotic, in February 1941, failed when their supplies

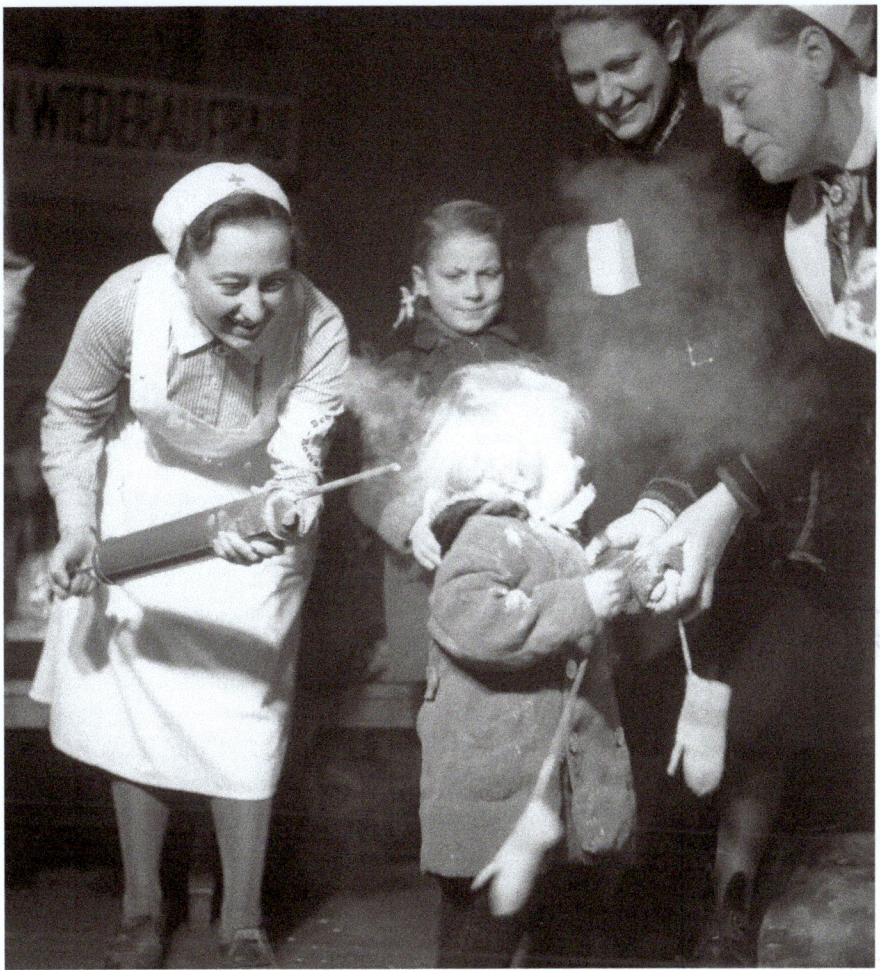

FIGURE 6.4 A child crying as she is sprayed with DDT delousing powder at the Cecilienschule, Nikolsburger Platz, Wilmersdorf, Berlin, Germany, October 1945. Photograph by George Konig/Keystone Features/Getty Images.

ran out before the patient was cured. It was clear that the successful production of penicillin could only be achieved by drawing on American expertise and production facilities. Heatley worked with the staff of the Northern Regional Laboratory of the US Department of Agriculture at Peoria, Illinois, to find not only a better strain of penicillium mold, but also to work out how to produce penicillin on a large scale. Significantly, the Peoria research facility was one of the four laboratories set up in 1938 as a result of lobbying by the chemurgic movement and its allies in the US Congress. The successful industrial production of penicillin by Pfizer and then by Merck was a military breakthrough not just for saving the lives of wounded servicemen, but also for preventing an epidemic of gonorrhea among otherwise healthy soldiers. Florey and Chain shared the 1945 Nobel Prize in physiology or medicine with Alexander Fleming, who had discovered penicillin by chance in 1928, thereby overlooking Heatley, who played an important role in its production.

The end of World War II was a low point for several leading chemical companies. IG Farben stood accused of helping the Nazis to wage war and for the use of slave labor. The firm's leading executives were tried by the United States at Nuremberg; several of them received prison sentences, but were released by 1952. There was an intensive program of technical information gathering and dismantlement of equipment by the Allies (Krammer 1981; Gimbel 1990). IG Farben was broken up and the fragments were put under the control of the occupying power in each of the four zones of occupation (Stokes 1988). The factories in the Soviet zone became the property of the German Democratic Republic, and remained so until 1990 (Karlsch 2000). In West Germany, three successor firms were set up in 1953 which took the names of the pre-1925 firms, namely, BASF, Bayer, and Hoechst. The leading firms in Italy and Japan were broken up in a similar way, although they would reappear in different guises from the 1950s. For instance, Montecatini was accused by the Allies of collaborating with the enemy, and Donegani was removed from the management of the firm. The US government insisted that the exchange agreement between Du Pont and ICI be terminated, thus closing off what had been one of most important sources of new products for both firms (Reader 1975). Despite these upheavals, many observers must have assumed, perhaps cynically, that things would continue much as before, given what had happened after World War I. However, there was in fact a radical change in the way that the chemical industry operated, which was largely the result of major changes in the raw materials used by the industry, and in the cost of energy.

MAKING PROFITS, 1945–1965

After 1945, coal faded in importance as the raw material for the chemical industry, although its use continued in West Germany for some time (Stokes 1994). It was largely replaced by so-called petrochemistry, which used the off-gases from petroleum refining and natural gas rather than petroleum directly. Petroleum refiners such as Esso and Shell moved into the chemical industry as a result of their prewar interest in synthetic fuels and the desire to profit from the growth of the chemical industry, especially in plastics. At the same time, established chemical companies began to see both technical and economic advantages of using petrochemistry. A few American companies such as Dow and Union Carbide had entered petrochemicals before World War II, but they did not gain any major advantage from their early entry into this field. The postwar shift was a significant cultural and psychological hurdle for the major

FIGURE 6.5 Synthetic rubber plant in the open air, 1950. Photograph by Andreas Feininger/The LIFE Picture Collection via Getty Images.

European companies such as BASF and ICI, and even for American firms like Du Pont. They had built up their technological prowess on the basis of classical organic chemistry as practiced in academic laboratories. Petrochemistry involved mastering an entirely new type of chemistry. It also meant building an entirely new generation of chemical factories based on petroleum refining rather than on classical industrial chemistry. Most notably, they were "open air" factories (and they were usually called plants rather than factories), rather than being enclosed in brick or concrete buildings. In the United States they were often erected in the southern states, which had hitherto lacked any significant chemical industry but were close to the oil and gas fields of Texas, Louisiana, and other regions surrounding the Gulf of Mexico.

Despite these barriers, petrochemistry swept through the chemical industry in the 1950s and early 1960s (Spitz 1988). This was partly the result of the rapid development on a huge scale of certain plastics such as polyethylene, PVC, and polypropylene, but mainly a consequence of the falling price of petroleum as new oil fields were opened up, above all in the Middle East. The price of oil remained both very stable and remarkably low at around $3 a barrel between 1948 and 1970 (Macrotrends 2018). An industry that had been hitherto based on an abundant but relatively expensive raw material, namely coal, was now using very cheap and even more abundant raw materials, petroleum and natural gas. The challenge was to find processes and products which could convert this cheap oil and gas into profits.

This seemingly easy route to riches turned out to be more difficult than chemical industry executives had assumed. While profits in the chemical industry were higher than in other industries, they were never very high for most chemical companies, certainly not on the scale of the interwar rayon industry. This was a result of the entry of new firms into the sector and the resulting downward push on prices as a result of this competition. In the end it was the customer who benefited rather than the shareholders, although most shareholders profited as well.

A major consequence of the transformation of the chemical industry was the rise of chemical engineering companies (van Rooij and Homburg, 2002). They had existed before World War II, but it was the development of the petrochemical industry and the modernization of the synthetic ammonia industry that revolutionized their roles in the chemical industry. The new academic discipline had grown in the United States as a result of the development of cracking processes, which converted crude petroleum into lighter fractions, and the rise of chemical engineering in technically oriented American universities, most notably under Warren K. Lewis at the Massachusetts Institute of Technology (Enos 1962; Furter 1982; Divall and Johnston 2000). The first process engineering firms were Universal Oil Products (UOP), founded to develop the cracking process developed by Carbon Petroleum Dubbs in 1914, and M.W. Kellogg, which moved into petroleum refining engineering in the 1930s. The British firms Humphreys and Glasgow (founded in 1892) and Power-Gas Corporation (founded in 1901 and merged with Davy-United in 1960) had their origins in gas production, respectively water gas and Mond producer (power) gas. Bradley Dewey and Charles Almy, two MIT-trained chemical engineers who had served in the US Chemical Warfare Service, founded Dewey and Almy in 1919. Dewey became the Rubber Director in charge of the USA's synthetic rubber program during World War II. In 1954 the firm was taken over by W.R. Grace & Co., itself a successor of a company which had been founded to ship nitrogen-rich guano from Peru in the mid-nineteenth century. In 1946, Ralph Landau, an MIT-trained chemical engineer at Kellogg, founded Scientific Design with two colleagues, Harry Rehnberg and Robert Egbert. In contrast to other firms in the field, Scientific Design concentrated on petrochemical processes, partly drawing on the information brought out of Germany by the Allied technical intelligence teams, for example the manufacture of ethylene oxide (Spitz 1988; Spitz 2019). In the 1960s, Landau decided to enter the chemical industry on the basis of Scientific Design's proprietary technology. Scientific Design became Halcon and collaborated with the petroleum firm Atlantic Richfield (ARCO). After the

Oil Crises of the 1970s, Landau pulled out of manufacturing and sold Halcon SD (its name at the time). In the late 2010s, it was jointly owned by the Saudi Arabian petroleum firm Sabic and Clariant (formerly Sandoz).

From the 1960s, the mature synthetic ammonia industry benefited from a number of major technical improvements (Greenberg et al. 1979; Smil 2001). Thermal efficiencies increased threefold, and energy costs dropped by almost two thirds. M.W. Kellogg in the United States introduced the first single train energy-integrated plant, for the American Oil Company, at Texas City (near Houston) in 1963. By the late 2010s, apart from China, which is still largely dependent on coal, hydrogen for ammonia synthesis was mainly prepared by steam reforming or partial oxidation of hydrocarbons. China, the world's leading manufacturer of ammonia since the early 1990s, accounts for one third of international production. The manufacture of synthetic ammonia continues to represent an important part of the global chemical industry, with around 80 percent consumed by the agricultural sector.

The chemical industry, hitherto divided between inorganic chemicals and organic chemicals, was now split between the mass production of a few chemicals – the major plastics and fertilizers – and chemicals produced on a relatively small scale – pesticides, drugs, and high-impact adhesives. One major aim of the industry was to prevent the scale of disease and hunger occurring in the aftermath of World War II. This involved the development of new drugs and production of cheap fertilizers, but also use of pesticides to kill insects and weeds with the aim of increasing crop yields (Russell 2001). In the 1970s the industry even attempted to develop synthetic foods, based on fermentation techniques first used for the production of penicillin.

Another aspect of the relationship between the chemical industry and population growth was the development of oral contraceptives in the late 1950s, drawing on cutting-edge academic research on the group of organic chemicals called steroids, and based on the conversion of natural products, most notably a substance found in the Mexican yam (Marks 2001). In the same period James Black at ICI Pharmaceuticals explored the idea of blocking specific biochemical pathways in the body to cure diseases. He developed propranolol, the first of the so-called beta-blockers, for high blood pressure and heart diseases (Quirke 2006). After ICI refused to allow him to work on similar targeted drugs for stomach ulcers, Black moved to Smith, Kline & French, and developed cimetidine, the first proton inhibitor for stomach acidity in the late 1960s. These new "blockbuster" drugs were very successful in both medical and economic terms. Nonetheless, few novel drugs were in the pipeline in the early twenty-first century. By contrast, the regulation of drug prices by national health care systems and lowering of drug prices through group purchases by health providers in the private sector created a good market for generics, at least as long as the pressure on prices was not too great.

The chemical industry also remained eager to produce polymers, namely plastics, synthetic fibers, and synthetic rubber. They were a potent example of how cheap hydrocarbon gases could be easily converted into products with significant added value. However, as more and more companies moved into this sector, prices and hence profits fell. Furthermore, the main producers of polymers, led by the example of Du Pont, were reluctant to be involved with the manufacture of the final products for consumer sale, preferring to concentrate on the production of the virgin material. They did this partly to avoid having their reputations damaged by inferior consumer products. By contrast in the field of synthetic fibers, the leading firms aimed to make their fibers a premium product, such as ICI's Terylene, Hoechst's Trevira, and Du Pont's Lycra. The final stage of manufacture was generally undertaken by small firms. These were initially based in Europe and the USA, and by the 1960s in Japan and Hong Kong, where millions of refugees from Communist China sought a new livelihood. While these activities doubtlessly generated both personal income and economic growth, the end products were often cheap and poorly made, thus harming the reputation of both plastics and the chemical industry. At the same time, the search for a "new nylon" by Du Pont and other chemical firms did not meet with success. The polyurethane rubbers and foams were a notable exception, although their discovery went back to wartime Germany.

Another breakthrough was the development of the stereospecific polymers – polypropylene, high-density polyethylene, and synthetic natural rubber – using the revolutionary catalysts developed by Karl Ziegler and Giulio Natta (McMillan 1979). Plastic bottles first made by Du Pont revolutionized the soft drinks industry, and provided new outlets for polyester and PVC. The use of polyester fiber as a mixture with wool was a marked success and allowed the continuing growth of synthetic fibers. However, ICI's attempts to make a wool substitute called Ardil from peanut meal (in an effort to support Britain's colonies in west Africa), and Du Pont's drive to make Corfam, a breathable leather substitute, were both complete failures (Kanigel 2007; Hall 2014). They failed because the products were not of the same quality as the natural product and were too expensive. As a result, the chemical industry's belief in R&D, based on the success of early products such as nylon and polyethylene, began to wane, and spending on research fell. Instead of attempting to make technological breakthroughs from pure research, the industry tried to improve the technology created between 1920 and 1960.

THE GATHERING STORM, 1965–1980

In a continuation of their "boosterism" of the interwar years, the chemical industry presented itself to broader society as an unalloyed force for good and the provider of higher living standards. In the 1960s, this optimistic view of the industry was increasingly met with skepticism or even rejection by the general public, especially in the United States and Germany. The concerns over harmful effects of chemical pesticides, especially DDT, expressed in Rachel Carson's exposé *Silent Spring* (1962) were first denied and then condemned by industry representatives, but her information proved correct, and helped spark a broad environmental movement (Morris 2019). About the same time, it was revealed that pregnant women who had been prescribed the popular sedative thalidomide gave birth to severely deformed babies. DDT became a cause célèbre in the USA, whereas thalidomide was more of an issue in Britain, where thousands of children were affected (Quirke 2013).

A series of environmental disasters in the mid-1970s and 1980s deepened the public's concerns about chemicals in the environment, notably the plant explosions at Flixborough, England (1974), Seveso, Italy (1976), and Bhopal, India (1984), and the discovery of severe ground water contamination at an abandoned factory site in Love Canal, New York State (1977). The Vietnam War not only fostered an anti-establishment mood among many young people in the United States and elsewhere, but also had a direct negative link with the chemical industry (and Dow in particular) through the Americans' broad-scale use of napalm (an antipersonnel incendiary), and the application of the defoliant Agent Orange, a mixture of weedkillers that also contained an extraordinarily deadly contaminant called dioxin, to millions of acres of Vietnamese forests and farmlands (Martini 2012).

In addition to these public relations setbacks, the industry was also beginning to suffer from overcapacity (Arora et al. 1998). There was only so much polyethylene, PVC, or polyester that the average consumer could use. The two Oil Crises of 1973 and 1979 took place against this darkening sky. The causes of these crises need not concern us here, for they were largely political, but their effect was dramatic. From $3.56 per barrel in June 1973, the price of oil trebled to $10.11 in March 1974 and then almost quadrupled to $39.50 in May 1980 (Macrotrends 2018). The price then slowly slid down to $25.43 in December 1984, partly because there was a recession, before spending the rest of the 1980s and the 1990s hovering between $10 and $20 a barrel. But the damage was done: the age of cheap oil was over, and the chemical industry never quite recovered its swagger of the late 1950s and 1960s.

As a reaction to the 1973 Oil Crisis, President Jimmy Carter launched an emergency program to make synthetic gasoline using the Fischer–Tropsch process, which had been used by apartheid-era South Africa to circumvent petroleum sanctions. While the program was largely successful in technical terms, it could not defeat the economic conundrum that whenever the price of oil rose, the price of coal was certain to follow, given the importance of oil as fuel. Hence, synthetic gasoline could never be competitive in price terms with the natural product, however expensive the latter might be (Crow et al. 1988; Morris 1990).

REARRANGING THE DECKCHAIRS ON THE TITANIC, 1980–2019

The chemical industry changed in several fundamental ways following the two Oil Crises (Spitz 2003; Chandler 2005; Galambos et al. 2007). The industry switched from expansion, to restructuring and diversification. Basic sectors such as polyethylene and PVC were consolidated, and then sold off. Because of ever-increasing environmental regulation and labor costs, some sectors, notably dyestuffs manufacture, largely migrated from the USA and Western Europe (with the partial exception of Germany) to India and China, and also Taiwan, South Korea, and Thailand. This shift was also driven by the growth of the textile industry in Asia, thereby keeping the industrial producers close to the consumers. Petroleum companies both in oil-producing and in oil-consuming nations continued to make petrochemicals. They were joined by entrepreneurs, notably Jon Huntsman, Sr., Gordon Cain, and more recently Jim Ratcliffe, who saw the opportunity to enter the sector at relatively low cost. Du Pont attempted to pursue both strategies, by acquiring the petroleum company Conoco while at the same time moving out of petrochemicals. This corporate marriage was not a happy one, and DuPont sold off Conoco in 1998.

In this period, many firms left petrochemicals and increasingly concentrated on high-value and small-volume products such as pharmaceuticals, agrochemicals, chemicals for the electronics industry, and other specialty chemicals. Pharmaceuticals were seen as particularly attractive because of aging populations and lifestyle changes on the one hand, and the rise of modeling of both drug interactions and organic synthesis on the other. As part of this diversification drive, some companies, notably Monsanto, moved into the field of genetically modified (GM) plants, and GM seeds in particular. A controversial aspect of this shift was the development of plants that were specifically resistant to one company's weedkiller, in particular Monsanto's widely used Roundup® (based on the chemical glyphosate).

Nonetheless, the overall global output of the industry worldwide remained largely unchanged in relative terms between 1975 and 2000. Since the 1990s there has been much talk of "green chemistry" – in which engineers incorporate environmental protection into the basic manufacturing process rather than removing the pollutants afterwards – and "sustainable chemistry," which seeks to save energy and avoid the use of scarce materials. However, the expected positive impact on the image of the industry as a whole has so far been relatively limited. Nonetheless, it is striking that immediately prior to its merger with Dow in 2017, DuPont concentrated on the development of GM seeds, solar cells, and alternatives to fossil fuels; it had spun off its performance chemicals into the Chemours subsidiary in 2015.

In terms of corporate identity, however, the chemical industry has been transformed. In 1975, the major chemical companies were still recognizably the same firms that had been dominant in 1950 and, as a result of the break-up of IG Farben, the same as in the early 1920s. These companies were also largely national in character. In the USA, there was Du Pont, Union Carbide, and Dow; in Germany, BASF, Bayer, and Hoechst (all successors of IG Farben); in Britain, ICI; in Italy, Montedison (the successor of Montecatini); in France, Rhône-Poulenc; and in Belgium, Solvay.

By contrast, the list of the ten largest firms in the world in 2016 according to *Chemical and Engineering News* included just three of these companies, BASF, Dow, and DuPont (Tullo 2017). Dow and DuPont merged on August 31, 2017 to form DowDuPont, the world's largest chemical corporation, with combined sales about 10 percent higher than the previous leader, BASF. However, as planned from the outset of the merger, the new behemoth was split by June 2019 into three separate companies: agriculture (Corteva), materials science (Dow), and specialty products (DuPont; Trager 2018). BASF alone has remained of the large firms of the twentieth century. Other old established firms in the top ten of 2016 are Mitsubishi Chemicals of Japan, and the oil company ExxonMobil, the successor of Standard Oil of New Jersey.

The relatively new entrants into the industry include the petrochemical giants of China (Sinopec) and Saudi Arabia (Sabic). Other rising powers in the industry are Formosa Plastics, originally founded in 1954 in Taiwan with American help, and Ineos, the former BP Petrochemicals business taken over by Ratcliffe in 1992 (as Inspec). Ineos and another top ten firm, LyondellBasell, are a result of the post-oil crisis restructuring of the chemical industry, both firms being constituted of production units (mostly petrochemicals) sold off by established firms in the late twentieth century. Basell was a joint petrochemicals venture of BASF and Shell (the name is a near-merger of the two parents) in 2000, which merged with Lyondell, a successor to ARCO Chemicals in the same way that Inspec was a successor to BP Chemicals, in 2007. Having briefly filed for bankruptcy in 2009, the firm bounced back after the inevitable financial restructuring.

As for the rest of the "old guard," the British firm ICI, under pressure from its shareholders, spun off its pharmaceutical business as Zeneca in 1993. Subsequently, it disposed of several historic parts of the company, including ammonia-soda, agricultural chemicals (sold to Norsk Hydro, which then spun off its own agricultural arm as Yara), and nylon (to DuPont). Having then tried to diversify by buying businesses from Unilever, ICI had to sell core businesses to pay off the resulting debt. In the end, the remains of this once great firm, basically just the coatings business, was taken over by AkzoNobel in 2007. Hoechst attempted in the 1980s to move into the USA, which it rightly perceived to be a key market. It acquired Celanese in 1987 and in an effort to break into the US pharmaceutical industry took over Marion Merrell Dow in 1995. Facing financial turbulence, the firm decided to convert itself into a pharmaceutical company. It spun off Celanese, abandoned its historic site near

Frankfurt, and under its new name of Hoechst Marion Roussel merged with the French firm Rhône-Poulenc to form Aventis (which in turn merged with Sanofi in 2004). Bayer moved its dyestuffs operations to a new company DyStar (formed in collaboration with Hoechst) in 1995, then spun off much of its chemicals and polymers business (including its historically important synthetic rubber unit) into Lanxess in 2004 and its specialty polymers (including its iconic polyurethanes and polycarbonates) into Covestro in 2015. It controversially merged with Monsanto in June 2018 to create a new major force in seeds and crop protection chemicals under the Bayer name (Trager 2018). The net result of this corporate whirlwind through the industry (which one suspects has been of more benefit to the big banks and venture capitalists, than the chemical industry) was globalization, the loss of historic firms, and a proliferation of contrived names for firms which are constantly merging (and sometimes de-merging).

CONCLUSION

The chemical industry experienced three fundamental technological and cultural shifts during the twentieth century. Initially, after World War I there was a period of intense innovation dominated by large national firms supported financially by national governments (a situation similar to the aerospace industry in the early twenty-first century). After World War II, the industry enjoyed a free-wheeling era of cheap petroleum and gas that was largely based on the technical innovations of the earlier period. Finally, after the Oil Crises of the 1970s, there has been an upheaval which has not seen much in the way of major radical innovations, but the now-globalized industry has been split into two parts: one associated with the petrochemicals of the second period and the other now largely devoted to pharmaceuticals, genetically modified seeds, crop protection products, and solar cells.

Since 1914, the industry has had a transforming impact on world society, its economy and its culture as a whole, delivering a better quality of life, vastly improved health, and sufficient food. At the same time the industry has also generated pollution on a global scale, from local incidents such as Love Canal, to the worldwide distribution of CFC gases. Plastics waste in particular has become justly controversial. The Pimentel report on "Opportunities in Chemistry" recognized the problem as far back as 1987: "each year in this country [the USA], you and I dump millions of tons of plastics into the environment. A high percentage of that is spewed directly into the oceans" (National Research Council 1987: 4). There is now the great Pacific garbage patch, consisting mostly of discarded plastics, and plastic microbeads in the Arctic ice. The world in the later twenty-first century will discover how the balance between the practical benefits and environmental costs of the chemical industry has tipped (Davis 1984; Crone 1986; Radkau 2008).

CHAPTER SEVEN

Learning and Institutions: *Global Developments since 1914*

JEFFREY ALLAN JOHNSON, YASU FURUKAWA,

AND LIJING JIANG

"Education is national, but science is international." This dictum by a German science policymaker in the 1920s highlights not only the dilemmas of science education in general (Harnack 1961: 182), but more particularly those of chemistry. Chemistry's economic, environmental, and military significance, manifested so clearly by World War I, made promoting chemical education a political priority in most nations. However, successfully pursuing this goal required the commitment of substantial resources, which were difficult to obtain in the postwar era. Intensifying international competition threw up the challenge of adapting national institutions to keep pace with those at the forefront of rapidly advancing international scientific standards. This was no small matter for the increasingly diverse nationalities who were coming to view chemistry as an indispensable element of a "modern" culture they were not always eager to embrace. Welcome or not, the century since 1914 brought exponential growth and fundamental changes in the discipline and its practitioners. Thus, the leading European and North American chemical institutions of 1914 would be dwarfed by their twenty-first century counterparts around the world, staffed by a diverse group scarcely recognizable by the white male chemists who dominated the profession in 1914. This successful growth notwithstanding, chemical educators faced continuing challenges, in particular an increasingly negative public image of chemistry (Schummer et al. 2007).

Chemists continued to share a common culture, derived from hands-on laboratory training in the manipulation of chemical substances that has characterized modern introductory chemistry courses everywhere. So along with considerable diversity in institutions and approaches, basic elements of chemical education remained the same. The common global culture of chemistry benefited both from enhanced international communications as well as from the restructuring of international politics since the 1980s, further enhanced to some extent by the efforts of international organizations to foster cooperation among national chemical societies.

CHEMICAL EDUCATION AND INSTITUTIONS DURING THE ERA OF THE WORLD WARS, 1914–1945

By 1914 chemistry had become an international scientific discipline and an integral part of European and American educational systems. Chemists in most independent countries had organized national societies by the end of the nineteenth century (cf. Table 7.1 for leading examples). By highlighting the military value of a sophisticated chemical industry, World War I brought chemistry even more recognition, while also complicating support for chemical education. If advanced nations now had greater incentive to promote chemical education at home, its potential military value gave them less motivation to teach chemistry to potentially rebellious natives in their Asian or African possessions. By contrast, independent Japan had developed institutions for chemical education as part of its broader strategy of Westernization, although leading Japanese chemists still tended to complete their research training in the West (Kikuchi 2018: 140–3). China also sought to develop its educational institutions, but continuing political chaos made this difficult. The independent nations of Latin America were just beginning to develop their scientific professions. Even in the British Dominions of Australia and New Zealand with their largely European populations, doctoral programs did not fully emerge until the late 1940s (Halton 2018: 267; Rae 2018: 44). Consequently, Europe and the United States, home to the majority of chemists in this period, will be the primary focus of this section, concluding with a brief discussion of the emerging international structure of chemistry.

The heightened national significance of chemical education brought institutional innovations. For example, the American Chemical Society (ACS) created a section (later, a division) of chemical education shortly after the war in 1921, followed by the founding of the *Journal of Chemical Education* in 1924 (Gordon 1924: 1). The Germans already had a specialty group for chemical and

TABLE 7.1 Numbers of chemists (in thousands – census figures) in the United States and Germany; memberships in principal professional associations in USA, Germany, and Japan (in thousands)

Year	USA		Germany		Japan
	Chemists (index)	ACS	Chemists (index)	VDCh/DChG [to 1940] GDCh/ChGDDR [1950–]	CSJ
1907			5.8 (100)		
1910	16 (100)	5.1		4.1/3.4	
1920	28 (175)	15.6		6.0/3.5	
1925			10.6 (183)	7.4/5.2	
1930	45 (281)	18.2		8.8/4.9	
1939			15.3 (264)	9.8/3.4	
1940	57 (356)	25.4			
1950	77	63.3		4.9	11.5
1960	84	92.2		11.0/1.6	22.6
1970	110	114.3		16.0/3.0	35.3
1980	128* (800)	120.4		17.8/3.2	31.8
1990/91	127*	142.6		21.4/4.5**	38.1
2000	82*	163.0		28.3	36.4
2010	80* (500)	163.1		28.8	30.9
2017	84* (525)	150.0		30.9	27.5

*Includes chemists, not biochemists or chemical engineers (US Census estimates, 1980, 1991 [but 1991 is not directly comparable with earlier years]; Bureau of Labor Statistics estimates of fully employed professionals, 2000 and later; these may not be comparable with the earlier US Census estimates, in particular by understating the total employment of chemists, but probably by no more than 10 percent in each case).
**The ChGDDR (*Chemische Gesellschaft der DDR* [Chemical Society of the GDR]) was disbanded at the end of 1990; 3,500 members joined the GDCh (*Gesellschaft Deutscher Chemiker* [Society of German Chemists]) in 1991 following German reunification (Osterath 2017: 1023).
Sources: Thackray et al. 1985: 247 [US census estimates to 1970], 250–2 [ACS (American Chemical Society) members to 1980]; US Bureau of the Census 1981: 402; US Bureau of the Census 1992: 392; US Bureau of Labor Statistics 2000, May 2010, May 2017 (national tables including occupational group 19–2031: Chemists); ACS 2018 [ACS membership, 2017]; Germany. Statistisches Reichsamt 1907: 130–1; Germany. Statistisches Reichsamt 1927; Germany. Statistisches Reichsamt 1941–44; Maier 2015: 19, 349, 362 [membership of VDCh (*Verein Deutscher Chemiker*, Association of German Chemists) and DChG (*Deutsche Chemische Gesellschaft*, German Chemical Society), 1910–1939]; GDCh 2018: 30 [GDCh membership 1950–2017]; Kiessling 2018 [ChGDDR (membership 1953–89); statistic data provided to one of the authors (YF) by the Chemical Society of Japan, May 2018, and to another of the authors (JAJ) by the American Chemical Society, April 2019.

technological education (*Fachgruppe für chemisch-technologischen Unterricht*) in the professionalizing Association of German Chemists (*Verein Deutscher Chemiker*). Since 1898 the Association of Laboratory Directors at German Colleges and Universities (*Verband der Laboratoriumsvorstände an deutschen Hochschulen*), an organization dominated by university professors, had set standards for predoctoral examinations. In the interwar period, it discussed how to improve doctoral training. At the same time, the powerful German chemical industry created funding organizations that expanded the smaller prewar network of industrial subsidies for German academic chemistry. The industrially supported Justus Liebig Society annually awarded around forty to seventy Justus Liebig postdoctoral fellowships, while the Adolf von Baeyer Society subsidized the journals and reference works of the German Chemical Society (*Deutsche Chemische Gesellschaft*). When subsidies for the Liebig Society decreased as the chemical industry cut their labor force and wages after 1930, the chemical combine I.G. Farben created its own program of postdoctoral fellowships for the retraining of unemployed industrial chemists (Johnson 2000: 28–36, 45).

At the same time, the chaos of revolution and civil war in the former Russian empire set back the development of chemical education for several years. After the Bolshevik triumph, Lenin began promoting technically useful science in the Soviet Union as the developmental path of "scientific socialism." Stalin intensified this process in the 1930s while also bringing scientific professions under Communist Party control and purging scientists (among many others) deemed ideologically unacceptable (Brooks 1997: 363–5). Given its relevance for industrial development, Soviet chemistry suffered fewer losses than physics or biology, apart from quantum chemistry, which appeared to impinge on the theory of dialectical materialism (Graham 1964). Certainly the Soviets had impressive numbers of chemistry professors and instructors in their institutions of higher education (examples in Lütke 1926a: 450–1; Lutke 1926b: 1000–27; Lutke 1926c: 1328–79).

On the level of secondary education, during this period most young people either left school in their early teens or pursued "vocational" training through relatively brief courses in technical schools. Only a small but growing minority attended secondary schools in preparation for higher education, for example in German Prussia, where the proportion of eighteen-year-old boys in such schools rose from under 3 percent of the total population in 1920 to 7.7 percent in 1937. The girls' share exceeded 1 percent only during the depths of the economic depression, in 1931 (Lundgreen 2000: 148). In 1914 elite schools still tended to be classically oriented, teaching Latin and Greek; while they incorporated some training in mathematics and the natural sciences (particularly physics), these were usually regarded with some disdain. To English gentlemen with an elite classical education, "stinks," i.e. chemistry, symbolized the social and cultural inferiority of modern science, chemists being equated with tradesmen, as the English poet W.H. Auden recalled (1948: 1). Even the German school reform of 1925 required only two hours of chemistry (less than 1 percent of the total teaching time) in the four upper grades of the classical secondary schools (*Gymnasium*), whereas the "modern" secondary schools devoted six to twelve hours (2–4 percent of the total) to chemistry (Bauer 1990: 10). Chemistry both gained and lost when the National Socialists – rejecting the "false" ideal of a classically cultured elite – in 1938 established an ideologically based standard curriculum for all schools, featuring racial biology, German studies, and physical training. Introduced on the eve of World War II, the new system designated four to seven hours for chemistry, generally taught in conjunction with physics, but also with an option for additional cooperative projects that would serve "national–political" purposes (REM 1938; Pine 2010: 44–8). Parallel versions of regimented secondary-level science education developed elsewhere, particularly in the Soviet Union and Fascist Italy.

The secondary schools of the United States enrolled a larger proportion of the population than in Europe, thus offering more students the opportunity to learn at least some chemistry. While overall enrollments in secondary schools roughly quadrupled over this period, a rate faster than overall population growth, the proportion enrolled in chemistry classes remained between 7 percent and 8 percent. Thus, the proportion of the population learning chemistry also roughly quadrupled, to more than a fifth in the 1940s (Thackray et al. 1985: 64–6 and tables 3.16–3.17, 293–4). This was more than twice the rate in Germany, where another generation of future chemists lost on the battlefield further undermined its former scientific preeminence.

Regarding higher education, most countries had two main types of establishments: the traditional academic universities offering study in the liberal arts and sciences (especially for teacher training) and professional training in the traditional faculties of medicine, law, and theology; and technical institutions, designed to train for careers in business and industry (Ringer 1979). Chemistry was found in both tracks, but the numbers of students involved, the types of certificates or degrees, the disciplinary focus, and the resulting level of expertise that they might attain varied widely. The British and American academic institutions normally distinguished between undergraduate training leading to a first degree and postgraduate training focused on research and leading to a master's or doctoral degree. The German universities had no undergraduate training in the Anglo-American sense, while in some countries, such as Italy, the universities apparently conferred only undergraduate degrees.

A central feature of the German approach to chemical instruction had gained general acceptance, at least in principle: a combination of lectures featuring experimental demonstrations, and laboratory training starting with an introduction to qualitative and quantitative analysis (Brock 1993). But the quality and quantity of laboratory training varied greatly; German university chemistry students received some of the most intensive training, spending entire days throughout the week in laboratories, while attending lectures in the evenings. Those countries in which the chemical industry had developed on the largest scale had the greatest demand for trained chemists. In 1914 the leaders were Germany, the United States, and the United Kingdom, whereas France and Italy offered more limited access to advanced research training.

Although American academic chemists may have prided themselves on their modern outlook, in elite universities such as Yale during the postwar era they still had to contend with a neo-medieval academic culture, reflected in the neo-Gothic university architecture that had become popular in the late nineteenth century. To deflect prejudices not unlike the English elite-school disdain for "stinks," Yale chemists found it prudent to avoid calling attention to the modernity of their discipline. They cloaked their otherwise ultra-modern chemistry teaching facility, the Sterling Chemical Laboratory constructed in 1923, in a strange, neo-Gothic exterior (compare Figure 7.1 and Figure 7.2).

FIGURE 7.1 In the student laboratory, Sterling Chemical Laboratory, Yale University, New Haven, Connecticut, 1926 (the laboratory was built in 1923). Soapstone table tops supported on metal standards provide working space. Note use of electric lighting and gas for burners. These were graduate students; Yale first admitted women undergraduates in 1969. Photograph by Print Collector/Getty Images.

FIGURE 7.2 Sterling Chemical Laboratory, Yale University, New Haven, Connecticut, 1926. Exterior view of the building constructed in 1923. Photograph by Print Collector/Getty Images.

During the period before 1945, national approaches to physico-chemical theories and methods reflected institutional differences. In the traditional, hierarchical German universities, organic chemists dominated the large teaching institutes and resisted even physical chemistry (ionic dissociation, solution theory, reaction kinetics, and thermodynamics), which had been promoted by Wilhelm Ostwald. Notably, physics – but not physical chemistry – was required for the preliminary examination, roughly equivalent to the knowledge expected of an American undergraduate degree in chemistry. In the less hierarchical departments of the colleges of technology (*Technische Hochschulen*), physical chemists played a more equal role, but their smaller numbers overall limited their influence on the educational policies of the Association of Laboratory Directors (Johnson 2013: 417–22; Weininger 2018).

American universities with their departmental structures appear to have integrated physical chemistry much more easily than did their German counterparts. Because students in American universities did not pay fees to lecturers, there was no incentive for the senior professors to teach large introductory courses, whereas in German universities, professors still earned a significant portion of their income from student fees. While American senior faculty taught smaller advanced undergraduate and graduate-level courses in their areas of specialization (then mainly classical organic and inorganic chemistry), junior faculty took the large introductory lecture courses in "general chemistry," which they could shape in innovative ways. This meant not only "classical" physical chemistry, but in the interwar years increasingly also more recent material such as the G.N. Lewis-Irving Langmuir octet theory of electron bonding, and by the end of the 1930s the quantum-related "chemical physics" notions of valence bond theory and molecular orbital theory. In 1937 Linus Pauling at Caltech began developing a highly influential modern textbook of general chemistry (Servos 1990: 248–50; Nye 2000b: 397–414).

During the years up to 1945, the periodic table gradually became an iconic feature of chemistry lecture halls, though the version initially most popular was a short form based on that of Dmitrii Mendeleev (Figure 7.3; Robinson 2018); the modern eighteen-column-long form did not become dominant until the 1950s (Bensaude-Vincent 2001: 133–61). Another iconic feature of lectures at this time, carried over from the nineteenth century, were the simple ball-and-stick molecular models, largely derived from older two-dimensional structural formulas with some three-dimensional features such as the tetrahedral carbon atom. Although three-dimensional space-filling molecular models based on quantum ideas of orbitals and bond angles emerged in the 1930s, they were expensive and somewhat controversial, and they remained relatively rare until after the war. The physical chemist Linus Pauling (Figure 7.4) became one of

FIGURE 7.3 Students in a large chemistry lecture, The University of Iowa, 1930s. Note the table and apparatus for demonstration experiments, the short-form periodic table, and also that the students appear to be all or nearly all male. Frederick W. Kent Collection of Photographs, University of Iowa Libraries, Iowa City, Iowa. Reproduced by kind permission of the University of Iowa Libraries.

FIGURE 7.4 The American physical chemist Linus Pauling (Nobel Prize in Chemistry, 1954; Nobel Peace Prize, 1962) presenting a lecture in 1967 using various types of molecular models. The largest ones, on the right, are space-filling molecular models based on quantum principles. Photograph by Universal History Archive/Universal Images Group via Getty Images.

their strongest advocates (Francoeur 1997; Johnson 2013; see also Chapter 2 in this volume).

The German universities at this time conferred only the doctorate and hence in principle trained all of their students for research, although between a third and a quarter of their students did not complete their degrees. The predoctoral "Association Examination" (*Verbandsexamen*) conferred no officially recognized degree. Although the German colleges of technology did award an intermediate *Diplom-Ingenieur* (diploma of engineering, roughly equivalent to an American master's degree), most of their chemistry students at this time went on to complete doctorates. In 1939 the National Socialist government dissolved the Association of Laboratory Directors and replaced its examination and the *Diplom-Ingenieur* examination for chemists by the first national certifying examination and associated professional title for chemists, the *Diplom-Chemiker* (certified chemist), with uniform requirements across all institutions, which became the German standard until phased out by the

Bologna Process after 1999 (Johnson 2013: 436–7, 448). Table 7.2 shows the figures for the numbers of doctorates awarded during this period, including those to women, who received fewer than 10 percent except during times of war (1918/1919) or mobilization (1937/1938 and later). The numbers of academic staff and postdoctoral assistants peaked before the Nazi regime took power in 1933, then fell steadily, only partly mitigated by increased numbers of women during World War II (Tables 7.3 and 7.4). Overall, National Socialism had a disastrous impact on the rising generation of German chemists.

By contrast, in the United States, although in 1914 chemistry was one of the most popular university subjects, relatively few chemistry students were trained in research. For the undergraduate degree, laboratory training took at most a few hours per day, supplementing lectures rather than playing a central role as in Germany. The US data (Table 7.5) clearly reflect the impact of wars and economic crisis. In 1918 and 1945, one sees much-reduced enrollments

TABLE 7.2 German doctorates in chemistry awarded by universities and other academic institutions 1914–1938, with percentage awarded to women.

Year	Ph.D.s	Percent to women
1913–14	335	1.8
1918–19	125	17.6
1922–23	864	4.5
1925–26	785	3.6
1930–31	496	7.7
1934–35	372	6 [est.]
1937–38	355	10–11 [est.]

Source: Verband der Laboratoriumsvorstände 1914–1939. Percentages of women were based on counts of names in institutional lists; due to occasional ambiguities, numbers are partly estimated.

TABLE 7.3 Academic chemists (excluding postdoctoral assistants) in German higher educational institutions, 1910–1938

Year	Regular, honorary professors	Index (1920 = 100)	% of total academic staff	Total academic staff	Index (1920 = 100)
1910	93	72	26	355	90
1920	130	100	33	393	100
1931	165	127	32	516	131
1938	148	114	31	473	120

Source: Compiled from Ferber 1956: 197.

TABLE 7.4 Assistants and doctoral students in German higher education, 1920–1943

Year	Assistants with Ph.D./ DrIng (% women)	Index	Doctoral students (% women)	Index
1920/21	233 (4.3)	100	1,959 (5.9)	100
1932/33	556 (3.2)	239	2,118 (6.4)	108
1936/37	415 (2.9)	178	1,612 (7.3)	82
1942/43*	328 (12.5)	141	717 (22.7)	37

*Numbers of assistants and students exclude men in military service.
Source: Adapted from Johnson 1998a: 11; Johnson 1998b: 67, 77.

TABLE 7.5 United States: Bachelor's degrees and doctorates granted in the natural sciences and chemistry

Year	Total numbers of bachelor's degrees and doctorates			
	In natural sciences		In chemistry (% of all natural science degrees)	
	BA/BS	Ph.D.	BA/BS	Ph.D.
1914	7,297	244	2,573 (35.3)	97 (39.8)
1915	7,095	268	2,397 (33.8)	107 (39.9)
1918	6,648	244	2,102 (31.6)	96 (39.3)
1920	8,312	268	2623 (31.6)	104 (38.8)
1925	12,371	533	3784 (30.6)	220 (41.3)
1930	15,297	945	4392 (28.7)	332 (35.1)
1935	15,911	1,140	4493 (28.2)	402 (35.3)
1940	22,976	1,459	6366 (27.7)	532 (36.5)
1945	15,361	764	4400 (28.6)	342 (44.8)
1947	23,273	1,154	6,344 (27.3)	418 (36.2)
1950	49,184	2,351	12272 (25.0)	967 (41.1)
1919–1950 (totals)		32,026		12,443 (38.9)

Source: Thackray 1985: 257–8, 265–6, 433.

during the final years of the world wars, as a result of the induction of much of the university-aged population into the military. It is also clear that despite the efforts of the leaders of the profession to popularize their role in the "chemists' war," the interwar period saw chemistry decreasing in relative popularity, which continued after a temporary increase during World War II. At the doctoral level, chemistry enjoyed a significantly higher level of popularity among the natural sciences, but the general trend was also declining (Thackray 1985: 257–8, 265–

6, 433). The American data unfortunately are not broken down by gender, but other sources suggest that, as in Germany, women chemistry students compensated for a dramatic decline in male student enrollments during both wars (Rossiter 1995: 19–20, 29–35, 76–7, 100). British undergraduate and graduate degrees resembled the American model, although British undergraduate chemists tended to receive more specialized and advanced training, especially in elite universities such as Oxford and Cambridge. British women too at least temporarily benefited from wartime educational opportunities during World War I (Rayner-Canham and Rayner-Canham 2008), but in the next war "an informal agreement" kept most universities from using "female students to compensate for the missing male undergraduates" (Horrocks 2011: 157).

As the International Association of Chemical Societies, founded in 1911, had collapsed during World War I, a new and more enduring organization emerged in 1919 as the International Union of Pure and Applied Chemistry (IUPAC) within the new framework of the International Research Council (IRC), which temporarily excluded Germany (blamed for instigating the war and for introducing chemical warfare) and Russia (as a result of the Bolshevik revolution). IUPAC remained based in Paris during this entire period, and until 1944 it had only one secretary general, the chemical engineer Jean Gérard. It continued the work of the prewar International Association by sponsoring commissions that held regular conferences to discuss various issues including nomenclature, standardization, and issues related to literature. After several years of often rancorous debate and negotiations, the German and Soviet chemists joined IUPAC in 1930 (subsequently known until after 1945 as the International Union of Chemistry or IUC). The IUC then coordinated the only International Congresses of Applied Chemistry held in the interwar period, in Madrid (1934) and Rome (1938). These conferences made for a stark contrast between the left-liberal ideals espoused by leading scientists in republican Spain, shortly before its destruction during the Spanish Civil War of 1936–1939, and the militant totalitarianism of Fascist Italy, which had employed chemical weapons in its recent conquest of Ethiopia and offered them to its anti-republican ally General Francisco Franco in Spain (Nieto-Galan 2019: 68–72, 84–92). With the advent of general European war in 1939, most of the IUC's work was suspended, leaving chemists to hope that internationalism would survive World War II (Fennell 1994: 47–75; Fauque 2016: 42–9; Fauque and Van Tiggelen 2019).

PART TWO: THE ERA OF DECOLONIZATION AND COLD WAR, 1945–1990

The aftermath of World War II brought the division of Europe into the Western and Soviet blocs, leading to four decades of Cold War. Ironically, despite the horrors of the National Socialist regime and the continuing division of Germany throughout this period, Germans achieved a faster reintegration into a relatively unchanged international scientific structure after 1945 than after the major reorganization of 1919. Most German scientific institutions revived rapidly during the economic recovery of the 1950s, those in the Soviet zone (from 1949, German Democratic Republic) rather more slowly than in the Western zones (from 1949, Federal Republic of Germany); many German scientists cooperated closely with their Western or Soviet counterparts (Stokes 2000: 15–54; Augustine 2007: 1–38; Crim 2018). However, the era of German dominance in chemical education had passed, as the Americans and the Soviets vied to take their place. Reconstruction in Western Europe proceeded more rapidly than in the Soviet bloc due to economic support from the United States, which had suffered virtually no damage during the war and comparatively few losses of professional manpower (Krige 2006). Indeed, the Americans had improved their scientific manpower in the form of prewar refugees who remained after the war, as well as a significant "brain drain" from Europe during the initial postwar period (Balmer et al. 2009). Similarly, more professionals fled eastern European countries after they came under Soviet control than moved in the other direction for political reasons (Dowty 1989: 122).

In the United States, the most successful of the older centers, there was an eightfold increase in the total number of chemists counted by the national census from 1910 to the high point of around 128,000 in 1980, while the membership of the ACS continued to increase, allowing it to claim the title of world's largest scientific society (cf. Table 7.1). The politically and economically weakened Western European powers gradually gave up most of their colonial possessions in the 1950s and 1960s. This era of decolonization created a large number of so-called "Third World" or "developing" nations in Asia and Africa, which tended to embrace science and education as an important tool for national development. Unfortunately, they suffered from perennial shortages of funds, which made it very difficult to appropriately develop the laboratory-based instruction expected in chemistry, and they also lacked qualified instructors (Sané 1979: 93–7). Many of these countries carried over institutional forms and even instructors from the previous colonial systems, which gave the former metropolitan powers continuing educational influence (cf. Adjangba 1979: 174; Pham et al. 2018: 434). During the Cold War, these newly independent nations became the targets for the competing attentions of the American and Soviet blocs, but the better-funded American universities attracted considerably more foreign students and postdoctoral researchers. Soviet universities, with intensive, standardized five-year programs for the first degree, were probably superior to the Americans for training ordinary chemists for industry (Slobodkin and Pickering 1988: 3–5). But advanced research training suffered from the over-centralization and bureaucratic inflexibility that were legacies of Stalinist planning. One American observer asserted that even the elite organic chemistry research institutes of the Soviet Academy lacked "the diversity, quantity, and quality of the instrumentation" in a good American university doctoral program (Wotiz 1971: 66).

There were, of course, other options. Following the victory of the Communist revolution under Mao Zedong in China in 1949, the Chinese also offered a model for development, albeit initially adopting the Soviet educational system, but political issues limited growth during most of this period. Wars also devastated and divided Korea and Vietnam in the 1950s, and Vietnam again in the 1960s and early 1970s. South Korea with United Nations and American support, and Vietnam with Soviet aid, went on to rebuild and expand significant facilities for chemical education. This allowed a growing proportion of their students to seek advanced degrees at home as well as in a variety of foreign nations (Do 2018: 190–9; Pham et al. 2018: 433–9). American occupation meanwhile forced a defeated and disarmed Japan to modify many of its institutions. Rebuilding their devastated cities, the Japanese achieved an "economic miracle" that also revived their chemical industry, beginning in the mid-1950s (Forsberg 2000). Japan was readmitted to IUPAC in 1951, at the same time as the Federal Republic of Germany (Kikuchi 2018: 150–1). These developments made it possible to rebuild and expand educational institutions and to resume the scientific development that had been interrupted by the war.

Postwar reconstruction, decolonization and Cold War competition brought global expansion in higher education. By the mid-1950s the number of institutions of higher education outside Europe had nearly doubled from an estimated 1,500 before the war, and many of the institutions had also grown substantially in size (Schuder 1956: vii). The postwar "baby boom" led to institutional expansion and innovation in the United States, particularly after the Soviets launched Sputnik in 1957 and panicky Americans demanded improvements in science education (Orna 2015: 1–9). This scare brought major changes in secondary-school chemistry curricula and teaching methods both in the USA and in Western Europe, especially in the UK, although there was not a clear consensus as to the best approach to take (see Chapter 2 in this volume).

In the absence of detailed and comparable figures for chemistry student enrollments in secondary schools during this period, one may note that in the case of the Federal Republic of Germany (but not in the Soviet-dominated German Democratic Republic) the postwar era reversed the National Socialist effort to create a uniform secondary-school system (Fenn 2012). Enrollments in academically oriented secondary schools rose from around 5 percent in 1952 to more than 20 percent of eighteen-year-olds in 1990 (Lundgreen 2000: 148). This would greatly increase the flow of students into the universities. An innovation that helped to generate secondary-student interest in chemistry during this period emerged from Eastern Europe in 1968, with the first

International Chemistry Olympiad held in Prague, Czechoslovakia. Initially confined to the Soviet bloc, by the 1980s the Olympiad began to meet also in Western Europe, English became its official language, and in ensuing decades it became fully global (Fung et al. 2018: 195–6).

Trends towards the end of this period also brought many more women into higher education. It has been estimated that whereas women's share of the total was nowhere greater than 10 percent in 1970, by 1980 it had doubled in Western nations (in the United States, particularly as a result of Affirmative Action legislation in the early 1970s), and by 1990 that share had doubled again, to around 40 percent. The proportion of women university students in other regions rose as well, albeit not to the same extent (Ramirez and Kwak 2015: 14, fig. 2.1).

How did these trends affect chemistry students? The figures for university degrees in chemistry in West Germany and the USA (see Table 7.6) suggest that after a period of rapid growth in the early 1950s, the early 1960s were a period of relative stability or decline in numbers. In the US, more women studied chemistry at both the undergraduate and graduate levels, compensating for a decline in the number of US men studying chemistry after 1985. By contrast, German doctoral candidates (including women) increased, so that by 1990, their numbers exceeded those of Americans (Table 7.6; Rossiter 2012: xv–xvii, 42, 96).

TABLE 7.6 Diplom certificates and doctorates awarded in German universities, 1956–1985; bachelor's and doctorates in US universities, 1950–1990

Year (acad./cal.)	Diplom certificates (% women)	Ger. Drs. (% women)	US bach. degrees (% women)	US Ph.D.s (% women)
1950			8,696 (15.0%)	967
1956–57	740 (8.2%)	581 (3.1%)		
1960			7,604 (19.8%)	
1965–66	568 (5.8%)	619 (3.1%)		1,379 (5.2%)
1970			11,617 (18.2%)	
1975–76	1,162 (8.1%)	1,079 (6.4%)		1,521 (10.1%)
1980–81	1,362 (18.4%)	971 (11.1%)	11,446 (28.6%)	1,269 (17.2%)
1985	1,913 (26.3%)	1,292 (16.6%)		1,432 (19.4%)
1990	2,200* (25%)*	1700* (22%)*	8,289 (40.1%)	1,497 (25.6%)

*Approximate.
Source: Roloff 1988: 54 [German data to 1985]; Schwarzl 2006: 77 [German data 1990]; Kuck 2006: 109, 113; Nelson and Rogers 2005: Appendix 2, Table 1 [US data for calendar years only (1966, 1975, 1980, 1985, 1990) and include only US citizens and permanent residents, excluding foreign students and temporary residents].

Continuing international differences in chemical education, including significant gaps between the relatively prosperous industrialized nations and the poorer nations throughout this period, became topics for a series of international conferences under various auspices. These became significant vehicles for the exchange of ideas about chemical education on a global scale. The IUPAC Committee on the Teaching of Chemistry took the initiative, organizing its first International Symposium on University Chemical Education in Italy in 1969, in conjunction with the Italian National Research Council (Chisman 1970). A second symposium followed in 1971 in Brazil, setting a pattern of biennial international conferences (Gómez-Ibáñez 1972). Their proceedings and related documents offer extremely useful sources for the challenges and achievements in the field during the period in a wide variety of countries, including eastern and southern Europe, Latin America, and Africa, which unfortunately cannot be discussed in detail (cf. Rao and Radhakrishna 1979; Brubacher et al. 1989).

Participants in the 1969 symposium, for example, raised several perennial issues in chemistry teaching: what was the value of lectures versus other forms of teaching? To what extent should physical chemistry be featured in introductory courses, and are physical chemists really physicists rather than chemists (Chisman 1970: 15, 23–36)? More fundamentally, was chemistry losing its popularity in comparison to other fields? For example, only 7 percent of secondary-school students in the UK were in chemistry classes at age seventeen and only 1 percent of the total went on to study chemistry in a university (most of the rest went into medicine). The underlying problem here appeared to be a lack of appreciation for chemistry among the general public. Were chemistry professors neglecting the task of general education (Halliwell 1970: 18)? Hence chemistry in the West appeared to be losing its attractiveness even before the notorious industrial disasters of the 1970s and 1980s (cf. Schummer 2007; Chapter 5 in this volume).

THE RISE OF ASIA: CHEMICAL EDUCATION AND INSTITUTIONS IN JAPAN AND CHINA

Although China had been one of the major sources of alchemical knowledge in the medieval world, and Japan had successfully incorporated modern chemistry into its educational system during the Meiji era, it was not until the 1980s that Asian chemists began to gain broader recognition, with the first Nobel Prizes awarded to a Japanese chemist (Kenichi Fukui, 1981) and a Taiwanese-American chemist (Yuan T. Lee, 1986). In 1978 UNESCO proposed a Federation of Asian Chemical Societies (FACS), parallel to the Federation of European Chemical Societies (FECS) that had been founded in 1970 (and ultimately became the European Chemical Society, see below). The FACS began with eleven members in 1979, grew to twenty-one by the end of the next decade, and reached thirty-one by 2019. Its original working groups included one on chemical education (Singh 1989: 2–3, 9; Patterson 2018: 220–1).

Before World War II, Japan's nine Imperial Universities and other private universities employed elite chemists in their faculties of science and engineering. In addition, a dozen national Technical High Schools in seven major cities (including Tokyo and Kyoto) provided engineering education to train chemical technologists and technicians for careers in industry. Some of these technical high schools were later awarded university status. After the war, Japan adopted a system consisting of six years of elementary school, three years of middle school, and three years of high school. The old universities were restructured, and new national and private universities were created under a new system, with a shift from the German style to the American style of higher education. The term of study extended from three years to four years for liberal arts education and professional education in specialized fields. The role of graduate schools was to further professional training.

A science and technology boom occurred during the late 1950s and the 1960s, which was a period of rapid economic recovery in Japan. As a governmental policy in response to the demand from industry for engineers, Japanese universities increased the admission of students to science and engineering courses by 20,000 in the 1960s. Although environmental problems caused by the chemical industry in the 1960s and 1970s harmed its public image, chemistry-major student numbers increased until the late 1990s (see Table 7.7). However, an increasing focus on university entrance examinations meant that science teaching at many high schools was increasingly oriented toward examination preparation while minimizing laboratory practice.

The number of chemists regularly conducting research in universities, institutes, and industry increased fourfold from the early 1960s to the late 1990s (Table 7.8), reflecting the output of higher education. The membership of the Chemical Society of Japan (CSJ) reached nearly 40,000 in the late 1990s (Table 7.1), reflecting the increase in the number of researchers. Founded in 1878, the CSJ is Japan's largest society for chemical sciences and technology, and during this period it had the second-largest membership of its kind in the world behind the ACS. The CSJ has a Division of Chemical Education, which has played an important role in exchanges among high-school teachers. The Division founded a journal, *Kagaku kyoiku sinpojiumu* (*Chemical Education Symposium*), in 1953. This was renamed *Kagaku kyoiku* (*Chemical Education*) in 1962, and then *Kagaku to kyoiku* (*Chemistry and Education*) in 1987.

One distinctive feature of Japan's chemical community is the establishment of many specialist societies relating to chemistry and chemical technology in the past 140 years. Twenty-six societies compassing a wide variety of specialties still exist today. In contrast to the ACS, which has thirty-three technical divisions under its umbrella, the CSJ until recently has had only five divisions, including

TABLE 7.7 Numbers of students majoring in chemical sciences* at universities and graduate schools in Japan; female students in parentheses

Year	Undergraduate students	Graduate students	Universities**
1960	21,445 (310; 1.4%)	985 (56; 5.7%)	245
1965	36,951 (1,248; 3.4%)	3,233 (135; 4.2%)	317
1970	53,086 (3,063; 5.8%)	5,035 (146; 2.9%)	382
1975	59, 973 (4,781; 8.0%)	5,661 (188; 3.3%)	420
1980	60,288 (6,130; 10.2%)	5,820 (256; 4.4%)	446
1985	64,241 (7,817; 12.2%)	7,210 (389; 5.4%)	460
1990	68,682 (9,467; 13.8%)	9,328 (691; 7.4%)	507
1995	76,492 (15,693; 20.5%)	9,654 (1,015; 10.5%)	565
2000	70,708 (13,591; 19.2%)	10,005 (1,331; 13.3%)	649
2005	63,462 (15,681; 24.7%)	10,100 (1,402; 13.9%)	726
2010	56,476 (13,981; 24.8%)	9,961 (1,301; 13.1%)	778
2017	50,511 (14,973; 29.6%)	9,017 (1,903; 21.1%)	780

*Chemical sciences include pure chemistry, applied chemistry, and agricultural chemistry.
**Total number of national, prefectural, municipal, and private universities in Japan.
Source: Compiled from Japan. Ministry of Education, Culture, Sports, Science and Technology 1960–2017.

the Division of Chemical Education. The other societies are all independent, publish their own journals, and seldom communicate with each other. This tendency appears to stem from specialists' liking for managing and controlling their own studies in isolated professional groups. In Japan, this specialized, inward-looking mentality may exist in other scientific and cultural circles. As a new step toward cooperation with other specialized societies and in response to the international movement, in 2018 the CSJ expanded its divisional system, creating sixteen new divisions.

Before World War II, it was common practice for Japanese elite chemists to travel to Germany to study under leading professors. After returning home, they introduced emerging chemical disciplines (such as physical chemistry, coordination chemistry, and polymer chemistry) and new techniques to Japan's chemical community. However, World War II deprived Japanese scholars of the opportunity to study abroad. Kenichi Fukui at Kyoto Imperial University, for example, had no option to study abroad as his student days and early career in chemistry coincided with the war years. Instead, he taught himself quantum mechanics by reading German papers and books, which opened his eyes to the new field of quantum chemistry. His frontier orbital theory would earn him Japan's first Nobel Prize in Chemistry in 1981 (Furukawa 2016: chap. 4; Furukawa 2018: 164). After the war, the Japanese undertook advanced studies

TABLE 7.8 Numbers of chemical researchers in Japan (women in parentheses)

Year	Chemical researchers*	Researchers**
1963	22,028	133,842
1966	29,544	162,428
1968	36,048	198,972
1972	43,582	247,309
1974	47,296	292,097
1976	47,280 (1,035; 2.2%)	316,860 (18,237; 5.8%)
1980	48,867 (1,013; 2.1%)	366,998 (23,161; 6.3%)
1982	50,939 (1196; 2.3%)	392,625 (23,448; 6.0%)
1984	54,504 (1,497; 2.7%)	435,340 (27,589; 6.3%)
1987	59,765 (2,157; 3.6%)	487,779 (32,650; 6.7%)
1988	61,241 (2,340; 3.8%)	513,267
1990	68,119 (2,986; 4.4%)	560,276
1992	71,088 (3,761; 5.3%)	598,333
1994	74,910 (4,317; 5.8%)	641,083
1996	75,277 (5,113; 6.8%)	673,421 (62,002; 9.2%)
1998	82,090 (6,663; 8.1%)	704,514
2000	74,604 (5,414; 7.3%)	739,504
2002	70,402 (5,711; 8.1%)	756,336
2004	68,317 (6,000; 8.8%)	787,264
2006	63,233 (6,200; 9.8%)	819,931 (99,193; 12.1%)
2008	62,232 (6,433; 10.3%)	827,291
2010	58,554 (6,697; 11.4%)	840,293
2012	60,275 (7,579; 12.6%)	844,430
2014	58,887 (8,085; 13.7%)	841,554
2016	61,161 (8,835; 14.4%)	847,093
2017	59,457 (8,929; 15.0%)	853,704 (137,100; 16.1%)

*Chemists who regularly carry out research in universities, research institutions, and industry. Assistants are not included. Chemical technologists and technicians whose major jobs are not research are also not included.
**Total of researchers in all areas.
Source: Compiled from Japan. Statistics Bureau, Ministry of Internal Affairs and Communications 1963–2017.

in science and chemistry at American universities. During the 1940s and 1950s, about 3,000 Japanese scientists and students went to study in the United States through either the GARIOA (Government Appropriation for Relief in Occupied Areas) program or the Fulbright Scholarship program (Nakayama et al. 2001).

Nobel laureates in chemistry, such as Hideki Shirakawa, Ryoji Noyori, Osamu Shimomura, and Ei-ichi Negishi, conducted research at American universities while on sabbatical leave at crucial stages of their careers. Some chemists, including Shimomura and Negishi, chose to remain in American universities and institutes, attracted (as were many other foreign scientists in the postwar decades) by the favorable research environment, advanced instrumentation, and high salaries.

Let us now turn to China. Although translations of European chemistry textbooks into Chinese began in the 1870s, led by John Fryer and Xu Shou, their influence in chemical education was limited. Few institutions taught chemistry, and Chinese intellectuals still deemed scientific studies inferior to the traditional ideal of a bureaucratic career after excelling in the Civil Examination (Reardon-Anderson 1991: 17–53). The situation changed at the beginning of the twentieth century, as chemistry began to be incorporated into general education in the new school system established in 1903. In the first two decades of the century, a number of schools adopted both textbooks and teaching methods from Japan, then heavily influenced by the German system. The quality of chemical education in China was considered low in this period, as most schools lacked facilities for hands-on experiments (Chen 1985: 328).

In 1917, the new republic started to reform the educational system and established a system of six years of elementary school, followed by three years of middle school, then three years of high school, and for a selected few, a four-year college education. Soon after this reform, chemistry became a required subject for general education for all science majors in both secondary and higher education. American philanthropic institutions such as the Rockefeller Foundation funded a number of Protestant mission schools in the 1920s. American and European chemists often staffed such foreign-funded schools. One in particular, Yenching University, took a leading role in training professional elites (Rosenbaum 2007: 21). As shown in Figure 7.5, it had a well-equipped laboratory for organic chemistry, and its students included women. The 1910s and 1920s also witnessed a large group of Chinese students studying science in the United States, some of whom chose chemistry as their major subject. There were ten chemistry doctorates granted to Chinese students in the United States in the 1920s alone, as chemistry became their most popular scientific discipline. Chinese students in the USA earned more than 400 chemistry doctorates in the period 1905–1960 (Mainz 2017: 110). An increasing number of chemistry students returned from the late 1920s to teach in better-equipped chemistry laboratories in the modernized educational system. They compiled textbooks based extensively on American models and helped to standardize curricula in higher and secondary education. They also promoted experimentation as crucial for higher education and started to establish graduate programs (Chen 1985: 329).

FIGURE 7.5 Organic chemistry teaching laboratory, Yenching University, China, 1929. Note that the laboratory appears to be open to both male and female students. Rockefeller Archive Center, RF Records Collection.

This successful development in chemical education, however, was disrupted by the Second Sino-Japanese War between 1937 and 1945, during which a number of universities had to relocate to southwestern regions of China due to the occupation of the major cities by Japanese troops. Although the quality of experimental work significantly declined, higher education in chemistry continued, often by combining resources and sharing equipment between relocated universities such as the case in the Southwestern Union University in Kunming. Interestingly, graduates in chemistry in Kunming topped the list of all science majors during the war (Israel 1999; Zhao 2003: 152).

Recovery from the Pacific War was soon followed by the takeover of the Chinese educational system by the Communists in 1949. One immediate change to both secondary and higher education in chemistry was the introduction of several textbooks from the Soviet Union, followed by the influential Thought Reform targeting intellectuals (1951) and changes in academic institutions (1952). During these reforms, many scientists were criticized because of their past experiences in Western countries, although – as in the Soviet Union – direct criticism based on specific scientific ideas in chemistry was minimal compared to biology and physics. More importantly, academic institutions in the summer of 1952 adopted the Soviet educational system wholesale. A clear distinction between universities and engineering schools led to the relocation of programs in chemistry and those in chemical engineering to different institutions. These adjustments increased the number of engineering programs, but diluted human resources for basic chemical education (Zhao 2003: 171). Moreover, adopting the highly specialized Soviet approach to education forced students to focus on increasingly limited and technically oriented subdisciplines of chemistry, especially in fields directly related to industry and agriculture.

Despite recurring criticism against foreign-trained intellectuals under the new regime, the generation of chemists who had returned from Western countries continued to be influential, and many made substantial contributions to Chinese chemical education (Wang 2010: 369–75). As importation of equipment and reagents from Western countries became difficult, Gao Chongxi (1901–1952), an inorganic chemist who obtained his Ph.D. from the University of Wisconsin–Madison, began to devote significant energy in developing China's own production of high-grade reagents and glassware to satisfy the needs of chemistry education (Zhao 2003: 162). As Sino-Soviet cooperation weakened in the late 1950s, educators moved away from Soviet models. Chemists redesigned curricula, combining basic theoretical education (the introduction of the periodic table, for example), specific knowledge of elements and substances, and applied chemistry.

In the Maoist period, chemistry benefited from its official status as a crucial discipline for industrial and agricultural self-reliance. This reduced the influence of political movements and promoted steadily increasing numbers of trained chemists and chemical engineers, except during the decade of the Cultural Revolution from 1966. However, at times the emphasis on practical applications skewed the focus of chemical education. During the Great Leap Forward Movement of 1958, for example, a significant waste of human and material resources resulted from the futile efforts of students and faculty members who, using textbooks focusing on indigenous steelmaking, went to remote areas to attempt to help local fruitless efforts to produce steel from scrap metals (Zhao 2003: 195). At the same time, the continuing positive portrayal of practical chemistry delayed until the 1980s the dissemination of a negative image that had become common in the West, associated with weaponry development and environmental pollution (see Chapters 5 and 8 in this volume).

The end of the Cultural Revolution and the economic reform starting in 1978, coupled with a revival of foreign training and visits, brought Chinese chemical education more closely in line with international norms. The increasing numbers of Chinese graduate students in chemistry in developed countries, especially the USA, were partly related to the declining interest in graduate work in the subject among students native to those countries. In 1980 the Chinese government began to develop chemistry doctoral programs in seven universities, moving away from the Soviet model in which all research had been concentrated in the Academy of Sciences in Beijing. In 1983 the first two students defended their doctoral dissertations in chemistry at Peking University. Others would follow in the 1980s, including seven chemistry doctorates awarded in 1984 to students trained by the Academy (Mainz 2018: 115, 120). Reflecting its increasingly international outlook, the Chinese Chemical Society joined the FACS in 1984 (in the same year as the Taiwanese Chemical Society; Singh 1989: 3). China was thereby laying the basis for dramatic growth to come in the next period.

PART THREE: THE ERA OF GLOBAL SCIENCE SINCE 1990

The 1990s inaugurated a new era with the collapse of the Soviet bloc in Eastern Europe, opening the way for a market-capitalist order and the expansion of the European Union. This produced greater investment in more modernized plants and technologies, with the potential for fostering more opportunities for chemists (although this was also to some extent negated by the losses of academic positions brought about by budget cutbacks in the aftermath of economic dislocations produced by the new era). Japan had grown rapidly in the preceding period, but in the 1990s suffered from relative economic stagnation that may have made it difficult to modernize educational institutions. In China, the ruling Communist Party violently suppressed demands for democratic reforms voiced by students in Tiananmen Square in 1989, while continuing to promote economic reforms that introduced elements of capitalism. The manufacturing sector grew dramatically, as the output of enterprises based on foreign investment, attracted by relatively cheap Chinese labor, rose from a negligible amount to more than a quarter of the country's total output within a decade after 1990. By 2010 China had passed the USA as the world's leading manufacturing nation; the government sought to maintain its position through strong support for research and development, technological innovation, and advanced education (Morrison 2018: 10–14). This offered greater opportunities for employment of Chinese chemists, thus stimulating a rapid growth in the number of students. The new era also brought greater opportunities for the flow of students, trained scientists, and scientific information across national boundaries. Initially this seemed to benefit the established European and American centers of chemical education, which could thereby attract more foreign students, while the leading national professional organizations (especially the American and British chemical societies) could attract more foreign members. At the same time, during this period nations throughout the world, and particularly in Latin America, Asia, and Africa, continued to promote their own scientific and technological development, albeit with continuing financial limitations for various reasons. On the global level, IUPAC celebrated its centennial in 2019, which was also the International Year of the Periodic Table, while undertaking a comprehensive strategic review to maintain the organization's relevance to the global chemical community and to enhance collaboration with other professional groups (IUPAC 2018; Fauque and Van Tiggelen 2019).

Technological innovations at the beginning of the twenty-first century created unprecedented opportunities for global interaction in education generally, and chemical education in particular. Electronic communications from email to video conferencing through the Internet increasingly transcended spatial and geographical limits. Arguably, this implied the end of the old dichotomy between "center" and "periphery" in the scientific world, as collaborators in widely separated laboratories could work simultaneously on related problems, hindered only by the time differential and the need to use a common language, which tended to be English.

Improved international communication was especially significant for American scientific institutions in the wake of the destruction of the World Trade Center in September 2001 and the ensuing American-led invasions of Afghanistan and Iraq. The ensuing political destabilization and wars in the Arab–Islamic world, culminating in floods of refugees seeking safety in other regions, added to global political tensions and disrupted international collaboration as so-called nationalist leaders with anti-immigrant policies took power in the United States and elsewhere. Moreover, the continuing dilemma of "dual use," symbolized most recently in the revival of large-scale chemical warfare against civilians by the Syrian dictatorship during the civil war from 2012 to 2018 (Schneider and Lütkefend 2019), would not enhance the reputation of chemistry as a source of peaceful progress. Despite these problems, as the following discussion will show, the profession continued to grow.

In Europe the process of international political integration that followed the collapse of the Eastern European communist bloc produced an analogous process in education. During the 1990s most of the former Soviet-bloc countries ratified the European Cultural Convention. Negotiations among the members led to the Bologna Declaration in 1999, initially signed by twenty-nine countries, which launched the Bologna Process, aimed at achieving uniform standards of quality and mutually recognized degrees. To enhance the implementation of the process, by 2010 the European Higher Education Area (EHEA) was formed with forty-eight full members, including almost every European state from Spain to Russia (EHEA 2018).

At the same time, in the wealthier nations it became necessary to incorporate much more sophisticated technologies and apparatus into chemical education. Partly this resulted from the now almost indispensable use of computers for graphics, modeling, and problem-solving, in many cases replacing the older mechanical models that had been popular in the postwar era (Orna 2015: 15–16). At the same time, technologies such as nuclear magnetic resonance had also become ubiquitous (Rovnyak and Stockland 2007). Thus, the traditional test-tube and balance chemistry of the nineteenth century, which had still been the basis of introductory training for chemists in the mid-twentieth century, seemed increasingly obsolescent. This created the potential danger of a widening gap in chemistry teaching facilities, as the wealthier advanced nations were able to steadily improve their facilities, while others could not keep pace.

At the same time, chemical educators in the advanced nations experienced an ominous sense of crisis as a result of declining shares in student enrollments, and made various suggestions to reverse the trend. These culminated in proposals advocating more systematic research into chemical education, by the Task Force on Chemical Education Research of the ACS in 1994, and in 1999 the Division of Chemical Education of the FECS (EuChemS 2018). Parallel to this process, there were increasing demands for greater international interactions in developing chemical education. Thus new journals began to emerge, also following the overall trend toward more web-based communications. The journal *Chemical Education Research and Practice in Europe* began publishing in 2000, soon became more global in focus, and then merged with *University Chemistry Education*, published by the Royal Society of Chemistry (RSC). One of the purposes of this journal was to provide a publication venue in English for research done by chemical educators whose language was not primarily English, in order to bring the results to the attention of English-speaking researchers, particularly in the United States (Tsaparlis 2003: 7–9).

On the secondary level, there were also efforts to improve journals for educators. The RSC's *Education in Chemistry*, founded in 1964 as a magazine for British secondary-school chemistry teachers, was revamped in 2015–2017 to make it more "technology-friendly," with the launching of an app and a colorful new website. The German equivalent, *Naturwissenschaften im Unterricht Chemie* [*Sciences in Instruction – Chemistry*], published since 1990, launched its new website around the same time (EIC 2018; Friedrich Verlag 2018).

The size and composition of the global population of chemists also changed during this period. The economic shock of German reunification and the shutting down of many obsolete factories in the former German Democratic Republic caused considerable unemployment during the mid-1990s, accompanied by a significant decrease in the number of beginning students in higher education. After 2000 this produced a sharp decline in the number of German doctorates awarded. Because the proportion of both foreign and female doctoral candidates rose after 2000, it may be inferred that German-born men in particular were turning away from chemistry (Table 7.9). In the United States, the chemical profession became more diversified after 1990, and after 2000 the total number of doctorates awarded once again exceeded that in Germany. The proportion of women degree-holders continued to grow, reaching approximate parity with men on the bachelor's and master's levels during the first decade of the new millennium, but women were slower to be accepted in chemistry faculties (Nelson and Rogers 2006: 2–3). Nevertheless, the proportion of women receiving US doctorates rose from less than a quarter of the total in 1990 to 40 percent in 2015 (Table 7.9), suggesting a trend toward gender parity in academic careers. Ethnic diversity increased as well, as the proportion of non-Hispanic white US citizens or permanent residents receiving chemistry doctorates fell from over 60 percent in 1991 to 45 percent in 2005, after which it leveled off. The rising proportion of foreign students in American universities compensated in part for a declining interest of white Americans in the field of chemistry during the 1990s; but this trend peaked in 2005 (Table 7.9).

After the terrorist attacks of September 11, 2001, the share of foreign male temporary visa-holders among US postdoctoral fellows fell slightly, although the proportion of women remained constant (Table 7.10; Borello et al. 2015: 137). Meanwhile, the proportion of foreign-born students obtaining chemistry doctorates in German universities rose significantly from 2000 to 2005, which may be related to the Bologna process of standardization in European degree programs. After 2005, however, the proportion of foreign-born doctorates began to fall again, while the proportion of women obtaining German doctorates began to approach 40 percent for the first time, a share close to that of US women (Table 7.9; Schwarzl 2006: 77; Gesellschaft Deutscher Chemiker 2010: 23; Gesellschaft Deutscher Chemiker 2016: 7). During this period, moreover, women chemists finally began to gain genuine professional recognition, as American, German, and Japanese women were finally elected presidents of their national societies (Koch 2011: 3; Rossiter 2012: 249–50; see below for Japan).

TABLE 7.9 Doctorates in German and US universities, 1990–2015, including nationality, ethnicity, and sex

Year	Ger. Drs.	(% foreign)	(% women)	US Ph.D.s	(% white)	(% foreign)	(% women)
1990	1700	n/a	(22)*	2100	(61.1)	(28.7)	(24.0)
1995	2079	(4.2)	(22)*	2162	(51.4)	(24.9)	(30.6)
2000	2011	(5.5)	(23)*	1989	(49.5)	(37.6)	(31.4)
2005	1331	(27.7)	(24)*	2126	(45.0)	(43.0)	(34.0)
2009	1513	(25.3)	(37.6)	2392	(44.6)	(41.8)	(37.4)
2015	1901	(19.7)	(37.3)	2675	(44.2)	(41.2)	(40.3)

Source: Schwarzl 2006: 77 (*approximations derived from graph); GDCh 2010: 6, 23; GDCh 2016: 7, 27; NSF-NCSES 1996: tables 2–4; NSF-NCSES 2006: table 5; NSF-NCSES 2011: table 17; NSF-NCSES 2017: tables 15, 18, 22, and 24. In the US statistics, "% white" refers to US citizens and permanent residents self-reporting as "non-Hispanic white"; "% foreign" refers to temporary visa holders or status unknown, not including permanent residents of foreign origin; "% women" refers to all women, regardless of ethnicity or citizenship status.

TABLE 7.10 Postdoctoral fellows in US universities 2001–2010, including percentage of foreign and women postdocs (chemistry and biochemistry)

Year	Total	No. (%) temp. visa holders	No. (%) temp. visa holders – women	No. (%) US men	No. (%) US women
2001	6,416	4,118 (64.2)	993 (15.5)	1,574 (24.5)	724 (11.3)
2010	6,774	3,928 (58.0)	1,049 (15.5)	1,912 (28.2)	934 (13.8)

Source: Borello et al. 2015: 137.

The major chemical societies continued to welcome foreign members, with important implications for the future of global chemistry. In particular, their presence in the British and American societies reflects the continuing centrality of English as the language of chemistry. The more than 40 percent share of presumably non-American chemists in the ACS in 2017 (considering the excess of ACS membership over US employment of chemists) was comparable to the proportion of foreign members in the Deutsche Chemische Gesellschaft (DCG, German Chemical Society) during its period of greatest relative size and global influence during the 1890s (Johnson 2017: 11,049). Just as a significant proportion of the foreign members of the DCG were Americans, who learned important lessons from the Germans on how to be productive chemists and organize a large national chemical society, in the early twenty-first century a significant proportion of the foreign members of the ACS are Indian or Chinese. Considering that the far larger population of the United States in the mid-twentieth century ensured that the ACS could outnumber any German society, one may expect that the Indian and Chinese chemical societies will eventually dwarf the ACS as well as the European organizations. In the early twenty-first century, however, European consolidation produced the largest single regional group. In 2004–2006 the FECS became the European Association for Chemical and Molecular Sciences, EuCheMS, and from 2018 the European Chemical Society, EuChemS, incorporating forty societies with 160,000 members, somewhat more than the ACS (EuChemS 2018; Table 7.1).

EDUCATION AND INSTITUTIONS IN ASIA SINCE 1990

In Japan, as shown in Table 7.7, from the mid-1990s onwards the number of undergraduate chemistry-major students dropped by 25,000, although the total number of enrollments in universities in general continued to increase. Several factors can help explain this decrease, including a science phobia among the youth, the collapse of the bubble economy at the end of the 1980s, changes in scholarship, and the reorganization of universities.

Chemistry departments at a number of universities merged with other science departments, such as biology and physics departments. However, the percentage of female undergraduates increased from 1.4 percent in 1960 to 29.6 percent in 2017, and the number of chemistry graduate students remained more stable than that of undergraduates. It should also be noted that Maki Kawai became the first female president of the Chemical Society of Japan in 2018.

The number of chemical researchers, as shown in Table 7.8, shows a similar pattern, with the figure dropping by 20,000 since 1998. Japan's population decreased during this period; however, the total number of researchers in all areas gradually increased, meaning the decrease in the number of chemical researchers cannot be explained simply by the population decrease. Instead, it appears to be related also to the movement of a large number of industrial researchers from chemistry to other growing fields, such as life sciences and information technology. In contrast, the percentage of female chemical researchers steadily increased, although the ratio (14.6 percent in 2017) was still less than half that in the United States and Germany (compare Tables 7.8 and 7.9).

In China, developments since the 1990s were marked by a dramatic increase in the numbers of chemists receiving advanced training in domestic institutions, although a significant number of Chinese students continued to study abroad. The increased numbers of Chinese chemists trained in research led to a corresponding increase in the number of Chinese publications in chemical journals. In 1996, China was seventh in the world in the total number of publications, but by 2009 China had passed all competitors to take first place, and their lead steadily grew, so that by 2017 – the latest year available – Chinese chemists were producing almost twice as many publications as American chemists (Figure 7.6). Despite this remarkable quantitative growth, the overall quality of Chinese chemical publications continued to lag behind the better-established chemical communities in North America, Europe, and Japan, as roughly indicated by relatively low citation averages per document and a relatively high proportion of self-citations (Table 7.11). Will Chinese research continue to improve? As of 2017 the Chinese government was investing heavily in "a long-term plan to develop world-class departments within world-class universities," intended to offer bilingual instruction (in English and Chinese) to European and American students (cf. Figure 7.7). Reflecting this investment, a Chinese institute that in the 1990s had only two NMR spectrometers had thirty in 2017, making it "better equipped than most of the US Ivy League universities." These and other prestigious American universities faced budget limitations and had to cut financial support for postdoctoral researchers, thus becoming increasingly less attractive to the best Chinese students (Tremblay 2017: 21–3).

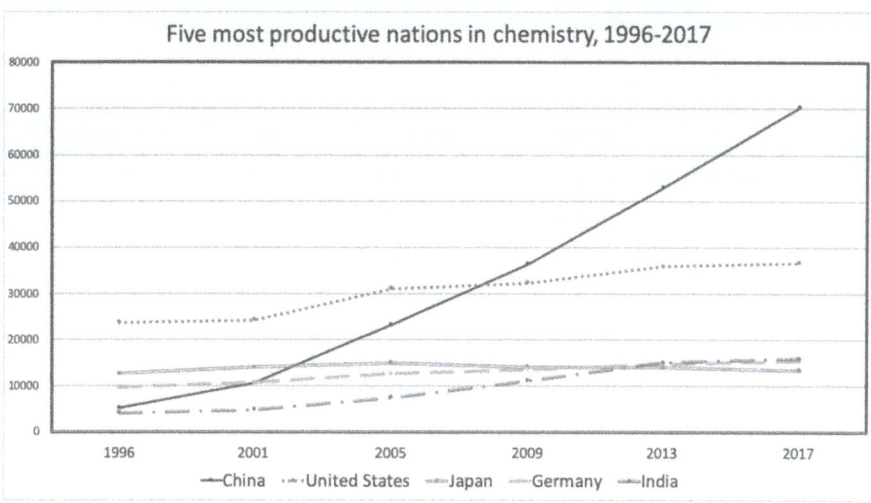

FIGURE 7.6 The five most productive nations in chemistry, 1996–2017, showing annual totals of publications in selected years. Drawn by the author (Johnson) from data for selected years shown, and for the entire period (SJR 2018). Note that this graph excludes countries that appeared in the top five during some of the selected years, but were not among the top five in cumulative publications.

FIGURE 7.7 Using electronic media in chemistry instruction in China, 2017. The instructor is an American chemist, Jay Siegel, Dean of the School of Pharmaceutical Science & Technology of Tianjin University. Tremblay 2017. Reproduced by kind permission of the School of Pharmaceutical Science and Technology, Tianjin University.

TABLE 7.11 The five most productive nations in chemistry, based on total publications, 1996–2017, ranked by average citations per published document

Rank	Country	Documents (000)	Citations (000)	Self-citations (000)	Citations per document
1	United States	672.8	21,710.8	7,678.6	32.3
2	Germany	280.0	7,026.1	1,827.1	25.1
3	Japan	304.0	6,544.4	1,889.7	21.5
4	China	687.3	10,042.2	5,775.5	14.6
5	India	210.7	2,875.2	1,068.9	13.7

Source: SJR 2018.

CONCLUSIONS AND OUTLOOK

In 1914 there could have been little doubt that Germany offered the world's best and largest facilities for advanced chemical education, so that the Germans had if not the largest professional organization, certainly the largest group of research-trained chemists. Accordingly, a reading knowledge of German was indispensable for a chemist anywhere in the world. More than a third of the members of the German Chemical Society were not Germans. It took only three decades – admittedly, decades of world wars, economic depression, and totalitarian dictatorship based on an extreme racial–nationalist ideology – for all of these advantages to be lost to the United States, which in 1945 was poised to dominate the world of chemistry for decades. Although Soviet Russia challenged that dominance for a time, their system ultimately collapsed. English thus became and in the second decade of the twenty-first century remained the central language of chemical research and chemical education, while American institutions still attracted students from all over the world, albeit in declining numbers. The ACS remained the world's leading national chemical organization, its main rivals the Japanese and in 2018 the newly renamed European Chemical Society. But for how much longer?

At the beginning of the UN-designated International Year of Chemistry (IYC) in 2011, the president of the Federation of Asian Chemical Societies was Bai Chunli of the Chinese Academy of Sciences. In his presidential address to FACS regarding IYC, he highlighted the "central role of chemistry in solving global challenges … through close international collaboration," and he emphasized the traditional importance of education to Asian culture. IYC thus offered to Asian chemists "a formidable opportunity" to improve the public image of chemistry, promote sustainable development, and, "most importantly, for attracting young talents with bright and original ideas to advance chemistry." Their success in the coming decades could lead to an "Asian century" in the world of chemistry (Bai 2010).

CHAPTER EIGHT

Art and Representation: *From the "Mad Scientist" to Poison Gas and Chemical Pollution*

JOACHIM SCHUMMER

INTRODUCTION

The task of providing a comprehensive survey of how chemistry was portrayed in twentieth-century media faces three main challenges: the tremendous multiplication of mass media, the immense growth and diversification of chemistry, and the lack of historical studies thus far.

No century before had created so many new media for a mass audience. Nineteenth-century writers and artists had still produced their works for an educated elite of frequently much less than 10 percent of the population. The new mass media of the twentieth century reached out to everyone, including illiterates and children, which made them economically extremely successful. Beginning with short motion pictures, screened in booths for a small price, the movie industry rapidly grew to national and international networks which performed full-length films in big theaters. Radio and television networks transmitted audible and visible entertainment, including films and series produced by television companies, into the most remote homes. Computer and video games, which frequently adopted movie themes, provided interactive tools to the story line, which at the end of the century became the economically most successful approach. At the same time, the Internet developed into a mass media that began to exceed anything else seen before in terms of content, omnipresence, and influence.

Despite all this competition, the print media continued to grow for most of the century. It did so mostly by targeting the same mass audience, not only in newspapers but also in fiction writing. For instance, novels were broken down into genres, such as romance, crime thriller, and fantasy, each with a general plot frame to be filled out with little variation. An army of writers produced simplistic short stories published in weekly or monthly magazines, out of which the US genre of science fiction emerged. Originally tailored to children, but soon equally popular with adults, comics became one of the most successful print media in many countries. Nonfiction writing also reached out far beyond its original audiences of students and specialists. Written in journalistic style, book-long reports on political scandals or environmental issues competed with novels on bestseller lists.

While the media moved toward content simplification, chemistry experienced an extraordinary sophistication, diversification, and growth during the twentieth century. The late-nineteenth-century triad of inorganic, organic, and physical chemistry, while still taught at universities, gave way to a multitude of specialties that resulted from interdisciplinary research, such as bio-, geo-, atmospheric, mineral, environmental, medical, quantum, mathematical, instrumental, bioinorganic, and physical organic chemistry; or from a specific interest in particular materials or processes, such as colloid, radio-, organometallic, polymer, petro-, surface, catalytic, supramolecular chemistry; or from a historical angle, such as cosmological chemistry, chemical evolution, and archeometry. Furthermore, the chemical industry which before World War I had been the domain of a few countries soon grew in all industrialized countries to one of the main manufacturing sectors, which, because of its unique demand for academically educated scientists, allowed chemistry to grow more than any other discipline. Following new material needs the chemical industry split up into many specialized branches, such as the dye, photographic film, pharmaceutical, petrochemical, polymer, biotechnology, and electronics materials industries.

Against the background of the tremendous growth and diversification of both the media and chemistry during the twentieth century, any attempt to cover chemistry's media representation seems futile. That may be one reason for the almost complete absence of previous studies on this subject, although chemists themselves have long complained about their bad public image.[1]

However, despite their diversification in form, the mass media frequently simplified and recycled themes developed already in nineteenth-century literature where the "mad scientist" had become the overarching theme with many variations (Schummer 2022). Thus, rather than providing a chronological overview of the media representation of chemistry, this chapter first explores how the nineteenth-century "mad scientist" was further developed in both Western mass media and original novels. Against the background of the diversity of the nineteenth century, the twentieth century favored extreme figures, particularly the "benevolent absent-minded scientist" and the "evil-minded supervillain" threatening the entire world in a drama reminiscent of the biblical battle of Armageddon.

The subsequent sections of this chapter will investigate the two topics that have probably more than anything else in that century shaped the public's view and debates about chemistry: poison gas in World War I, and environmental pollution. However, in both cases the impact on the representation of chemistry is complex, because writers and artists struggled long and hard in their own traditions. Chemical warfare challenged first of all heroic war poetry and became emblematic of the modern war probably only in the second half of the century, but it provoked themes of chemical apocalypse already in the interwar period. The creeping chemical pollution, about which the general public was most concerned, could neither be painted nor would it easily fit into the disaster framework that had emerged out of radicalizing the "mad scientist." Moreover, among environmental activists who pointed out the issues were many scientists, including chemists, which for a while confounded the received black-and-white scheme.

THE "MAD SCIENTIST"

Adaption of Nineteenth-Century Themes

During the entire twentieth century the "mad scientist" theme, created by nineteenth-century writers (Schummer 2006), continued to play a dominant role in the public representation of chemistry. Originally that trope resulted from a transformation of the "mad alchemist," a popular theme in late medieval and early modern fiction and painting in order to satirically denounce human folly and greediness. Seduced by a fiendish tempter, the "mad alchemist" ruins his health, wealth, and reputation through his unsuccessful obsession with gold-making. In its original form, the nineteenth-century "mad scientist" still resembles that character, but he (and he was invariably male) now has a variety of ambitious goals, including the making of diamonds, elixirs of longevity, or the creation of human-like beings, and he is typically a chemist or physician conducting chemical experiments. While he now works mostly for altruistic reasons or "in the name of science," he still does so in social seclusion with the madness of obsession. However, unlike his alchemical forerunner, he is successful in his deeds but disastrous in its effects because he produces unforeseen runaway developments. That served writers to point to the combination of scientific power, hubris, and moral naivety by using the religious template of "playing God." Many associated chemistry even with materialism, nihilism, and atheism.

Almost from the start of the twentieth century, movies recycled nineteenth-century "mad scientist" classics and then developed the theme further. Particularly popular were *Dr Jekyll and Mr Hyde* and *Frankenstein*, of which each several hundred productions have appeared since 1908 and 1910, respectively. Whereas the literary characters usually had philanthropic motives but failed out of moral naivety and hubris, their movie counterparts increasingly appeared as evil madmen who only pretend to act in the name of science (Toumey 1992). For instance, Mary Shelley's original composition of alternating monologues by Victor Frankenstein and his alter-ego-like creature, which are both full of sophisticated moral deliberation, increasingly turned into simplistic plots of an evil-minded scientist who creates a mass-murdering monster, from the Universal Picture film *Frankenstein* (1931) to its numerous sequels.

Already in the nineteenth century, writers had transformed the standard "scientific madness" of chemists into a variety of forms (Schummer 2022), one of which was the "morally perverted scientist" that now became particularly popular in novel-adapting Hollywood movies. Another form of madness had been "criminal intent," which through film adaption of *The Invisible Man* and original plots, including *The Cabinet of Dr. Caligari* (1920), *Dr. Mabuse* (1922), *Maniac* (1934), and *The Mad Ghoul* (1943), enjoyed similar popularity for most of the twentieth century and thereafter. For instance, in the enormously successful US TV series *Breaking Bad* (2008–2013), a chemistry teacher engages himself and his student in the illegal production of synthetic drugs. However, the most important further development of the "criminal intent" theme was the invention of the supervillain, which will be dealt with in the next section.

A third, and much more radical, transformation in the opposite direction was the detective-turned-scientist, whom Arthur Conan Doyle had invented in 1887 with the character of Sherlock Holmes. This smart and benevolent hero might appear as the opposite of the "mad scientist," but he still bears some traits in common, particularly his eccentric character and his occasional obsession with chemical experimentation, which only Arthur Benjamin Reeve in 1910 stripped off in his chemist-detective Craig Kennedy. Both figures enjoyed great popularity in numerous movie and TV series adaptions, but later crime serials replaced the scientist-detective by the policeman as hero. In this genre science returned to the fore only with the US TV series *CSI: Crime Scene Investigation* (2000–2015).

The "benevolent but absent-minded scientist," while rare in the nineteenth century, became fully developed at first in two genres of comic strips that soon were adapted in animated cartoons, movies, and TV series. One is the adventure travel genre, from which it originated and in which a sometimes eerie scientist helps the hero overcome obstacles by his profound knowledge, as for instance in *Flash Gordon* (since 1934). The second one is the fable-like comic strips for children after World War II, most of which remarkably included a scientist-engineer (see below). In the 1960s the "absent-minded scientist" also turned into a source of public amusement in comedy science fiction movies such as *The Absent-Minded Professor* (1961) and *The Nutty Professor* (1963).

Whereas the nineteenth-century "mad scientist" was mostly a chemist or a physician carrying out chemical experiments, his twentieth-century counterparts represented various disciplines, including psychology, biology, physics, engineering, and eventually computer science, or increasingly science and engineering together. Nonetheless, chemistry featured prominently in "mad scientist" stories but went far beyond the earlier focus on the making of gold, diamonds, or elixirs of life. On one hand, writers employed a more diverse and fanciful repertoire of chemistry, thereby establishing "science-fiction." On the other hand, and more important here, they envisioned global chemical hazards. While the medieval "mad alchemist" first of all harms himself and his family, and the nineteenth-century "mad scientist" causes damage to his local surroundings, the twentieth-century chemist is frequently a threat to the entire world. Before the advent of nuclear physics, chemistry played an exceptional role in global disaster stories because it was more than any other science suspected to have control over all environmental media that sustain life (air, water, soil, and nutrition) and over destruction by poisons and weapons (see below). Global chemical threats occurred in such diverse genres as apocalyptic and post-apocalyptic science fiction, disaster, espionage, war, and environmental thrillers. Under the impact of the two world wars and the Cold War, which posed the new threat of a nuclear war, Western popular culture showed a hitherto unseen fascination with doomsday scenarios.

I will first have a look at the two extreme, and most important, stereotypes into which the "mad scientist" was transformed in the twentieth century: the supervillain and the benevolent scientists, both originating from comics strips. Then I will investigate the origin of the disaster theme in writings about poison gas in World War I, and finally ask how writers dealt with the actual issues of environmental pollution.

Superheroes and Supervillains

Several factors contributed to the invention of the supervillain in the US. First, the rise of new media for mass entertainment, comics strips and movies, required the plots to be adapted to the target audience that was assumed to want simple moral messages. Second, the moral simplification of plots, which for instance tended to exclude cases of good intentions with bad outcomes, followed a quasi-religious (Manichean) scheme of dividing up the world into intrinsically good and evil parts. Rather than fights over local issues, the world as a whole was now at stake. Mass entertainment thus was now modeled after the battle of

Armageddon between angelic and diabolical forces, between God and Satan, to be represented by a superhero and a supervillain. Third, because the superhero/supervillain genre started at the beginning of World War II and temporarily faded afterwards, it probably resonated well with the black/white schemes of wartime articulated in public and political debates at the time. Indeed, American and British writers frequently related their "mad scientists" to the Nazi regime up to the 1980s, either by direct reference or by choosing an obviously German name. Because Germany also had a reputation in chemistry, the old link between the "mad scientist" and chemistry could thereby be reinforced. Indeed, many of the supervillains are associated with chemistry.

In 1938 Detective Comics Inc. (later renamed DC Comics; Gabilliet 2010: 16) introduced in their debut issue of *Action Comics* the first superhero, Superman, as an extraterrestrial angelic being in the disguise of an ordinary man. His counterpart and archenemy, the supervillain Lex Luthor, appeared only two years later as a diabolical maverick with extraordinary scientific and engineering skills, who was later developed into a powerful industrialist. Based on the success of the comic books, Fleischer studios soon produced a series of Superman cartoon films, starting with *The Mad Scientist* (1941), in which the angelic world-saver prevents the evil-minded scientist from destroying the fictional "universe" Metropolis with a ray gun.

In 1939, Detective Comics increased their output by introducing the superhero Batman, who first fought Doctor Death's world-poisoning ambitions, before in 1940 The Joker became his regular archenemy. The Joker owes both his criminal activity and character to chemistry. He poisons his victims with his chemical invention "Joker venom" or Smilex, which leaves the dead with a grim smile on their face, and is said to have turned evil-minded after he fell into a tank of chemical waste. Two years later the same publisher introduced the superheroine Wonder Woman who fights, among other supervillains, Doctor Poison, alias Princess Maru, head of a Nazi team that tries to poison the US.

Public interest in the superhero genre faded immediately after World War II and revived only after Marvel Comics launched *Spider-Man* in 1962 (Gabilliet 2010: 34, 54). Two years later its main supervillain became Norman Orman, head of Oscorp (later called Alchemax), a chemical company that he has cofounded with his former chemistry professor Dr. Mendel Stromm. Rather than poisoning or pollution, their main business is working on chemical formulas for the enhancement of human strength and power and for controlling people's daily life. Political and economic issues clearly dominate over health and environmental issues, but the focus is on chemistry.

All these superhero/supervillain comics stories later became very popular worldwide, and increasingly so through numerous Hollywood movie adaptions and countless imitations in cinema and TV productions well into the twenty-first century. The original enemy, Nazi Germany, was during the Cold War increasingly replaced by the Soviet Union, and then by extraterrestrial diabolical forces, but the supervillains' associations with chemistry remained for several decades. With the science – or better, weapons technology – turning more fanciful, the superhero also makes use of it and increasingly wins through his technological advancement rather than by his inherent angelic capacities, which is reminiscent of the Manhattan Project and in accordance with the official arms race policy. Thereby the originally science-critical attitude of the genre turned into the reverse.

The Absent-Minded, Benevolent Scientist in Comics for Children

The fascination of US publishers and producers with the "mad scientist" theme, of which only a small fraction is sketched above, also infected Walt Disney who, together with his brother, had founded a firm in Hollywood in 1923 that soon specialized in animated films for children featuring animal characters, as in traditional fables. Their first successful series was *Mickey Mouse*, starting in 1928, for which they produced the short film *The Mad Doctor* (1933), in which Dr. XXX kidnaps Mickey's dog Pluto in order to conduct a hybrid experiment that should connect the dog's head with the body of a chicken. However, the story shocked children, many US theaters refused to show the film, and other countries even banned it, so that the character and the theme disappeared from entertainment for children.

The figure was replaced by the "absent-minded, benevolent scientist", who probably first appeared in the illustrated short stories for children about Professor Branestawm (1933–1937, 1970–1983) by British author Norman Hunter, and then as Dr. Hans Zarkov in *Flash Gordon*, a US space opera adventure comic strip published since 1934. Both the figure and the storyline of *Flash Gordon* borrowed from Jule Verne's classic *Off on a Comet* (1877), featuring Professor Rosette. In 1944 the Belgian comic strip *The Adventures of Tintin* (published since 1929) introduced Professor Calculus, a largely benevolent inventor and genius in all scientific and engineering matters but clumsy at best in social affairs, frequently absent-minded and secluded, and at times revealing an eccentric or irascible character.

With little variation that figure was soon incorporated as a secondary character in many famous Western comics for children. In the first competing Belgian comics *Suske en Wiske* (French: *Bob et Bobette*, British: *Spike and Suzy*, American: *Willy and Wanda*) he appeared as Professor Barabas or Barnabas since 1945; in the second one, *Jommeke*, as Professor Gobelijn since 1955. Walt Disney's *Donald Duck* adapted him in 1952 as Gyro Gearloose. The German imitation *Fix & Foxi* called him Professor Knox (since 1953). The most famous French comics, *Astérix*, even made him in 1961 its founding figure in the form of the Druid Panoramix (English: Getafix). This pre-modern chemist, whose white coat resembles the modern lab coat, concocts, besides numerous elixirs, the magic potion that makes the people of a small Celtic village invincible against the troops of Julius Caesar.

The "absent-minded, benevolent scientist," rather than having a disciplinary specialty, is a universal scientist, engineer, and inventor, and chemistry is one of his main areas (Carter 1988). For instance, Gyro Gearloose creates a large variety of chemical preparations, from more realistic inventions, such as a super strong sealant (1954), a tranquilizer for animals (1955), high-speed gasoline (1956), and a bear repellant (1959), to more fanciful things, such as a liquid that turns inanimate objects into living beings (1959) and a drug that shrinks people (1960). However, not all these products work as the helpful inventor expects, and some run out of control as in the classic "mad scientist" stories.

Only a few comics went further and stripped off all the remaining traits of the "mad scientist." One example is the Croatian animated television series *Professor Balthazar* (1967–1978). This thoroughly benevolent, social, and open-minded chemist develops a solution to any problem of people who consult him, by the help of his magical machine for chemical concoctions.

Thus, while scientists play a marginal role in the overall literature for adults with the exception of science fiction, they remarkably became a regular figure of the set of characters in comics for children, unlike for instance criminals, policemen, lawyers, or politicians, who are regularly employed for the entertainment of adults. Real scientists might also appreciate that most of the traits of the mad scientist have been removed from the comic characters to the point that the origin is hardly recognizable any more. Chemistry is largely portrayed as the art of producing wondrous concoctions, mostly for useful purposes.

The extraordinary importance of the "benevolent scientist" in comics for children, compared to literature for adults, calls for explanation. If one compares comics for children with traditional fairy tales, the scientist, by his wondrous inventions that seem to make anything possible, corresponds to the fairy. And because, unlike fairy tales, comics were produced as serials with the same characters, there was a particular need for variations that a character with surprising capacities could supply. In addition, the general attitude toward science and technology was quite favorable in the 1950s (Miller et al. 1997), in stark contrast to the period following World War I.

FROM POISON GAS IN WORLD WAR I TO CHEMICAL APOCALYPSE

It would seem that chemical warfare in World War I was the first real-life confirmation of the large number of "mad scientist" stories written since the early nineteenth century without any reference to actual science. Hundreds of chemists, such as Fritz Haber in Germany, Charles Moureu in France, William

Jackson Pope in the UK, James B. Conant in the US, and Fritz Pregl in Austria, put all their "scientific" efforts on developing and improving or even overseeing the production and deployment of a new kind of devastating weapon that could kill up to 100,000 people (Freemantle 2014; Schummer 2018). Poison gases, which were actually mostly liquids, were not the only new weaponry invented. Tanks, machine guns, aircraft, submarines, and shells filled with high explosives, developed by various engineers and scientists including chemists, also had their first big stage entrance on the battlefields of World War I. They all posed new and difficult to foresee threats to soldiers, and confounded the received military routines. However, the use of "asphyxiating gases" and poisons was prohibited by the Hague Conventions of 1899 and 1907, signed by all major belligerent parties, of which all chemists of course were aware, at the latest from the worldwide (including German) newspapers reports of Haber's first deployment of chlorine gas on April 22, 1915. How did fiction writers and artists respond to that new kind of "scientific madness"?

Poetic Self-Reflection

Chemical warfare plays only a very small quantitative role in the myriad poems and numerous novels of World War I written both during and after the war, and similarly in the later movies, such that most literary scholars have ignored it.[2] For instance, one of the most famous English war poems, *In Flanders Fields*, written by John McCrae on the battlefield of Ypres where Haber had directed the first poison gas attack just ten days earlier, does not even allude to chemical warfare. Poets and writers probably encountered difficulties in capturing the invisible and silent threat of poison gases by traditional metaphorical resources (Löschnigg 1994: 152). Chemical warfare's first impact on poetry was, strangely enough, an urge for poets to reflect on their own profession.

Almost from the beginning of the war, poetry became a powerful tool for psychological war mobilization on all belligerent sides. In England, France, and particularly in Germany, people submitted hundreds of thousands of poems to newspapers that called for patriotic engagement in a heroic and glorious war; according to one probably overrated estimate, Germans alone submitted as many as 50,000 war poems per day (!) to their newspapers during the first month (Marsland 1991: 1–2). Encouraged by nearly all the well-known poets, the "poetical mobilization" turned into a mass movement – a "poetry slam" – about who could best express hatred for the enemy and pathos for the unconditional willingness to sacrifice one's life. For instance, Gerhart Hauptmann, who had received the Nobel Prize for Literature in 1912, published his influential poem *Reiterlied* (Cavalry Song) on August 12, 1914, which offered a call to arms against France, Russia, and England, and employed the outdated but romantic theme of the sword-fighting chivalrous cavalry.

Chemical warfare became a remedy to romantic war poetry for poets who regretted their earlier war enthusiasm and later wrote antiwar poems, usually published only after the war because of censorship. For instance, Wilfred Owen, who in 1914 had written "But sweeter still and far more meet / To die in war for brothers" (*Fragment 126*), radically changed his mind in the battlefield and wrote his famous *Dulce et Decorum est* (1917). By contrasting the "old lie" of how sweet it is to die for the fatherland – the poem title refers to a quote from an ode by the Roman poet Horace – with the cruel death of a gas-poisoned soldier, Owen denounced the abuse of poetry for war propaganda altogether: "If you could hear, at every jolt, the blood / Come gargling from the froth-corrupted lungs / … My friend, you would not tell with such high zest / To children ardent for some desperate glory, / The old Lie: Dulce et decorum est / Pro patria mori."

Many other poets, however, contrasted the glory of the traditional war, which they still seemed to uphold, with the poison gas of the modern war. For instance, British writer Gilbert Frankau set "fighting cleanly" with traditional weapons versus "murdering" with gas in his poem *Poison* (1919). Austrian writer Karl Kraus invented the term "chlorreich" in opposition to "glorreich" (chlorine-rich versus glory-rich), which he frequently used after 1918. The German poet Klabund, who once had been a prolific writer of propaganda poetry but spent the war in neutral Switzerland working for the German intelligence service, wrote afterwards in his *Die Ballade des Vergessens* (1926, The Ballad of Forgetting): "Once in the battle, head against head counted / And man against man – however / Today the chemist pushes the button / And the hero is forgotten, forgotten. … The new war … comes with charges of poison and gases, / Brewed in the devil's hearths."

Chemical warfare killed not only hundreds of thousands of soldiers, it also killed the dreams of heroic war, the martial myth, that was so dear to the tens of thousands of professional and would-be poets during the first years. That made the few poison gas poems famous in the first place. Later, in the interwar period and after World War II, when people were tired of war, they became models of pacifist poetry, even though many people still upheld the received ideas of glorious warfare in contrast to chemical warfare.

The Gas Mask Becomes a Deferred Emblem of the War

Strange to say, the gas mask became a worldwide visual symbol of World War I in general (Spear & Summersgill 1991: 321; Smith et al. 2003: 88f.; Münkler 2013, chap. 4.10; Skrebels 2014). In poems they hardly appear; a very rare exception is the one by Owen mentioned above. However, most World War I novels that mention chemical warfare – no matter at what time and from what country they originated, but particularly so those from the late 1920s – include short passages about the fear of soldiers in the trenches of putting on their gas masks at the right time and properly, to be protected in cases of gas alarm.[3] They describe the difficulties of wearing a gas mask more often than they mention casualties: the uncomfortable feeling and constricted breathing, the hampered communication and limited vision as well as the alienated, animal-like faces of their comrades. However, that alone would hardly justify making the gas mask an emblem of the war. Instead, the issue of what image best symbolizes the war – and by that, the role of chemical warfare in general – has been negotiated over many decades, as the following example illustrates.

Probably the most famous World War I novel is Erich Maria Remarque's *Im Westen nichts Neues* (1928, *All Quiet on the Western Front*) that soon after its publication outnumbered by far the sales of any other World War I novel in Germany, the UK, and many other countries (Owen 1989). It was translated into fifty languages, sold in total at least 20 million copies by 2007, and was twice adapted into award-winning movies of US (1930) and US/UK productions (1979). This novel by Remarque, who himself served only for less than two months on the Western Front before he was injured, is famous for describing first-hand experience of the atrocities of the war. However, poison gas attacks are only briefly mentioned three times (in chapters 4 and 6). Another mention of gas occurs in chapter 7, where the main figure, Paul, on a short leave, meets total misunderstanding and incomprehension about the war by people in his hometown. Among them is his mother, who worries that it is "terrible out there now, with the gas and all the rest of it," upon which Paul indeed remembers cruel gas casualties. In the final chapter Paul is said to have "swallowed some gas," which allows him to take a break from the war and, lying in the sun, reflect on the miserable condition of his generation.

As if by tacit censorship, the US 1930 movie adaptation removed all mention of gas, at a time when the implications of the Geneva Protocol of 1925 were still heavily debated in the USA (Ede 2002), and instead focused on shells and trench fighting. By contrast, the 1979 movie adaption put more emphasis on gas than the novel. For instance, it has Paul saying: "Gas – the most feared, the most obscene weapon of all!" (Skrebels 2014: 85), whereas in the novel Paul argues that "tanks ... more than anything else embody for us the horror of war" (chapter 11). Note that Robert Graves, in his autobiographical novel *Goodbye to All That* (1929), reckoned that British soldiers regarded the use of German bowie knives particularly atrocious, while the Germans most feared the British Mark VII rifle bullet.

Early pictorial representations (paintings and drawings) of gas warfare are similarly rare and diverse. For instance, of the 316 drawings by famous Dutch artist Louis Raemaekers (1869–1956) in *Raemaekers' Cartoon History of the War* (1919) only four refer to chemical warfare – a soldier carrying poison, a gas attack symbolized by a dragon and gas-wounded soldiers, and a German chemist producing weapons in his laboratory – but not a single one shows a gas mask. The international online exhibition "Art of the First World War"

(1998), with 110 works from the major six European museums of war history, included six images on chemical warfare.[4] The three British and American contributions represent the effect of poison gas in the form of blinded soldiers, indicated by blindfolds or hands covering their eyes (William Roberts, *The First German Gas Attack at Ypres*, 1918; Eric Kennington, *Gassed and Wounded*, 1918; John Singer Sargent, *Gassed*, 1918–1919). The French painting places a hardly recognizable poison gas cloud in a burning landscape full of colorful beams (Félix Vallotton, *Verdun, tableau de guerre interprété ...* 1917). In both the Belgian and German contribution, poison gas is only alluded to by soldiers wearing gas masks (Henri de Groux, *Masques à gaz* [Gas masks] 1915; Otto Dix, *Sturmtruppe geht unter Gas vor* [Assault under Gas] 1924). It seems that this kind of image would eventually dominate the visual culture of chemical warfare because neither the blindfold nor a colored fog conveyed a simple and strong message.

It might appear cynical that neither the weapon nor its disastrous effects, but its protective device, the gas mask, eventually became the strongest emblem of chemical warfare, and, along with the helmet, an emblem of World War I as a whole. However, the gas mask came to epitomize the technological sophistication, the anonymity, and the horror of the war, because it covers and distorts the human face with a technological apparatus, and because its wide eye windows convey the impression of fear. Probably much more than poetry, novels, and even paintings, photographs of soldiers wearing gas masks, which in the course of the twentieth century were increasingly used as illustrations in magazines and books, shaped the visual image of the war.

Chemical Apocalypse

An irony of history, the first apocalyptic chemical warfare novel was published in early 1914 "written under the immediate shadow of the Great War," as the author later remarked in the preface to the 1921 edition. Already in 1913, when he wrote *The World Set Free*, H.G. Wells saw Europe willing to engage in an all-devastating war, and chemistry ready to provide the appropriate weapon: not poison gas but nuclear fission, discovered by the fictional chemist Holston in 1933, rather than by chemists Otto Hahn and Fritz Strassmann in 1938. The novel at first narrates the history of alchemy and chemistry ("Prelude") and then the discovery that makes true the dreams of the alchemists, elemental transmutation to gold, as well as that of the military, eternal fire (chapter 1), which is then used in the "great war" to clean the world from all evil (chapters 2–3). After the war a "World Society" emerges that is based on a new integrated science to provide health and love to everyone (chapters 4–5). In stark contrast to his many "mad scientist" stories, on the eve of World War I Wells envisioned chemistry as a purgatory tool to reach a Golden Age.

Such millennialist war utopias – which later came to be known as post-apocalyptic science fiction – faded during the first years of the actual war.

However, apocalyptic themes soon arose anew, among others from devastated battlefields to which the effects of poison gases added further bizarre impressions. An early example is Arnold Ulitz's poem *Gasangriff* (1916, Gas Assault), written from the point of view of the aggressor who unleashes the "gas predator" that, to his own horror, kills not only the enemy but all living nature around, turning spring into eternal winter. One of his next poems, *Frühling in Litauen* (1917, Spring in Lithuania), describes the hoped-for end of the war in metaphors of the biblical flood, while his futuristic war novel *Ararat* (1920) is completely framed around the Noah myth. As late as 1932, the German painter Otto Dix depicted an apocalyptic war landscape featuring a gas mask in his famous tryptichon *Der Krieg* (The War), reminiscent of sixteenth-century paintings by Hieronymus Bosch.

The first genuine work of End Times literature on World War I is probably *Die letzten Tage der Menschheit* (*The Last Days of Humanity*), which the famous Austrian writer Karl Kraus wrote and published in pieces between 1915 and 1922, satirically commenting on various episodes of the war in verse. In the epilogue (written in 1917) the Berlin (chemical) engineer Abendrot (meaning "evening glow") proudly presents the latest art of chemical warfare to be used by "pushing the button" from the distance, shortly after which the Martians execute the "Last Judgment" by destroying humanity. That may have provoked the German writer Erich Kästner to write his sarcastic poem *Das letzte Kapitel* (The Last Chapter, 1930) in which neither the Martians nor God but the human world government, in a last effort to bring peace to earth, decides in 2003 to destroy all human beings by pouring poison gas from airplanes around the globe. Poison gas (chlorine) became the first weapon of mass destruction in fiction because, due to its high density, it flows and spreads like a fluid over the entire surface and into every corner, reminiscent of the biblical flood.

Once established as a literary trope, "chemical apocalypse" was soon employed for various purposes in different genres. It featured prominently in warnings of the next world war, for instance in the mentioned poem by Kästner, Stephen Southwold's *The Gas War of 1940* (1931), and H.G. Wells' *The Shape of Things to Come* (1933). When combined with socialist critiques of capitalism, the apocalyptic threat comes from chemical companies, as in Georg Kaiser's play *Gas II* (1920), Johannes R. Becher's *Levisite oder der einzig gerechte Krieg* (1926, Lewisite or the Only Just War), or Michail Dubson's silent movie *Giftgas* (1929, Poison Gas). Equip the chemical company, or its leader, with some traits of the "mad scientist," willing to destroy the whole earth with some poison for some weird reason, and you have the basic ingredients for popular thriller movies that were produced in the late twentieth century, perhaps the most famous one being the James Bond movie *Moonraker* (1979).

After the US deployment of atomic bombs on Japan in 1945, nuclear apocalypses dominated the genre for decades, to be followed by global threats of runaway bacteria or viruses and climate change. However, all those topics were originally inspired by fictional ideas about chemistry, which quite recently moved from the apocalyptic theme to the eco-thriller genre (see below).

CHEMICAL POLLUTION

Before we deal with the representation of chemical pollution in fiction, it is useful to first have a look at its history and representation in nonfiction. As was pointed out in the Introduction, nonfiction writing also changed drastically during the twentieth century and produced new genres such as environmental reports written by journalists, science writers, and non-governmental organizations, which were increasingly targeted at a mass audience and out of which the environmental movement developed. That required the development of new styles of writing, and it shaped new narratives that fiction writers later adapted. In the last third of the century, chemistry when mentioned in public media was probably most often associated with environmental pollution.

Three Kinds of Chemical Threats

During the twentieth century the public increasingly perceived the chemical industry as a threefold threat: production accidents, pollution, and the adverse side effects of its products (Koch and Vahrenholt 1978).

There were several industrial production disasters. In terms of immediate casualties, the three worst accidents, or disasters in the narrow sense, were the plant explosions at the Union Carbide subsidiary in Bhopal, India, in 1984 with several thousand casualties, and at the BASF factory at Oppau near Ludwigshafen, Germany in 1921 and 1948, which caused 561 and 207 casualties, respectively. Because the last two cases are not well known, it is unlikely that industrial accidents were perceived as the main threat. Moreover, disasters in other industries greatly outnumbered those of the chemical industry both in terms of frequency and casualties, such as dam failures, mining tragedies, and aircraft accidents.

However, in addition to the immediate deaths, the damage to health is important in terms of judging hazards, which in the case of Bhopal alone affected hundreds of thousands of people (Eckerman and Børsen 2018). The main threat of most chemicals, as perceived by the public, is the slowly creeping pollution that develops its poisonous effect from either one-time or long-term exposure, and which affects not only humans but all living beings by polluting the entire environment. Public concerns about air, water, and soil pollution from the chemical industries are almost as old as these industries themselves (Tarr 1996; Homburg et al. 1998: 121–201; Tarr 2002; Bernhardt 2004). In the second half of the twentieth century, widely noticed chemical disasters without immediate human casualties included the release of methylmercury in waste water over several decades into Minamata Bay in Japan since 1932, causing mercury poisoning of humans and wildlife discovered only as late as 1956; the release of the extremely toxic dioxin in Seveso, Italy in 1976; and the Sandoz chemical spill in Switzerland in 1986, releasing various toxic agrochemicals into the air and the Rhine River and killing aquatic wildlife.

A third chemical threat increasingly raised public concerns during the second half of the twentieth century: the toxic and eco-toxic effects of the chemical products themselves. The worst case happened in late 1961, when thalidomide, after having been on the European market without clinical tests for more than four years, was found to be severely teratogenic. Recommended as a side effect-free sleeping pill and widely used against nausea by pregnant women, the drug had caused more than 10,000 stillborn children and between 5,000 and 10,000 surviving babies with malformations before the German company Grünenthal and its international licensees could no longer dismiss warnings and were eventually forced to withdraw it (Ruthenberg 2016).

Since the 1960s, hundreds of chemical products have been shown to have adverse effects on human health and other living beings by their regular use and release to the environment, from the insecticide DDT, to chlorofluorocarbons (CFCs), to polymer additives which may cause endocrine disruption. Over time, many people suspected all chemical products are more harmful than useful and began to prefer "natural" products instead.

In summary, it was not industrial accidents but pollution through plant leakages and waste disposal, and the use of chemical products that were perceived as the main threat to humans and the natural environment, and heavily debated in public media.

Industrial Pollution Narratives in Non-Fiction

Less than seven months after the shock of the thalidomide scandal broke, the American marine biologist and by then already famous nature writer Rachel Carson serialized her fourth book, *Silent Spring* (1962), in three parts in *The New Yorker* magazine. Unlike her previous books, which described the history, diversity, and beauty of marine wildlife, *Silent Spring* accused the chemical industry of indiscriminately and severely threatening life on earth, from insects to fish, birds, and mammals, including humans. Based on previous research and her own field investigations, she argued that insecticides such as DDT (called "Elixirs of Death" by Carson), which since the 1940s had widely been used to fight insect pests and insect-borne diseases like malaria, accumulate in the food chain and damage organisms even in remote areas. The American chemical industry started a counter-campaign, denying the threat, denigrating Carson, and suing the magazine. However, Carson had made her case carefully and convincingly for many, before she died in 1964. Although various other political factors favored the US ban of DDT as late as 1973 (Morris 2019), she is widely considered the person who started the environmentalist fight against insecticides.

During the second half of the twentieth century, there was a recurrent pattern in the chemical industry's responses to environmental and health issues. Even if the historical details did not always accurately match this pattern, journalistic narratives highlighted those aspects that corresponded to the pattern, as if the same story had to be reenacted and/or narrated over and over again, like a classical myth. In this narrative, a chemical company faced with the accusation of causing a health or environmental hazard first ignores and downplays the hazard, then counteracts with campaigns that highlights their merits, denigrates and sues the accusers, pays for "scientific" counter-reports, lobbies policymakers, and changes its behavior only when a better alternative is in its pipeline or when required to address the issue by legal regulation or a court judgement. In full or in part, we have this narrative, besides in numerous journalistic descriptions of local cases, also in scholarly accounts: Minamata, thalidomide, DDT, Seveso, Bhopal, CFCs, and endocrine disruptors, as well as other industries, such as nuclear energy production, waste disposal, and for global climate change and tobacco smoke (Oreskes and Conway 2010). In several cases, like that of DDT and that of Love Canal – a suburb in New York state built on a chemical dumping site that caused health impairments in the 1970s (Fjelland 2016) – the narrative includes a heroine who fights the powerful industrial–governmental complex, reminiscent of the Joan of Arc theme.

In the present context, it does not matter if the narrative was always correct or not; it became an essential part of the public view of the chemical industry, and frequently that of chemistry altogether. Unlike the "mad scientist" of the nineteenth century, the "chemical-industry-as-villain" stereotype was not developed and shaped by fiction writers. Rather, it was created in real life interactions between the chemical industry and its critics, including victims, environmental activists, scientists, and journalists.

Environmental Doomsday Narratives in Fiction

While activists, including scientists and journalists, fought for a safer environment, artists and fiction writers long faced difficulties to include the topic in their works. In the late nineteenth century, when industrial pollution was already publicly debated, the smokestack was still employed as a symbol of economic progress in landscape paintings. Wilhelm Raabe's pioneering environmental novel *Pfisters Mühle* (1884, *Pfister's Mill*), in which industrial pollution destroys a holiday resort near Berlin, remained for many decades the only one of its kind worth mentioning (Schummer 2022). When the environmental movement began in the 1970s, the formative arts hardly engaged with it, with few notable exceptions. For instance, in his art performance of planting "1,000 oaks" (1982) the German artist Joseph Beuys referred to a popular phrase attributed to Martin Luther, who would have planted a new tree even on the day before Doomsday.

Writers probably lacked interest in pollution because many had long since also embarked on the doomsday theme, which dwarfed any local, creeping, and unspectacular pollution. Moreover, it was difficult to feature an environmental activist-scientist in plots that still recycled the nineteenth-century "mad scientist." However, writers were extremely inventive in attributing diverse kinds of global disasters, first of all to chemistry including famine, nuclear energy, run-away bugs, global warming, and the removal (not pollution!) of the Earth's drinkable water and breathable oxygen. And they remained steady in their reference to biblical themes.

Probably the first twentieth-century chemical disaster novel was written in 1904 by H.G. Wells. In *The Food of the Gods and How It Came to Earth*, chemists invent a substance that let children grow endlessly, resulting in a bitter war between the normal and the giant people about food. This was not just an interesting story about biochemistry, but a variation of the biblical myth of the fallen angels or devils: under the leadership of Satan they build a godless kingdom on earth and mate with human females to beget giants who eat all food causing a human famine and insurgency. Wells thus put chemists and devils in parallel. Ten years later, on the eve of World War I, he published the already mentioned post-apocalyptic novel *The World Set Free* (1914), in which a chemist manages to achieve chain reactions of nuclear fission that is soon employed as a weapon in a devastating world war. That set the stage for both chemical End Times stories after World War I and for the nuclear disaster theme that became popular after World War II.

British authors continued to dominate the genre with original ideas, which American directors would much later turn into movies. Scottish chemistry professor Alfred Walter Stewart, who wrote novels under the pen name J.J. Connington, published the first runaway bacteria drama, which became the favorite disaster theme towards the end of the twentieth century. In *Nordenholt's Million* (1923) the amateur scientist Wotherspoon has recently extended his chemical research to include "denitrifying bacteria" when ball lightning happens to enter his laboratory still full of "flasks, retorts, test-tube racks." The electric discharge modifies his bacteria such that they proliferate at a tremendous speed. Spread within weeks by airplanes to all continents, they stop the growths of plants worldwide and ruin agriculture by turning fertile soils into denitrified sandy soil, causing a global famine. Replace bacteria with locusts, and you have a disaster similar to the eighth Plague of Egypt from the Book of Exodus.

Global warming, which would become a popular literary and movie theme only in the twenty-first century, was attributed to chemistry as early as in 1964. In *The Burning World* British novelist J.G. Ballard envisioned polymer waste from the chemical industry flushed into the oceans where it forms a semipermeable layer on the surface that hinders the vaporization of seawater, resulting in a global drought and heat wave. The novel is also one of the earliest examples that focuses on industrial waste and thus is rightly considered to be a classic of the eco-novel genre. But note that the point is not the harm to marine wildlife as in Carson's book two years before, but a disaster reminiscent of the fourth End Times plague from the Book of Revelation, a burning heat wave.

In another science-fiction classic, *Cat's Cradle* (1963) by American author Kurt Vonnegut, the chemist/physicist Felix Hoenikker (supposedly modeled after Irving Langmuir) invents ice-nine, a modification of water that is solid at room temperature. A seed crystal can turn ordinary water into ice. Originally created for the military purpose of crossing swamps, a piece falls by accident into the sea, solidifies the water of the oceans, and kills almost all life. Ice-nine is, so to speak, the doomsday equivalent for contemporary water pollution.[5]

In the aforementioned disaster stories, "mad chemists" destroy the life-sustaining media of nutrition, soil, water, and climate, or threaten life with weapons and poisons. Pollution of the air other than by poison gas was even harder to adjust to the disaster theme. Rather than by pollution, the spoiling of air was imagined by de-oxygenation, as for instance in *Dalkey Archive* (1964) by Irish novelist Flann O'Brien, which features the evil-minded "mad scientist" De Selby who tries to destroy all life on earth through his invention of an oxygen-absorbing substance called DMP.

The chemical industry in the form of an evil-minded company, with malicious intent to pollute the environment, appeared – apart from post-World War I poison gas-producing companies – very late in novels. In post-World War II America, like in post-World War I Germany, this trope was the product of an anti-capitalist movement. Left-wing US science fiction authors developed dystopian views of transnational corporations taking over governmental power and abusing society for their corporate interest, such as in Frederik Pohl and Cyril M. Kornbluth's *The Space Merchants* (1953), as well as in the *Superman* and *Spider-Man* series. However, the first environmental activist-hero in a novel who fights pollution (and a corrupt US government in the far future) is probably the Savior-like character Austin Train in the dystopian novel *The Sheep Look Up* (1972), written by British science fiction author John Brunner. Note that in the same year the Club of Rome had already identified pollution as one of five limiting factors of human civilization in their *Limits to Growth*.

It would be several more years before the standard narrative from non-fiction – a scientist-hero fighting the chemical industry as villain – was adopted by fiction writers. One of the earliest novels of that kind was written by German author Michail Krausnick, who published a series of books with chemical companies featuring as villains. In *Die Paracana-Affäre* (1975, *The Paracana Affair*) they intentionally poison part of humanity; in *Lautlos kommt der Tod* (1982, *Death Comes Silently*) they secretly develop biochemical weapons;

and in *Im Schatten der Wolke* (1980, *In the Shadow of the Cloud*) they are responsible for a mysterious epidemic. Only in the last one does the scientist-hero enter the stage, in the form of Professor Kovacs, who traces the cause of the epidemic to a chemical plant leakage.

Neal Stephenson developed this trope further in his novel *Zodiac* (1988). Here, the chemist Sangamon Taylor works for an environmentalist group and fights against the fictional chemical company Basco. The firm has been illegally dumping toxic waste in Boston Harbor and releasing genetically modified bacteria; once confronted with evidence of its crimes, it tries to denigrate Taylor and accuses him of terrorism. Eventually also the female heroine combating industrial pollution received a vehicle in Steven Soderbergh's movie *Erin Brockovich* (2000), based on actual events.

At the turn of the century, a temporary twist of the eco-thriller genre by two famous US authors briefly shocked environmental activists. In T.C. Boyle's *A Friend of the Earth* (2000) they are portrayed as miserable figures. And Michael Crichton, who had in *Prey* (2002) still employed the classical "mad scientist" framework for the fictitious nanotechnology of software-controlled molecular machines running amok, now displayed environmentalists as terrorists and villains in his *State of Fear* (2004). They unscrupulously fabricate the deadly disasters they had previously warned about. However, the genre soon returned to its original scheme and has produced many more works ever since.

CONCLUSION

The mostly negative image of chemistry in public representations, as described in this chapter, is to some extent influenced by the selection of topics. A history of the many representations sponsored by chemical industries and societies, from advertisement campaigns to paid publications and donated museum exhibits, might tell a different story. However, it is difficult to dispute that the three topics highlighted here – the recycling and further development of the nineteenth-century "mad scientist" theme, the impact of poison gas in World War I, and environmental pollution – strongly influenced the public image of chemistry. There would be many other negative stories to tell, which, for instance, explain the public opposition between chemistry and nature, as in the aversion to chemical medicines, food additives, cosmetics, plastics, and fertilizers.

In addition to the benevolent scientist in comics for children, there are also some independent, more nuanced or favorable representations of chemistry, particularly at the end of the century (Ball 2007), by writers who did not feel obliged to cater to a mass audience and who objected to the business of standardized genre writing. However, the twentieth century is the first era in which the publication and entertainment industry reached out to a mass audience for both media-technological and economic reasons. The most successful way to do so was to reproduce and exaggerate simple stereotypes, mostly developed in the previous century.

Most Western stories about chemistry are framed after biblical themes, including the classical "mad scientist" who commits the sin of "playing God"; the fallen angels who bring evil to the world; the biblical flood of the Noah myth by global poisoning; the fourth End Times plague of global warming; and the battle of Armageddon in the fight between the angelic superhero and the diabolical chemical supervillain. That suggests that for the targeted mass audience, or their most influential writers, the Bible provided the dominating framework for entertaining drama and for formulating moral positions about chemistry throughout the twentieth century and beyond.

The religious approach could draw on a much older tradition that associated the precursors of chemistry with diabolical work (Schummer 2017). It already began with the apocryphal *Book of Enoch* (*ca.* 200 BCE) that for the first time introduced the concept of devils or fallen angels who reveal chemical crafts to humans, including the secrets of Creation. From the late medieval "mad alchemist" through the literary trope of the Faust myth to the classical "mad scientist" of the nineteenth century, all of which employ a fiendish tempter, alchemy and chemistry were portrayed as the devil's work. That religious trope seems to be so deeply rooted in Western culture that it can be, and somehow must be, narrated over and again for both entertaining and moral purposes, like a classical myth.

Two topics in this chapter – poison gas in World War I and chemical pollution – both illustrate the process and the obstacles of incorporating important historical events into the religious framework of literary narratives. Although many writers were for a period after World War I busy with reflecting on their earlier enthusiasm for heroic war, some soon developed from the theme of chemical warfare the first-ever notion of apocalypse by technological means, a global poison gas cloud reminiscent of the biblical flood. Once the trope was established, chemistry, much more so than any other science or technology, became firmly associated with the total destruction of the world as we know it. Indeed, as was shown above, all the global disaster themes in fiction that evolved during the twentieth century – from poisoning or suffocating, nuclear energy, and run-away bugs, to famine, drought, and global warming – each began with a novel in which the cataclysm was attributed to fictional chemists.

Although environmental pollution was already a severe problem in the nineteenth century, fiction writers had difficulties dealing with the issue for most of the twentieth century because those mostly local and insidious problems did not fit well their preferred plot, apocalyptic disasters of global dimensions. Instead, many scientists, including chemists, worked on these issues and tried to raise public awareness. They thereby became actors in a non-fiction narrative, the scientifically minded environmentalist fighting the polluting industry, told in numerous journalistic and scholarly accounts. The late adoption of the theme in fiction, including the temporary twist of "eco-terrorism," suggests that the scientist-as-hero was difficult to reconcile with the received "mad scientist" theme and its religious connotations. The most successful way of re-adoption was by attributing elements of the "mad scientist" to the chemical industry.

Oddly, the chemical profession encountered similar difficulties to identify environmental issues as part of their own core field. Environmental chemistry did not easily match the dominant self-image of chemists that highlighted making things in the laboratory rather than understanding the natural environment and moral engagement. If chemists were asked to articulate the goals of their research, many just pointed to the chemical industry, which, in the public view, dealt with environmental issues largely by denying, downplaying, campaigning, suing, and lobbying. While the chemical industries underwent radical reforms since the late 1970s (for instance, in waste treatment, labor safety, risk management, public communication, and acquiring non-chemical expertise), chemical societies largely missed the opportunity to develop a modern profile of their science that corresponded to the diversity of research fields that had emerged over the century. They particularly neglected to emphasize their important contributions to environmental issues such as in environmental analysis, monitoring, and cleanup, "green chemistry" production, materials recycling, biodegradable materials, and renewable energy.

The literary and media representations of chemistry during the twentieth century, dominated by the "mad scientist" theme, apocalyptic threats, and environmental pollution, probably contributed to a social isolation of chemists, which is incidentally a feature of the "mad scientist." Chemists did comparatively well with physicists, biologists, and engineers in interdisciplinary work. But there have been huge barriers to the humanities and social sciences, even to those that could help chemists locate their field within the broader cultural context, overcome their societal isolation, and correct the public image. For instance, physics, mathematics, astronomy, biology, medicine, and engineering all actively contributed to the academic establishment of the history, philosophy, and sometimes ethics of their respective field. In contrast, the history of chemistry has remained marginal within the history of science, and has met with disinterest by most working chemists, while the small field of philosophy of chemistry has largely remained beyond their horizon. Ironically, for a discipline that has, much more than any other, faced moral accusations in public debates and media representations, not least through the hubris motif of the "mad scientist," the ethics of chemistry hardly exists (Schummer and Børsen 2021).

NOTES

CHAPTER 1

1. I gratefully thank Bretislav Friedrich, Peter J.T. Morris, Robert A. Nye, Alan J. Rocke, and Stephen J. Weininger for their comments and suggestions.

CHAPTER 8

1. The most detailed treatment of science in literature is still Haynes (1994). Shorter and more specific works include Stocker (1998), Labinger (2011), Ziolkowski (2015), and various chapters in Schummer et al. (2007), particularly Haynes (2007), Weingart (2007), and Ball (2007).
2. The topic is missing from most of the many monolingual anthologies and monographs about literature on World War I, and hardly appears in comparative studies such as Riegel (1978) (which discusses some 200 French, English, American, and German novels) and Marsland (1991). The international bibliography of the journal *Krieg und Literatur/War and Literature* 1989–1994 includes nothing related to chemical warfare among its about 700 index terms. Earlier studies include Haber (1986: 230–8), Spear and Summersgill (1991), Löschnigg (1994: 150–63), and Kaufmann (2017). On the poor representation of gas warfare in later movies, see Skrebels (2014).
3. For instance, Henri Barbusse's *Le Feu* (1916, chapter 19); Roland Dorgelès' *Les Croix de bois* (1919: 286–7, 303); John Dos Passos' *One Man's Initiation: 1917* (1920, chapter 6); Erich Maria Remarque's *Im Westen Nichts Neues* (1928, chapters 4 and 6); Robert Graves' *Goodbye to All That* (1929: 90–1, 164–5, 220).
4. Parts of the exhibition are still temporarily available at http://www.memorial-caen.fr/10EVENT/EXPO1418/.
5. Ironically, the story inspired the polywater media hype in the US in the late 1960s, based on the suspicion that the Russians could have invented something like ice-nine. That in turn inspired the post-apocalyptic science fiction novel *Year of the Cloud* (1970) by Kate Wilhelm and Theodore L. Thomas, where all water turns into a gel through the impact of a cosmic cloud. Scientists are eventually able to solve the problem, which illustrates that chemists can also play the hero in disaster stories.

BIBLIOGRAPHY

Abelshauser, Werner, Wolfgang von Hippel, Jeffrey Allan Johnson, and Raymond G. Stokes. 2004. *German Industry and Global Enterprise. BASF: The History of a Company*. Cambridge: Cambridge University Press.
Achilladelis, Basil. 1970. "A Study in Technological History, Part I: The Manufacture of 'Perlon' (nylon 6) and Caprolactam by I.G. Farbenindustrie." *Chemistry and Industry*, 107: 1549–54.
Adjangba, Messan. 1979. "Position Paper: Togo." In Peter E. Childs and James E. Gowan (eds), *International Conference on Chemical Education, Proceedings: The Teaching of Chemistry – Interaction between Secondary and Tertiary Levels*. Dublin: Organising Committee, ICCE.
Aftalion, Fred. 2001. *A History of the International Chemical Industry: From the "Early Days" to 2000*. 2nd ed. Philadelphia, PA: Chemical Heritage Press.
Allen, Barbara L. 2003. *Uneasy Alchemy: Citizens and Experts in Louisiana's Chemical Corridor Disputes*. Cambridge, MA: MIT Press.
American Chemical Society. 2000. *ChemCensus*. Washington, DC: American Chemical Society.
American Chemical Society. 2018. "Comparison of ACS with Other Scientific Societies – 2017." In *ACS Financial Information – FAQs*. Available online: https://www.acs.org/content/acs/en/about/aboutacs/financial/faq.html.
American Chemical Society. 2019. "Fast Facts about ACS." Available online: https://www.acs.org/content/acs/en/about/aboutacs.html.
Anderson, Philip W. 1972. "More Is Different." *Science*, 177: 393–6.
Anderson, Philip W. 2011. *More and Different: Notes from a Thoughtful Curmudgeon*. London: World Scientific.
Angier, Natalie. 2017. "Mildred Dresselhaus, the Queen of Carbon, Dies at 86." *New York Times* (February 23). Available online: https://www.nytimes.com/2017/02/23/science/mildred-dresselhaus-dead-queen-of-carbon.html.
Arora, Ashish, Ralph Landau, and Nathan Rosenberg (eds). 1998. *Chemicals and Long-term Economic Growth: Insights from the Chemical Industry*. New York: John Wiley.

Ashby, Eric, and Mary Anderson. 1981. *The Politics of Clean Air*. Oxford: Oxford University Press.

Atkins, Peter W. 2010. *Liquid Materialities: A History of Milk, Science and the Law*. Farnham: Ashgate.

Auden, W.H. 1948. "Editor's Introduction." In W.H. Auden (ed.), *The Portable Greek Reader*. New York: Viking Penguin.

Augustine, Dolores L. 2007. *Red Prometheus: Engineering and Dictatorship in East Germany, 1945–1990*. Cambridge, MA: MIT Press.

Austen, Kat F. 2016. "Theory Choice in Chemistry: Attitudes to Computer Modelling in Chemistry." In Emma Tobin and Chiara Ambrosio (eds), *Theory Choice in the History of Chemical Practices*. Cham: Springer.

Authier, André. 2015. *Early Days of X-ray Crystallography*. Oxford: Oxford University Press.

Bächi, Beat. 2009. *Vitamin C für alle. Pharmazeutische Produktion, Vermarktung und Gesundheitspolitik (1933–1953)*. Zürich: Chronos Verlag.

Bai Chunli. 2010. "Message from the President of FACS (2009–2011): IYC and our Mission for a Sustainable Society." Available online: http://www.facs-as.org/c12.

Baird, Davis. 1993. "Analytical Chemistry and the 'Big' Scientific Instrumentation Revolution." *Annals of Science*, 50: 267–90.

Ball, Philip. 2007. "Chemistry and Power in Recent American Fiction." In Joachim Schummer, Bernadette Bensaude-Vincent, and Brigitte Van Tiggelen (eds), *The Public Image of Chemistry*. Singapore: World Scientific.

Ballard J.G. 1964. *The Burning World*. New York: Berkley.

Balmer, Brian, Matthew Godwin, and Jane Gregory. 2009. "The Royal Society and the 'Brain Drain': Natural Scientists Meet Social Science." *Notes and Records*, 63: 339–54.

Barbusse, Henri. 1916. *Le Feu*. Paris: Flammarion.

Barrow, Richard F., and C. John Danby. 1991. *The Physical Chemistry Laboratory: The First Fifty Years; A History of the Physical Chemistry Laboratory, University of Oxford, 1941–1991*. Oxford: Physical Chemistry Laboratory. Available online: http://ptcl.chem.ox.ac.uk/history/ptcl2.html.

Bauer, Helmut. 1990. "Zur Geschichte des Chemieunterrichts an allgemeinbildenden Schulen im deutschen Sprachraum." *Mitteilungen, Gesellschaft Deutscher Chemiker – Fachgruppe Geschichte der Chemie*, 4: 3–12.

Becher, Johannes R. 1926. *Levisite oder der einzig gerechte Krieg*. Vienna and Berlin: Agis.

Bell, Michelle, Devra L. Davis, and Tony Fletcher. 2004. "A Retrospective Assessment of Mortality from the London Smog Episode on 1952: The Role of Influenza and Pollution." *Environmental Health Perspectives*, 112: 6–8.

Benfey, Otto Theodor, and Peter J.T. Morris. 2001. *Robert Burns Woodward: Architect and Artist in the World of Molecules*. Philadelphia, PA: Chemical Heritage Foundation.

Bennett, Stuart. 1991. "'The Industrial Instrument – Master of Industry, Servant of Management': Automatic Control in the Process Industries, 1900–1940." *Technology and Culture*, 32: 69–81.

Bennett, Stuart. 2002. "Production Control Instruments in the Chemical and Process Industries." In Peter J.T. Morris (ed.), *From Classical to Modern Chemistry: The Instrumental Revolution*. Cambridge: Royal Society of Chemistry.

Bensaude-Vincent, Bernadette. 1998. *Eloge du mixte: Matériaux nouveaux et philosophie ancienne*. Paris: Hachette.

Bensaude-Vincent, Bernadette. 2001. "Graphic Representations of the Periodic System of Chemical Elements." In Ursula Klein (ed.), *Tools and Modes of Representation in the Laboratory Sciences*. Dordrecht: Springer.

Bensaude-Vincent, Bernadette. 2009. *Les vertiges de la technoscience: façonner le monde atome par atome*. Paris: La Découverte.

Bensaude-Vincent, Bernadette. 2018. "Chemists without Borders." *Isis*, 109: 597–607.

Bensaude-Vincent, Bernadette, and Jonathan Simon. 2008. *Chemistry: The Impure Science*. London: Imperial College Press.

Bensaude-Vincent, Bernadette, and Isabelle Stengers. 1996. *A History of Chemistry*. Cambridge, MA: Harvard University Press.

Bensaude-Vincent, Bernadette, Antonio García-Belmar, and José R. Bertomeu-Sanchez. 2003. *L'émergence d'une science des manuels. Les livres de chimie en France (1789–1852)*. Paris: Editions des Archives Contemporaines.

Bensaude-Vincent, Bernadette, Sacha Loeve, Alfred Nordmann, and Astrid Schwarz (eds). 2017. *Research Objects in their Technological Settings*. Abingdon: Routledge.

Bernal, J. Desmond. 1968. "X-ray Diffraction in the Last Fifty Years of Physics." *Physics Bulletin*, 19: 137–40.

Bernhardt, Christoph (ed.). 2004. *Environmental Problems in European Cities in the 19th and 20th Century*. Münster: Waxmann.

Bertazzi, Pier Alberto. 1991. "Long-Term Effects of Chemical Disasters. Lessons and Results from Seveso." *The Science of the Total Environment*, 106: 5–20.

Bertomeu-Sánchez, José R. 2013. "Managing Uncertainty in the Academy and the Courtroom: Normal Arsenic and Nineteenth-century Toxicology." *Isis*, 104: 197–225.

Bertomeu-Sánchez, José R. 2015. "Chemistry, Microscopy and Smell: Bloodstains and Nineteenth-century Legal Medicine." *Annals of Science*, 72: 490–516.

Bertomeu-Sánchez, José R., Duncan Thorburn Burns, and Brigitte van Tiggelen (eds). 2008. *Neighbours and Territories: The Evolving Identity of Chemistry*. Louvain-la-neuve: Mémosciences.

Bertrams, Kenneth, Nicolas Coupain, and Ernst Homburg. 2013. *Solvay: History of a Multinational Family Firm*. Cambridge: Cambridge University Press.

Bigg, Charlotte, 2002. "Adam Hilger, Ltd. and the Development of Spectrochemical Analysis." In Peter J.T. Morris (ed.), *From Classical to Modern Chemistry: The Instrumental Revolution*. Cambridge: RSC Publications.

Bigg, Charlotte. 2014. "Representing the Experimental Atom." In Carsten Reinhardt and Ursula Klein (eds), *Objects of Chemical Inquiry*. Sagamore Beach, MA: Science History Publications.

Birkenfeld, Wolfgang. 1964. *Der synthetische Treibstoff, 1933–1945: ein Beitrag zur nationalsozialistischen Wirtschafts- und Rüstungspolitik*. Göttingen: Musterschmidt.

Bohn, John L., Ana Maria Rey, and Jun Ye. 2017. "Cold Molecules: Progress in Quantum Engineering of Chemistry and Quantum Matter." *Science*, 357: 1002–10.

Blanc, Paul David. 2007. *How Everyday Products make People Sick: Toxins at Home and in the Workplace*. Berkeley: University of California Press.

Blanc, Paul David. 2016. *Fake Silk: The Lethal History of Viscose Rayon*. New Haven, CT: Yale University Press.

Bonner, James. 1989. *Arie Jan Haagen-Smit (1900–1977): A Biographical Memoir*. Washington, DC: National Academy of Sciences.

Bonneuil, Christophe, and Jean-Baptiste Fressoz (eds). 2016. *The Shock of the Anthropocene. The Earth, History and Us*. London: Verso.

Børsen, Tom, and Joachim Schummer (eds). 2016–2018. *Ethical Case Studies of Chemistry*. Available online: http://www.hyle.org/journal/issues/special/ethical-cases.html.

Borello, Lisa J., Robert Lichter, Willie Pearson, and Janet L. Bryant. 2015. "International Status of Women in the Chemical Sciences." In Willie Pearson, Jr., Lisa M. Frehill, and Connie L. McNeely (eds), *Advancing Women in Science: An International Perspective*. Cham: Springer.

Borman, Stu. 2016. "What Type of Research Wins the Nobel Prize in Chemistry?" graphic in "Notable Chemists Who Should Have Won the Nobel." *Chemical and Engineering News*, 94(11 April): 19–20. Available online: https://cen.acs.org/content/cen/articles/94/i15/Five-chemists-should-won-Nobel.html.

Bottle, Robert T. 1962. "Background Information." In Robert T. Bottle (ed.), *Use of the Chemical Literature*. London: Butterworth.

Boudia, Soraya, and Nathalie Jas (eds). 2014. *Powerless Science? Science and Politics in a Toxic World*. New York: Berghahn Books.

Boudia, Soraya, Angela Creager, Scott Frickel, Emmanuel Henry, Nathalie Jas, Carsten Reinhardt, and Jody Roberts. 2018. "Residues. Rethinking Chemical Environments." *Engaging Science and Technology*, 4: 165–78.

Boyle, T.C. 2000. *A Friend of the Earth*. New York: Viking Books.

Bragg, William Lawrence. 1975. *The Development of X-ray Analysis*. London: Bell.

Brickman, Ronald, Sheila Jasanoff, and Thomas Ilgen. 1985. *Controlling Chemicals. The Politics of Regulation in Europe and the United States*. Ithaca, NY: Cornell University Press.

Brock, William H. (ed.). 1973. *H.E. Armstrong and the Teaching of Science, 1880–1930*. Cambridge: Cambridge University Press.

Brock, William H. 1975. "From Liebig to Nuffield. A Bibliography in the History of Science Education, 1839–1974." *Studies in Science Education*, 2: 67–99.

Brock, William H. 1993. *The Norton History of Chemistry*. New York: Norton.

Brock, William H. 2017. "British School Chemistry Laboratories, 1830–1920." *Ambix*, 64: 43–65.

Brooks, Nathan M. 1997. "Chemistry in War, Revolution, and Upheaval: Russia and the Soviet Union, 1900–1929." *Centaurus*, 39: 349–67.

Broughton, Edward. 2005. "The Bhopal Disaster and its Aftermath: A Review." *Environmental Health: A Global Access Science Source*, 4: 1–6.

Brown, Andrew. 2005. *J. D. Bernal: The Sage of Science*. Oxford: Oxford University Press.

Brown, Mary Jean, and Stephen Margolis. 2012. *Lead in Drinking Water and Human Blood Lead Levels in the United States*. Atlanta, GA: Department of Health and Human Services. Available online: https://www.cdc.gov/mmwr/pdf/other/su6104.pdf.

Brown, Phil. 1992. "Popular Epidemiology and Toxic Waste Concentration: Lay and Professional Ways of Knowing." *Journal of Health and Social Behavior*, 33: 267–81.

Brubacher, Lewis J., Chung Chieh, Donald E. Irish, and A. Donald Maynes (eds). 1989. *Proceedings of the Tenth International Conference on Chemical Education, August 20 to 25, 1989, University of Waterloo, Ontario, Canada*. Waterloo: Tenth International Conference on Chemical Education.

Brunner, John. 1972. *The Sheep Look Up*. New York: Harper & Row.

Bruno, Anne-Sophie, Eric Geerkens, Nicolas Hatzfeld, and Catherine Omnès (eds). 2011. *La santé au travail, entre savoirs et pouvoirs (19e–20e siècle)*. Rennes: Presses universitaires de Rennes.

Brush, Stephen G. 1999. "Dynamics of Theory Change in Chemistry: Part 2. Benzene and Molecular Orbitals, 1945–1980." *Studies in the History and Philosophy of Science*, 30: 263–302.

Bud, Robert F. 2007. *Penicillin: Triumph and Tragedy*. New York: Oxford University Press
Bud, Robert F. 2013. "Life, DNA and the Model." *British Journal for the History of Science*, 46: 311–34.
Bud, Robert F. 2018. "Modernity and the Ambivalent Significance of Applied Science: Motors, Wireless, Telephones and Poison Gas." In Robert Bud, Paul Greenhalgh, Frank James, and Morag Shiach (eds), *Being Modern: The Impact of Science on Culture in the Early Twentieth Century*. London: UCL Press.
Cahan, David. 1997. "The Zeiss Werke and the Ultramicroscope: The Creation of a Scientific Instrument in Context." In Jed Z. Buchwald (ed.), *Scientific Credibility and Technical Standards in 19th and Early 20th Century Germany and Britain*. Dordrecht: Kluwer.
Cahn, Robert W. (ed.). 2001. *The Coming of Materials Science*. Oxford: Elsevier Science.
California Air Resources Board. 2017. *History*. Available online: https://ww2.arb.ca.gov/about/history.
Carson, Rachel. 1962. *Silent Spring*. Boston: Houghton Mifflin.
Carter, Henry A. 1988. "Chemistry in the Comics: Part 1. A Survey of the Comic Book Literature." *Journal of Chemical Education*, 65: 1029–35.
Chandler, Alfred D., Jr. 1990. *Scale and Scope: The Dynamics of Industrial Capitalism*. Cambridge, MA: Belknap Press.
Chandler, Jr., Alfred D. 2005. *Shaping the Industrial Century: The Remarkable Story of the Evolution of the Modern Chemical and Pharmaceutical Industries*. Cambridge, MA: Harvard University Press.
Chandler, Alfred D., Jr., and Stephen Salsbury. 1971. *Pierre S. du Pont and the Making of the Modern Corporation*. New York: Harper & Row.
Channel, David F. (ed.). 2017. *A History of Technoscience: Erasing the Boundaries between Science and Technology*. Abingdon: Routledge.
Chen, Guangxu. 1985. "Zhongguo huaxue jiaoyu de fazhan [The development of chemistry education in China]" in *Zhongguo huaxue wushi nian, 1932–1982* [Fifty Years of Chemistry in China, 1932–1982]. Beijing: Science Press.
Chisman, Dennis G. (ed.) 1970. *University Chemical Education: Proceedings of the International Symposium on University Chemical Education held in Frascati (Rome), Italy, 16–19 October 1969*. London: Butterworths.
Clavin, Whitney. 2016. "Garnet Chan Talks Quantum Chemistry and Chinese Food." *Caltech News*, December 9. Available online: http://www.caltech.edu/news/garnet-chan-talks-quantum-chemistry-and-chinese-food-53248.
Cohen, Clive. 1996. "The Early History of Chemical Engineering: A Reassessment." *British Journal for the History of Science*, 29: 171–94.
Coleman, Donald C. 1969. *Courtaulds: An Economic and Social History*. 2 vols. Oxford: Clarendon Press.
Connington, J.J. 1923. *Nordenholt's Million*. London: Constable & Co.
Craighead, John E., and Brooke T. Mossman. 1982. "The Pathogenesis of Asbestos-Associated Diseases." *New England Journal of Medicine*, 306: 1446–55.
Creager, Angela N.H. 2013. *Life Atomic: A History of Radioisotopes in Science and Medicine*. Chicago, IL: University of Chicago Press.
Crichton, Michael. *State of Fear*. London: HarperCollins.
Crichton, Michael. 2002. *Prey*. New York: HarperCollins.
Crim, Brian E. 2018. *Our Germans: Project Paperclip and the National Security State*. Baltimore, MD: Johns Hopkins University Press.

Crone, Hugh D. 1986. *Chemicals and Society: A Guide to the New Chemical Age*. Cambridge: Cambridge University Press.

Crow, Michael, Barry Bozeman, Walter Meyer, and Ralph Shangraw, Jr. 1988. *Synthetic Fuel Technology Development in the United States: A Retrospective Assessment*. Westport, CT: Praeger.

Davies, Frederick Rowe. 2014. *Banned. A History of Pesticides and the Science of Toxicology*. New Haven, CT: Yale University Press.

Davis, Lee Niedringhaus. 1984. *The Corporate Alchemists: Profit Takers and Problem Makers in the Chemical Industry*. New York: William Morrow.

DeBoer, George E. 1991. *A History of Ideas in Science Education: Implications for Practice*. New York: Teachers College Press.

DeLillo, Don. 1984. *White Noise*. New York: Viking.

Depalma, Anthony. 2004. "Love Canal Declared Clean, Ending Toxic Horror." *New York Times*, March 18, 1.

Divall, Colin, and Sean F. Johnston. 2000. *Scaling Up: The Institution of Chemical Engineers and the Rise of a New Profession*. Dordrecht: Kluwer.

Djerassi, Carl. 1981. *The Politics of Contraception*. New York: W.H. Freeman

Djerassi, Carl. 1990. *Steroids Made it Possible*. Washington, DC: American Chemical Society.

Djerassi, Carl. 1992. *The Pill, Pygmy Chimps and Degas' Horse: The Autobiography of Carl Djerassi*. New York: Basic Books.

Djerassi, Carl. 2015. *In Retrospect: From the Pill to the Pen*. London: Imperial College Press.

Do, Choon H. 2018. "A Short Story of Chemistry in South Korea." In Seth C. Rasmussen (ed.), *Igniting the Chemical Ring of Fire: Historical Evolution of the Chemical Communities of the Pacific Rim*. London: World Scientific.

Doogab, Yi. 2015. *The Recombinant University: Genetic Engineering and the Emergence of Stanford Biotechnology*. Chicago, IL: University of Chicago Press.

Dorgelès, Roland. 1919. *Les Croix de bois*. Paris: Albin Michel.

Dos Passos, John. 1920. *One Man's Initiation: 1917*. London: George Allen & Unwin.

Downie, David, and Jessica Templeton. 2014. "Persistent Organic Pollutants." In Paul G. Harris (ed.), *Routledge Handbook of Global Environmental Politics*. Abingdon: Routledge.

Dowty, Alan. 1989. *Closed Borders: The Contemporary Assault on Freedom of Movement*. New Haven, CT: Yale University Press.

Doyle, John M., Bretislav Friedrich, and Edwardas Narevicius. 2016. "Physics and Chemistry with Cold Molecules." Introduction to Special Issue. *ChemPhysChem*, 17: 3581–2.

Dunlap, Thomas R. 1981. *DDT: Scientists, Citizens and Public Policy*. Princeton, NJ: Princeton University Press, 1981.

Dyer, Davis, and David B. Sicilia. 1990. *Labors of a Modern Hercules: The Evolution of a Chemical Company*. Boston, MA: Harvard Business School Press.

Eckerman, Ingrid, and Tom Børsen. 2018. "Corporate and Governmental Responsibilities for Preventing Chemical Disasters: Lessons from Bhopal." *Hyle: International Journal for Philosophy of Chemistry*, 24: 29–53. Available online: http://www.hyle.org/journal/issues/24-1/eckerman.htm.

Ede, Andrew. 2002. "The Natural Defense of a Scientific People: The Public Debate Over Chemical Warfare in Post-WWI America." *Bulletin for the History of Chemistry*, 27: 128–35.

Ede, Andrew. 2007a. "Abraham Cressy Morrison in the Agora: Bringing Chemistry to the Public." In Joachim Schummer, Bernadette Bensaude-Vincent, and Brigitte Van Tiggelen (eds), *The Public Image of Chemistry*. Singapore: World Scientific.

Ede, Andrew. 2007b. *The Rise and Decline of Colloid Science in North America. 1900–1935*. Aldershot: Ashgate.

EIC: *Education in Chemistry*. 2018. Available online: https://eic.rsc.org/.

Emsley, John. 2015. *Chemistry at Home: Exploring the Ingredients in Everyday Products*. Cambridge: Royal Society of Chemistry.

Enos, John L. 1962. *Petroleum, Progress and Profits: A History of Process Innovation*. Cambridge, MA: MIT Press.

EPA (Environmental Protection Agency). 2008. *Ground Water Rule: A Quick Reference Guide*. Washington, DC: EPA. Available online: https://nepis.epa.gov/Exe/ZyPDF.cgi?Dockey=P100156H.txt.

EPA (Environmental Protection Agency). 2016. *The Frank R. Lautenberg Chemical Safety for the 21st Century Act*. Available online: https://www.epa.gov/assessing-and-managing-chemicals-under-tsca/frank-r-lautenberg-chemical-safety-21st-century-act.

Eriksen, Marcus, Laurent C.M. Lebreton, Henry S. Carson, Martin Thiel, Charles J. Moore, Jose C. Borerro, Francois Galgini, Peter G. Ryan, and Julia Reisser. 2014. "Plastic Pollution in the World's Oceans: More than 5 Trillion Plastic Pieces Weighing over 250,000 Tons Afloat at Sea." *PLoS ONE*, 9 (12): 1–15. Available online: https://doi.org/10.1371/journal.pone.0111913.

Ernst, Richard R. 2010. "Zurich's Contributions to 50 Years Development of Bruker." *Angewandte Chemie International Edition*, 49: 8310–15. Available online: http://www.ncbi.nlm.nih.gov.

Ertl, Gerard. 2008. "Reactions at Surfaces: From Atoms to Complexity (Nobel Lecture)." *Angewandte Chemie International Edition*, 47: 3524–35.

Ettre, Leslie S. 1977. "Gas Chromatography." In Herbert A. Laitinen and Galen W. Ewing (eds), *A History of Analytical Chemistry*. Washington, DC: Division of Analytical Chemistry of the American Chemical Society.

Ettre, Leslie S. 2001. "The Birth of Partition Chromatography." *LC-GC North America*, 19: 506–12.

Ettre, Leslie S. 2008. *Chapters in the Evolution of Chromatography*. London: Imperial College Press.

Ettre, Leslie S., and Albert Zlatkis (eds). 1979. *75 Years of Chromatography: A Historical Dialogue*. Amsterdam: Elsevier.

EuChemS (European Chemical Society). 2018. "What is EuChemS?" Available online: https://www.euchems.eu/about-us/what-is-euchems/.

European Commission. 2013. "Women in Science: Gender Equality is Not Yet Fulfilled." *Horizon: The EU Research & Innovation Magazine*, March 18. Available online: https://horizon-magazine.eu/article/women-science-gender-equality-not-yet-fulfilled.html

European Higher Education Area. 2018. "How does the Bologna Process Work?" Available online: http://www.ehea.info/pid34247/how-does-the-bologna-process-work.html.

Evert, Sarah. 2015. "A Brief History of Chemical War." *Distillations*, May 11. Available online: https://www.sciencehistory.org/distillations/a-brief-history-of-chemical-war

Ewald, Paul P. (ed.). 1962. *Fifty Years of X-Ray Diffraction: Dedicated to the International Union of Crystallography on the Occasion of the Commemorative Meeting in Munich, July 1962*. Utrecht: Oosthoek.

Fast, Kenneth V. 1963. "The Role of Laboratory Experiences in the CHEM Study Program." *School Science and Mathematics*, 63: 147–56.

Fauque, Danielle M. 2016. "Jean Gérard, Secretary General and Driving Force of the International Chemical Conferences between the Wars." In Masanori Kaji, Yasu Furukawa, Hiroaki Tanaka, and Yoshiyuki Kikuchi (eds), *International Workshop on the History of Chemistry 2015, Tokyo. Transformation of Chemistry from the 1920s to the 1960s: Proceedings*. Tokyo: Japanese Society for History of Chemistry. Available online: http://kagakushi.org/iwhc2015/proceedings.

Fauque, Danielle M., and Brigitte Van Tiggelen (eds). 2019. "Special IUPAC 100: A Glance at the Union's History." *Chemistry International*, 41(3): 1–59. Available online: https://www.degruyter.com/view/j/ci.2019.41.issue-3/issue-files/ci.2019.41.issue-3.xml.

Fenn, Monika. 2012. "Schulwesen (nach 1945)." In *Historisches Lexikon Bayerns*. Available online: http://www.historisches-lexikon-bayerns.de/Lexikon/Schulwesen_(nach_1945).

Fennell, Roger. 1994. *History of IUPAC, 1919–1987*. Oxford: Blackwell.

Ferber, Christian von. 1956. *Untersuchungen zur Lage der deutschen Hochschullehrer. 3: Die Entwicklung des Lehrkörpers der deutschen Universitäten und Hochschulen 1864–1954*. Göttingen: Vandenhoeck & Ruprecht.

Ferry, Georgina. 1998. *Dorothy Hodgkin: A Life*. London: Granta Books.

Ferry, Georgina, 2007. *Max Perutz and the Secret of Life*. London: Chatto and Windus.

Finlay, Mark R. 2003. "Old Efforts at New Uses: A Brief History of Chemurgy and the American Search for Biobased Materials." *Journal of Industrial Ecology*, 7: 33–46.

Finlay, Mark R. 2009. *Growing American Rubber: Strategic Plants and the Politics of National Security*. New Brunswick, NJ: Rutgers University Press

Fjelland, Ragnar. 2016. "When Laypeople are Right and Experts are Wrong: Lessons from Love Canal." *Hyle: International Journal for Philosophy of Chemistry*, 22: 105–25. Available online: http://www.hyle.org/journal/issues/22-1/fjelland.htm.

Fleck, Ludwik. 1979. *Genesis and Development of a Scientific Fact*. Chicago, IL: University of Chicago Press.

Fleming, James Rodger. 1998. *Historical Perspectives on Climate Change*. Oxford: Oxford University Press.

Forsberg, Aaron. 2000. *America and the Japanese Miracle: The Cold War Context of Japan's Postwar Economic Revival, 1950–1960*. Chapel Hill: University of North Carolina Press.

Francoeur, Eric. 1997. "The Forgotten Tool: The Design and Use of Molecular Models." *Social Studies of Science*, 27: 7–40.

Fredga, Arne. 1964. "1947 Nobel Prize in Chemistry Presentation Speech." In *Nobel Lectures, Chemistry 1942–1962*. Amsterdam: Elsevier. Available online: https://www.nobelprize.org/nobel_prizes/chemistry/laureates/1947/press.html.

Freeman, Ray, and Gareth A. Morris. 2015. "The Varian Story." *Journal of Magnetic Resonance*, 250: 80–4.

Freemantle, Michael. 2014. *The Chemists' War 1914–1918*. Cambridge: Royal Society of Chemistry.

Frenz, Bert. 1988. "Computers and Crystallography." *Computers in Physics*, 2: 42–8.

Friedman, Robert Marc. 2001. *The Politics of Excellence: Behind the Nobel Prize in Science*. New York: W.H. Freeman.

Friedrich, Bretislav. 2016. "How Did the Tree of Knowledge Get Its Blossom? The Rise of Physical and Theoretical Chemistry with an Eye on Berlin and Leipzig." *Angewandte Chemie International Edition*, 55: 5378–92.

Friedrich, Bretislav, Dieter Hoffmann, Jürgen Renn, Florian Schmaltz, and Martin Wolf (eds). 2017. *One Hundred Years of Chemical Warfare: Research, Deployment, Consequences*. Cham: Springer. Available on Open Access Online: https://link.springer.com/book/10.1007%2F978-3-319-51664-6.

Friedrich Verlag. 2018. *Naturwissenschaft im Unterricht Chemie*. Available online: https://www.friedrich-verlag.de/sekundarstufe/naturwissenschaften/chemie/unterricht-chemie/.

Frohlich, Xaq. 2017. "The informational turn in food politics: The US FDA's nutrition label as information infrastructure." *Social Studies of Science*, 47: 245–71.

Fung, Fun Man, Martin Putala, Petr Holzhauser, Ekasith Somsook, Cecilia Hernandez, and I-Jy Chang. 2018. "Celebrating the Golden Jubilee of the International Chemistry Olympiad: Back to Where It All Began." *Journal of Chemical Education*, 95: 193–6.

Furter, William F. (ed.). 1982. *A Century of Chemical Engineering*. New York: Plenum.

Furukawa, Yasu. 1998. *Inventing Polymer Science. Staudinger, Carothers, and the Emergence of Macromolecular Chemistry*. Philadelphia: University of Pennsylvania Press.

Furukawa, Yasu. 2016. "From Fuel Chemistry to Quantum Chemistry: Kenichi Fukui and the Rise of the Kyoto School." In Masanori Kaji, Yasu Furukawa, Hiroaki Tanaka, and Yoshiyuki Kikuchi (eds), *International Workshop on the History of Chemistry 2015, Tokyo. Transformation of Chemistry from the 1920s to the 1960s: Proceedings*. Tokyo: Japanese Society for History of Chemistry. Available online: http://kagakushi.org/iwhc2015/proceedings.

Furukawa, Yasu. 2018. "Gen-itsu Kita and the Kyoto School's Formation." In Seth C. Rasmussen (ed.), *Igniting the Chemical Ring of Fire: Historical Evolution of the Chemical Communities of the Pacific Rim*. London: World Scientific.

Gabilliet, Jean-Paul. 2010. *Of Comics and Men: A Cultural History of American Comic Books*. Jackson: University Press of Mississippi.

Galambos, Louis, Takashi Hikino, and Vera Zamagni (eds). 2007. *The Global Chemical Industry in the Age of the Petrochemical Revolution*. Cambridge: Cambridge University Press.

Galan, Leo de. 2003. Review of Peter J.T. Morris (ed.). 2002. *From Classical to Modern Chemistry: The Instrumental Revolution. Ambix*, 50: 234–6.

Galan, Leo de. 2012. "The Four Players in the Analytical Performance." *Journal of Analytical Atomic Spectrometry*, 27: 1173–6.

Galison, Peter (ed.). 1992. *Big Science: The Growth of Large Scale Research*. Stanford, CA: Stanford University Press.

Galison, Peter, and Andrew Warwick. 1998. "Introduction: Cultures of Theory." *Studies in History and Philosophy of Science*, 29B: 287–94.

Ganz, Cheryl R. 2008. *The 1933 Chicago World's Fair: A Century of Progress*. Urbana: University of Illinois Press.

García-Belmar, Antonio. 2006. The Didactic Uses of Experiment: Louis Jacques Thenard's Lectures at the Collège de France." In José R. Bertomeu-Sánchez and Agustí Nieto-Galan (eds), *Chemistry, Medicine and Crime: Mateu J.B. Orfila (1787–1853) and His Times*. Sagamore Beach, MA: Science History Publications.

García-Belmar, Antonio, and José R. Bertomeu-Sánchez. 1999. *Nombrar la materia. Una introducción histórica a la terminología química*. Barcelona: Serbal.

García-Belmar, Antonio, José R. Bertomeu-Sánchez, and Bernadette Bensaude-Vincent. 2005. "The Power of Didactic Writings: French Chemistry Textbooks of the Nineteenth Century." In David Kaiser (ed.), *Pedagogy and the Practice of Science: Historical and Contemporary Perspectives*. Cambridge, MA: MIT Press.

Garrigós-Oltra, Luis, Carles Millán-Verdú, and Georgina Blanes-Nadal. 2006. *El color líquido. Instrumentos y útiles de la colorimetría en el siglo XIX*. Alicante: Aguaclara.

Gaudillière, Jean-Paul, and Volker Hess (eds). 2013. *Ways of Regulating Drugs in the 19th and 20th Centuries*. Basingstoke: Palgrave Macmillan.

Gavroglu, Kostas (ed.). 2000. *Theoretical Chemistry in the Making*. Special Issue of *Studies in the History and Philosophy of Modern Physics*, 31: 429–609.

Gavroglu, Kostas, and Ana Simões. 2012. *Neither Physics nor Chemistry: A History of Quantum Chemistry*. Cambridge, MA: MIT Press.

Gay, Hannah. 2012. "Before and After *Silent Spring*: From Chemical Pesticides to Biological Control and Integrated Pest Management – Britain." *Ambix*, 59: 88–108.

Gedeon, Andras. 2006. *Science and Technology in Medicine: An Illustrated Account Based on Ninety-Nine Landmark Publications from Five Centuries*. New York: Springer.

Germany. Statistisches Reichsamt. 1907. *Statistik des Deutschen Reichs*, N.S., 221: *Zusammenfassende Übersichten für die Gewerbliche Betriebszählung von 1907*. Berlin: Verlag für Sozialpolitik, Wirtschaft und Statistik.

Germany. Statistisches Reichsamt. 1927. *Statistik des Deutschen Reichs*, N.S., 402 (1–2): *Volks-, Berufs- und Betriebszählung vom 16. Juni 1925*. Berlin: Verlag für Sozialpolitik, Wirtschaft und Statistik.

Germany. Statistisches Reichsamt. 1941–1944. *Statistik des Deutschen Reichs*: *Volks-, Berufs- und Betriebszählumg vom 17. Mai 1939*. Berlin: Verlag für Sozialpolitik, Wirtschaft und Statistik.

Gerontas, Apostolos. 2014. "Creating New Technologists of Research in the 1960s: The Case of the Reproduction of Automated Chromatography Specialists and Practitioners." *Science & Education*, 23: 1681–700.

Gesellschaft Deutscher Chemiker. 2010. *Chemiestudiengänge in Deutschland: Statistische Daten 2009*. Frankfurt am Main: Gesellschaft Deutscher Chemiker. Available online: https://www.chemanager-online.com/sites/chemanager-online.com/files/GDCh_Chemiestudienanfaenger_Statistik_2009.pdf.

Gesellschaft Deutscher Chemiker. 2016. *Chemiestudiengänge in Deutschland: Statistische Daten 2015*. Frankfurt am Main: Gesellschaft Deutscher Chemiker. Previously available online: https://www.gdch.de/fileadmin/downloads/Ausbildung_und_Karriere/Karriere/Statistik/GDCh_2015Statistikweb.pdf.

Gesellschaft Deutscher Chemiker. 2018. *Jahresbericht 2017*. Frankfurt am Main: Gesellschaft Deutscher Chemiker. Available online: https://www.gdch.de/fileadmin/downloads/Service_und_Informationen/Downloads/Jahresberichte/2017_Jahresbericht.pdf.

Gilmour, C. Stewart. 2004. *Fred Terman at Stanford: Building a Discipline, a University, and Silicon Valley*. Stanford, CA: Stanford University Press.

Gimbel, John. 1990. *Science, Technology and Reparations: Exploitation and Plunder in Postwar Germany*. Stanford, CA: Stanford University Press.

Glasstone, Samuel, Keith J. Laidler, and Henry Eyring. 1941. *The Theory of Rate Processes: The Kinetics of Chemical Reactions, Viscosity, Diffusion and Electrochemical Phenomena*. New York: McGraw-Hill.

Glazer, A. Michael. 2016. *Crystallography: A Very Short Introduction*. Oxford: Oxford University Press.
Glazer A. Michael, and Patience Thomson. 2015. *Crystal Clear: The Autobiographies of Sir Lawrence and Lady Bragg*. Oxford: Oxford University Press.
Gómez-Ibáñez, José D. (ed.). 1972. *Chemical Education; Invited Lectures Presented at the International Symposium on Chemical Education held at São Paulo, Brazil, 30 August–3 September 1971*. London: Butterworths.
Gordin, Michael D. 2015. *Scientific Babel*. Chicago, IL: University of Chicago Press.
Gordon, Neil E. 1924. "Editors' Outlook." *Journal of Chemical Education*, 1: 1.
Gortler, Leon. 1985. "The Physical Organic Community in the United States, 1925–1950: An Emerging Network." *Journal of Chemical Education*, 62: 753–7.
Gortler, Leon, and Stephen J. Weininger. 2017. "Private Philanthropy and Basic Research in Mid-Twentieth Century America: The Hickrill Chemical Research Foundation." *Ambix*, 64: 66–94.
Graham, Loren R. 1964. "A Soviet Marxist View of Structural Chemistry: The Theory of Resonance Controversy." *Isis*, 55: 20–31.
Grayson, Michael A. (ed.). 2002. *Measuring Mass: From Positive Rays to Proteins*. Philadelphia, PA: Chemical Heritage Press.
Grayson, Michael A. 2011. "John Bennett Fenn: A Curious Road to the Prize. *Journal of The American Society for Mass Spectrometry*, 22: 1301–8.
Greenaway, Frank, Robert G.W. Anderson, Susan E. Messham, Ann M. Newmark, and Derek A. Robinson. 1978. "The Chemical Industry." In Trevor I. Williams (ed.), *A History of Technology*, vol. VI, *The Twentieth Century, c.1900–1950*, part I. Oxford: Clarendon Press.
Greenberg, Edward, Christopher T. Hill, and David J. Newburger. 1979. *Regulation, Market Prices, and Process Innovation: The Case of the Ammonia Industry*. Boulder, CO: Westview Press.
Gribbin, John. 1993. *The Hole in the Sky: Man's Threat to the Ozone Layer*, rev. ed. New York: Bantam.
Guerrini, Anita. 2003. *Experimenting with Humans and Animals: From Galen to Animal Rights*. Baltimore, MD: Johns Hopkins University Press.
Haber, Fritz. 1924. *Fünf Vorträge aus den Jahren 1920–23*. Berlin: Springer.
Haber, Ludwig F. 1971. *The Chemical Industry, 1900–1930: International Growth and Technological Change*. Oxford: Clarendon Press.
Haber, Ludwig F. 1986. *The Poisonous Cloud: Chemical Warfare in the First World War*. Oxford: Clarendon Press.
Hager, Thomas. 1995. *Force of Nature: The Life of Linus Pauling*. New York: Simon and Schuster.
Hager, Thomas. 1998. *Linus Pauling and the Chemistry of Life*. Oxford: Oxford University Press.
Hall, Kersten T. 2014. *The Man in the Monkeynut Coat: William Astbury and the Forgotten Road to the Double-Helix*. Oxford: Oxford University Press.
Halliwell, H. Frank. 1970. "The Problem of Conflict between Specialism and Generalism in Chemical Education." In Dennis G. Chisman (ed.), *University Chemical Education: Proceedings of the International Symposium on University Chemical Education held in Frascati (Rome), Italy, 16–19 October 1969*. London: Butterworths.
Halton, Brian. 2018. "The Development of Chemistry in New Zealand." In Seth C. Rasmussen (ed.), *Igniting the Chemical Ring of Fire: Historical Evolution of the Chemical Communities of the Pacific Rim*. London: World Scientific.

Hamill, Sean D. 2008. "Unveiling a Museum, a Pennsylvania Town Remembers the Smog That Killed 20." *The New York Times* (November 1), A22.

Handler, Philip. 1970. *The Physical Sciences. Report of the National Science Board submitted to the Congress*. Washington, DC: National Science Foundation.

Harikumar, K.R., Iain R. McNab, John C. Polanyi, Amir Zabet-Khosousi, and Werner A. Hofer. 2011. "Imprinting Self-Assembled Patterns of Lines at a Semiconductor Surface, Using Heat, Light or Electrons." In John T. Yates and Charles T. Campbell (eds), *PNAS Surface Chemistry Special Issue "Frontiers in Surface Chemistry."* PNAS, 108: 950–5.

Harnack, Adolf von. 1961. "Aus einer Tischrede des Präsidenten v. Harnack auf der 12. Hauptversammlung, 16. Dezember 1926." In *50 Jahre Kaiser-Wilhelm-Gesellschaft und Max-Planck-Gesellschaft zur Förderung der Wissenschaften, 1911–1961: Beiträge und Dokumente*. Göttingen: Generalverwaltung der Max-Planck-Gesellschaft z.F.d.W.e.V.

Harré, Rom. 2013. "Preface." In Jean-Pierre Llored (ed.), *The Philosophy of Chemistry: Practices, Methodologies, and Concepts*. Cambridge: Cambridge University Press.

Harris, Richard, and Jeremy Paxman. 2002. *A Higher Form of Killing: The Secret Story of Gas and Germ Warfare*. London: Arrow.

Hayes, Peter. 1987. *Industry and Ideology: IG Farben in the Nazi Era*. Cambridge: Cambridge University Press.

Haynes, Roslynn D. 1994. *From Faust to Strangelove: Representations of Scientists in Western Literature*. Baltimore, MD: Johns Hopkins University Press.

Haynes, Roslynn D. 2007. "The Alchemists in Fiction: The Master Narrative." In Joachim Schummer, Bernadette Bensaude-Vincent, and Brigitte Van Tiggelen (eds), *The Public Image of Chemistry*. Singapore: World Scientific.

Health and Safety Executive. 1975. *The Flixborough Disaster: Report of the Court of Inquiry*. London: HMSO. Available online: http://www.hse.gov.uk/comah/sragtech/caseflixboroug74.htm.

Heartney, Eleanor. 2013. *Kenneth Snelson's Art and Ideas*. New York, Marlborough Gallery. Only available free online: http://kennethsnelson.net/KennethSnelson_Art_And_Ideas.pdf.

Heilbron, John L. 1974. *H.G.J. Moseley: The Life and Letters of an English Physicist, 1887–1915*. Berkeley: University of California Press.

Hentschel, Klaus. 2002. *Mapping the Spectrum: Techniques of Visual Representation in Research and Teaching*. Oxford: Oxford University Press.

Hentschel, Klaus. 2014. *Visual Cultures in Science and Technology: A Comparative History*. Oxford: Oxford University Press.

Herbert, Vernon, and Attilio Bisio. 1985. *Synthetic Rubber: A Project That Had to Succeed*. Westport, CT: Greenwood Press.

Hermes, Matthew E. 1996. *Enough for One Lifetime: Wallace Carothers, Inventor of Nylon*. Washington, DC: American Chemical Society.

Hepler-Smith, Evan. 2015. "'Just as the Structural Formula Does:' Names, Diagrams, and the Structure of Organic Chemistry at the 1892 Geneva Nomenclature Congress." *Ambix*, 62: 1–28.

Hepler-Smith, Evan. 2018. "'A Way of Thinking Backwards': Computing and Method in Synthetic Organic Chemistry." *Historical Studies in the Natural Sciences*, 48: 300–37.

Herschbach, Dudley. 2012. "Interview: Interleaved Excerpts from Interviews of Dudley Herschbach (DH) by John Rigden (JR) on May 21–22, 2003 and Bretislav Friedrich

(BF) on March 5–9, 2012." *Molecular Physics: An International Journal at the Interface between Chemistry and Physics*, 110: 1549–90.

Herschbach, Dudley. 2017. "Michael Polanyi: Patriarch of Chemical Dynamics and Tacit Knowing." *Angewandte Chemie International Edition*, 56: 3434–44.

Hinkle, Amber S., and Jody A. Kocsis (eds). 2005. *Successful Women in Chemistry: Corporate America's Contribution to Science*. Washington, DC: American Chemical Society.

Hirota, Noboru. 2016. *A History of Modern Chemistry*. Kyoto: Kyoto University Press and Melbourne: Trans Pacific Press, 2016. Originally published in Japanese by Kyoto University Press, 2013.

Hobsbawm, Eric. 1994. *Age of Extremes. A History of the World, 1914–1991*. London: Michael Joseph.

Hochheiser, Sheldon. 1986. *Rohm and Haas: History of a Chemical Company*. Philadelphia: University of Pennsylvania Press.

Hoffmann. Roald. 1988. "Under the Surface of the Chemical Article." *Angewandte Chemie. International Edition in English*, 27: 1593–602.

Hoffmann, Roald. 2012a. "What Might Philosophy of Science Look Like If Chemists Built It?" In Jeffrey Kovac and Michael Weisberg (eds), *Roald Hoffmann on the Philosophy, Art, and Science of Chemistry*. Oxford: Oxford University Press.

Hoffmann, Roald. 2012b. "Qualitative Thinking in the Age of Modern Computational Chemistry, or What Lionel Salem Knows." In Jeffrey Kovac and Michael Weisberg (eds), *Roald Hoffmann on the Philosophy, Art, and Science of Chemistry*. Oxford: Oxford University Press.

Hoffmann, Roald, Sason Shaik, and Philippe C. Hiberty. 2003. "A Conversation on VB vs MO Theory: A Never Ending Rivalry?" *Accounts of Chemical Research*, 36: 750–6.

Holmes, Frederic L. 1989. "The Complementarity of Teaching and Research in Liebig's Laboratory." *Osiris*, 5: 121–64.

Holmes, Frederic L. 1991. *Hans Krebs*, vol. 1, *Formation of a Scientific Life, 1900–1933*. New York: Oxford University Press.

Holmes, Frederic L. 1993. *Hans Krebs*, vol. 2, *Architect of Intermediary Metabolism, 1933–1937*. New York: Oxford University Press.

Holmes, Frederic L. 2004. *Investigative Pathways: Patterns and Stages in the Careers of Experimental Scientists*. New Haven, CT: Yale University Press.

Homburg, Ernst. 1999. "The Rise of Analytical Chemistry and its Consequences for the Development of the German Chemical Profession (1780–1860)." *Ambix*, 46: 2–32.

Homburg, Ernst. 2018. "Chemistry and Industry: A Tale of Two Moving Targets." *Isis*, 109: 565–76.

Homburg, Ernst, Anthony S. Travis, and Harm G. Schröter (eds). 1998. *The Chemical Industry in Europe, 1850–1914: Industrial Growth, Pollution, and Professionalization*. Dordrecht: Kluwer.

Homburg, Ernst, and Elisabeth Vaupel (eds). 2019. *Hazardous Chemicals. Agents of Risk and Change, 1800–2000*. New York: Berghahn.

Horrocks, Sally M. 2000. "A promising pioneer profession? Women in industrial chemistry in inter-war Britain." *British Journal for the History of Science*, 33: 351–67.

Horrocks, Sally M. 2011. "World War II, Post-war Reconstruction and British Women Chemists." *Ambix*, 58: 150–70.

Hounshell, David A., and John K. Smith, Jr. 1988. *Science and Corporate Strategy: Du Pont R&D, 1902–1980*. Cambridge: Cambridge University Press.

Hughes, Jeff. 2003. *The Manhattan Project: Big Science and the Atomic Bomb*. New York: Columbia University Press.

Hüntelmann, Axel C. 2011. *Paul Ehrlich. Leben, Forschung, Ökonomien, Netzwerke*. Göttingen: Wallstein.

Hunter, Graeme K. 2004. *Light is a Messenger: The Life and Science of William Lawrence Bragg*. Oxford: Oxford University Press.

Hutchinson, Eric. 1977. *The Department of Chemistry Stanford University, 1891–1976*. Stanford, CA: Department of Chemistry, Stanford University.

IG Farbenindustrie Proko-Büro. 1938. *Erzeugnisse unserer Arbeit*. Frankfurt am Main: IG Farbenindustrie.

Ihde, Aaron J. 1964. *The Development of Modern Chemistry*. New York: Harper and Row.

IUPAC (International Union of Pure and Applied Chemistry). 2018. "Strategic Plan." Available online: https://iupac.org/who-we-are/strategic-plan/.

Israel, John. 1999. *Lianda: A Chinese University in War and Revolution*. Stanford, CA: Stanford University Press.

Jackson, Catherine M. 2015. "The Wonderful Properties of Glass: Liebig's Kaliapparat and the Practice of Chemistry in Glass." *Isis*, 106: 43–69.

Jambeck, Jenna R., Roland Geyer, Chris Wilcox, Theodore R. Siegler, Miriam Perryman, Anthony Andrady, Ramani Narayan, and Kara Lavender Law. 2015. "Plastic Waste Inputs from Land into the Ocean." *Science*, 347(6223): 768–71. Available online: www.sciencemag.org.

James, Frank A.J.L. 1983. "The Establishment of Spectro-chemical Analysis as a Practical Method of Qualitative Analysis, 1854–1861." *Ambix*, 30: 30–53.

James, Frank A.J.L. 1988. "The Practical Problems of 'New' Experimental Science: Spectro-Chemistry and the Search for Hitherto Unknown Chemical Elements in Britain, 1860–1869." *British Journal for the History of Science*, 21: 281–94.

Japan. Ministry of Education, Culture, Sports, Science and Technology. 1960–2017. "Report on School Basic Survey" [in Japanese]). Tokyo: Ministry of Education, Culture, Sports, Science and Technology.

Japan. Statistics Bureau, Ministry of Internal Affairs and Communications. 1963–2017. "Survey of Research and Development" [in Japanese]. Tokyo: Statistics Bureau, Ministry of Internal Affairs and Communications.

Jas, Nathalie. 2014. "Gouverner les substances chimiques dangereuses dans les espaces Internationaux." In Dominique Pestre (ed.), *Le Gouvernement des Technosciences: Gouverner le progrès et ses dégâts depuis 1945*. Paris: Découverte.

Jenkin, John. 2008. *William and Lawrence Bragg, Father and Son: The Most Extraordinary Collaboration in Science*. Oxford: Oxford University Press.

Johnson, Jeffrey Allan. 1998a. "German Women in Chemistry, 1895–1925 (Part 1)." *NTM: International Journal of History and Ethics of Natural Sciences, Technology and Medicine*, N.S. 6: 1–21.

Johnson, Jeffrey Allan. 1998b. "German Women in Chemistry, 1925–45." *NTM: International Journal of History and Ethics of Natural Sciences, Technology and Medicine*, N.S. 6: 65–90.

Johnson, Jeffrey Allan. 2000. "The Academic–Industrial Symbiosis in German Chemical Research, 1905–1939." In John E. Lesch (ed.), *The German Chemical Industry in the Twentieth Century*. Dordrecht: Kluwer.

Johnson, Jeffrey Allan. 2008. "Germany: Discipline – Industry – Profession. German Chemical Organizations, 1867–1914." In Anita Kildebæk Nielsen and Soná Strbánová (eds), *Creating Networks in Chemistry: The Founding and Early History of Chemical Societies in Europe*. Cambridge: Royal Society of Chemistry.

Johnson, Jeffrey Allan. 2010. *The Kaiser's Chemists: Science and Modernization in Imperial Germany*. Chapel Hill: University of North Carolina Press.

Johnson, Jeffrey Allan. 2013. "The Case of the Missing German Quantum Chemists: On Molecular Models, Mobilization, and the Paradoxes of Modernizing Chemistry in Nazi Germany." *Historical Studies in the Natural Sciences*, 43: 391–452.

Johnson, Jeffrey Allan. 2017. "Between Nationalism and Internationalism: The German Chemical Society In Comparative Perspective, 1867–1945." *Angewandte Chemie International Edition*, 56: 11044–58.

Johnson, Louise N. 1998c. "The Early History of Lysozyme." *Nature Structural Biology*, 5: 942–4.

Johnston, Hamish. 2010. "Quantum Computer Takes on Quantum Chemistry." *Physics World*, January 19. Available online: http://physicsworld.com/cws/article/news/2010/jan/19/quantum-computer-takes-on-quantum-chemistry.

Johnstone, Alex. H. 1993. "The Development of Chemistry Teaching: A Changing Response to Changing Demand." *Journal of Chemical Education*, 70: 701–5.

Jones, James H. 1993. *Bad Blood. The Tuskegee Syphilis Experiment*. New and expanded edition. New York: Free Press.

Jones, Jerry L., and Konrad T. Semrau. 1984. "Wood Hydrolysis for Ethanol Production – Previous Experience and the Economics of Selected Processes." *Biomass*, 5: 109–35.

Jonker, Joost, and Jan L. van Zanden. 2007. *The History of Royal Dutch Shell*, vol. 1, *From Challenger to Joint Industry Leader, 1890–1939*. Oxford: Oxford University Press.

Jump, Henry D., and John M. Cruice. 1904. "Chronic Poisoning from Bisulphide of Carbon." *University of Pennsylvania Medical Bulletin*, 17: 193–6. Available online: https://babel.hathitrust.org/cgi/pt?id=mdp.39015018403884&view=1up&seq=201.

Kaiser, David. 2005. *Drawing Theories Apart: The Dispersion of Feynman Diagrams in Postwar Physics*. Chicago, IL: University of Chicago Press.

Kaiser, Georg. 1920. *Gas: Zweiter Teil; Schauspiel in drei Akten*. Potsdam: Kiepenheuer.

Kanigel, Robert. 2007. *Faux Real: Genuine Leather and 200 Years of Inspired Fakes*. Washington, DC: Joseph Henry Press.

Karachalios, Andreas. 2001. *I chimici di fronte al fascismo: il caso di Giovanni Battista Bonino (1899–1985)*. Palermo: Istituto Gramsci siciliano.

Karlsch, Rainer. 2000. "Capacity Losses, Reconstruction, and Unfinished Modernization: The Chemical Industry in the Soviet Zone of Occupation (SBZ)/GDR, 1945–1965." In John E. Lesch (ed.), *The German Chemical Industry in the Twentieth Century*. Dordrecht: Kluwer.

Kaufman, Morris. 1969. *The History of PVC: The Chemistry and Industrial Production of Polyvinyl Chloride*. London: Maclaren and Sons.

Kaufmann, Doris. 2017. "'Gas, Gas, Gaas!' The Poison Gas War in the Literature and Visual Arts of Interwar Europe." In Bretislav Friedrich, Dieter Hoffmann, Jürgen Renn, Florian Schmaltz, and Martin Wolf (eds), *One Hundred Years of Chemical Warfare: Research, Deployment, Consequences*. Cham: Springer. Available on Open Access Online: https://link.springer.com/book/10.1007%2F978-3-319-51664-6.

Kiechle, Melanie A. 2017. *Smell Detectives: An Olfactory History of Nineteenth-Century Urban America*. Seattle, WA: University of Washington Press.

Kiessling, Renate. 2018. "Chemische Gesellschaft: Mitgliederentwicklung." [Table of membership of the Chemical Society of the German Democratic Republic (ChGDDR), 1953–89, compiled from annual reports]. Personal communication to Jeffrey A. Johnson, August 13, 2018).

Kikuchi, Yoshiyuki. 2013. *Anglo-American Connections in Japanese Chemistry: The Lab as Contact Zone*. Basingstoke: Palgrave Macmillan.

Kikuchi, Yoshiyuki. 2018. "International Relations of the Japanese Chemical Community." In Seth C. Rasmussen (ed.), *Igniting the Chemical Ring of Fire: Historical Evolution of the Chemical Communities of the Pacific Rim*. London: World Scientific.

King, Donna. 2012. "New Perspectives on Context-Based Chemistry Education: Using a Dialectical Sociocultural Approach to View Teaching and Learning." *Studies in Science Education*, 48: 51–87.

Kinkela, David. 2011. *DDT and the American Century: Global Health, Environmental Politics, and the Pesticide That Changed the World*. Chapel Hill: University of North Carolina Press.

Klein, Ursula. 2001. "Paper Tools in Experimental Cultures." *Studies in History and Philosophy of Science*, 32: 265–302.

Klein, Ursula. 2003. *Experiments, Models, Paper Tools: Cultures of Organic Chemistry in the Nineteenth Century*. Stanford, CA: Stanford University Press.

Klein, Ursula. 2016. *Nützliches Wissen: Die Erfindung der Technikwissenschaften*. Göttingen: Wallstein.

Klein, Ursula, and Wolfgang Lefèvre. 2007. *Materials in Eighteenth-Century Science: A Historical Ontology*. Cambridge, MA: MIT Press.

Koch, Egmont R., and Fritz Vahrenholt. 1978. *Seveso ist überall. Die tödlichen Risiken der Chemie*. Cologne: Kiepenheuer und Witsch.

Koch, Wolfram. 2011. "First female GDCh President: Barbara Albert." *EuCheMS Newsletter*, November. Available online: https://www.euchems.eu/wp-content/uploads/Euchems11_2011.pdf.

Kohler, Robert E. 1971. "The Origin of G.N. Lewis's Theory of the Shared Pair Bond." *Historical Studies in the Physical Sciences* 3: 343–76.

Kohler, Robert E. 1975. The Lewis–Langmuir Theory of Valence and the Chemical Community, 1920–1928." *Historical Studies in the Physical Sciences*, 6: 431–68.

Kohler, Robert E. 1982. *From Medical Chemistry to Biochemistry: The Making of a Biomedical Discipline*. Cambridge: Cambridge University Press.

Kohn, Walther. 1999. "An Essay on Condensed Matter Physics in the Twentieth Century." *Reviews of Modern Physics*, 71: S59–77.

Kornhauser, Aleksandra, C.N.R. Rao, and David J. Waddington. 1980. *Chemical Education in the Seventies*. Oxford: Pergamon Press.

Kragh, Helge. 2000. "The Chemistry of the Universe: Historical Roots of Modern Cosmochemistry." *Annals of Science*, 57: 353–68.

Kragh, Helge. 2001. "From Geochemistry to Cosmochemistry: The Origin of a Scientific Discipline, 1915–1955." In Carsten Reinhardt (ed.), *Chemical Sciences in the 20th Century. Bridging Boundaries*. Weinheim: Wiley-VCH.

Krammer, Arnold. 1978. "Fueling the Third Reich." *Technology and Culture*, 19: 394–422.

Krammer, Arnold. 1981. "Technology Transfer as War Booty: The U.S. Technical Mission in Europe, 1945." *Technology and Culture*, 22: 68–103.

Kraus, Karl. 1922. *Die letzten Tage der Menschheit*. Vienna and Leipzig: Die Fackel

Krausnick, Michail.1975. *Die Paracana-Affäre*. Würzburg: Arena.
Krausnick, Michail. 1980. *Im Schatten der Wolke*. München and Wien: Pestum.
Krausnick, Michail. 1982. *Lautlos kommt der Tod*. Munich: F. Schneider.
Krige, John. 2006. *American Hegemony and the Postwar Reconstruction of Science in Europe*. Cambridge, MA: MIT Press.
Kuck, Valerie J. 2006. "Women in Academe: An Examination of the Education and Hiring of Chemists." In Cecilia H. Marzabadi, Valerie J. Kuck, Susan A. Nolan, and Janine P. Buckner (eds), *Are Women Achieving Equity in Chemistry? Dissolving Disparity and Catalyzing Change*. Washington, DC: American Chemical Society.
Labinger, Jay. 2011. "Chemistry." In Bruce Clarke and Manuela Rossini (eds), *The Routledge Companion to Literature and Science*. London: Routledge.
Labinger, Jay A. 2013. *Up From Generality: How Inorganic Chemistry Finally Became a Respectable Field*. Heidelberg: Springer.
Laeter, John R. de, and Mark D. Kurz. 2006. "Alfred Nier and the Sector Field Mass Spectrometer." *Journal of Mass Spectrometry*, 41: 847–54.
Laidler, Keith. 1993. *The World of Physical Chemistry*. Oxford: Oxford University Press.
Langston, Nancy. 2010. *Toxic Bodies: Hormone Disruptors and the Legacy of DES*. New Haven, CT: Yale University Press.
Law, John. 1973. "The Development of Specialties in Science: The Case of X-Ray Protein Crystallography." *Science Studies*, 3: 275–303.
Lear, Linda. 1997. *Rachel Carson: Witness for Nature*. New York: Henry Holt.
Lenoir, Timothy. 1988. "A Magic Bullet: Research for Profit and the Growth of Knowledge in Germany Around 1900." *Minerva*, 26: 66–88.
Lenoir, Timothy, and Christophe Lécuyer. 1995. "Instrument Makers and Discipline Builders: The Case of Nuclear Magnetic Resonance." *Perspectives on Science*, 3: 276–345.
Lesch, John E. (ed.). 2000. *The German Chemical Industry in the Twentieth Century*. Dordrecht: Kluwer.
Lesch, John E. 2007. *The First Miracle Drugs: How the Sulfa Drugs Transformed Medicine*. New York: Oxford University Press.
Leslie, Stuart W. 1994. *The Cold War and American Science: The Military–Industrial–Academic Complex at MIT and Stanford*. New York: Columbia University Press.
Levere, Trevor H. 1971. *Affinity and Matter: Elements of Chemical Philosophy 1800–1865*. Oxford: Clarendon Press.
Levere, Trevor H. 2001. *Transforming Matter: A History of Chemistry from Alchemy to the Buckyball*. Baltimore, MD: Johns Hopkins University Press.
Lewis, Jack. 1985. "The Birth of the EPA." *EPA Journal* (November). Available online: https://archive.epa.gov/epa/aboutepa/birth-epa.html.
Lodge, Timothy. 2017. "The Promise of Polymers." *Physics Today*, 70(12): 10–12.
Lönngren, Rune. 1992. *International Approaches to Chemicals Control: a Historical Overview*. Stockholm: Kemi.
Löschnigg, Martin. 1994. *Der Erste Weltkrieg in deutscher und englischer Dichtung*, Heidelberg: Winter.
Lundgreen, Peter. 2000. "Schule im 20. Jahrhundert. Institutionelle Differenzierung und expansive Bildungsbeteiligung." In Dietrich Benner and Heinz-Elmar Tenorth (eds), *Bildungsprozesse und Erziehungsverhältnisse im 20. Jahrhundert*. Weinheim: Beltz.
Lütke, Gerhard (ed.). 1926a. "Charkhov." *Minerva: Jahrbuch der gelehrten Welt*, 28: 450–1.
Lütke, Gerhard (ed.). 1926b. "Leningrad." *Minerva: Jahrbuch der gelehrten Welt*, 28: 1000–27.

Lütke, Gerhard (ed.). 1926c. "Moskva." *Minerva: Jahrbuch der gelehrten Welt*, 28: 1328–79.

Maestrutti, Marina. 2011. *Imaginaires des nanotechnologies. Mythes et fictions de l'infiniment petit*. Paris: Vuibert.

MacLeod, Roy, and Jeffrey Allan Johnson (eds). 2006. *Frontline and Factory: Comparative Perspectives on the Chemical Industry at War, 1914–1924*. Dordrecht: Springer.

Macrotrends. 2018. "Crude Oil Prices – 70 Year Historical Chart." Available online: http://www.macrotrends.net/1369/crude-oil-price-history-chart.

Mai Pham, Thi Ngoc, Thi Anh Huong Nguyen, Tien Duc Pham, Quoc Anh Hoang, and Thi Thao Ta. 2018. "History of Vietnamese Chemistry from Decolonization to the 21st Century." In Seth C. Rasmussen (ed.), *Igniting the Chemical Ring of Fire: Historical Evolution of the Chemical Communities of the Pacific Rim*. London: World Scientific.

Maier, Helmut. 2007. *Forschung als Waffe. Rüstungsforschung in der Kaiser-Wilhelm-Gesellschaft und das Kaiser-Wilhelm-Institut für Metallforschung 1900–1945/48*. Göttingen: Wallstein Verlag.

Maier, Helmut. 2015. *Chemiker im "Dritten Reich": Die Deutsche Chemische Gesellschaft und der Verein Deutscher Chemiker im NS-Herrschaftsapparat*. Weinheim: Wiley-VCH.

Mainz, Vera V. 2017. "List of Chinese Students Receiving a Ph.D. in Chemistry between 1905 and 1964." University of Illinois at Urbana-Champaign. Available online: https://doi.org/10.13012/B2IDB-0064468_V2.

Mainz, Vera V. 2018. "History of the Modern Chemistry Doctoral Program in Mainland China." In Seth C. Rasmussen (ed.), *Igniting the Chemical Ring of Fire: Historical Evolution of the Chemical Communities of the Pacific Rim*. London: World Scientific.

Manafu, Alexandru. 2013. "Concepts of Emergence in Chemistry." In Jean-Pierre Llored (ed.), *The Philosophy of Chemistry: Practices, Methodologies, and Concepts*. Cambridge: Cambridge University Press.

Marcovich, Anne, and Terry Shinn. 2014. *Toward a New Dimension: Exploring the Nanoscale*. Oxford: Oxford University Press.

Markowitz, Gerald, and David Rosner. 2002. *Deceit and Denial. The Deadly Politics of Industrial Pollution*. Berkeley: University of California Press.

Markowitz, Gerald, and David Rosner. 2013. *Lead Wars: The Politics of Science and the Fate of America's Children*. Berkeley: University of California Press.

Marks, Lara V. 2001. *Sexual Chemistry: A History of the Contraceptive Pill*. New Haven, CT: Yale University Press.

Marks, Lara V. 2015. *The Lock and Key of Medicine: Monoclonal Antibodies and the Transformation of Healthcare*. New Haven, CT: Yale University Press.

Marschall, Luitgard. 2000. *Im Schatten der chemischen Synthese: Industrielle Biotechnologie in Deutschland (1900–1970)*. Frankfurt: Campus.

Marsland, Elizabeth A. 1991. *The Nation's Cause: French, British and German Poetry of the First World War*. London: Routledge.

Martin, Joseph D. 2015. "Fundamental Disputations: The Philosophical Debates That Governed American Physics, 1939–1993." *Historical Studies in the Natural Sciences*, 45: 703–57.

Martin, Joseph D., and Michel Janssen. 2015. "Beyond the Crystal Maze: Twentieth-Century Physics from the Vantage Point of Solid State Physics." *Historical Studies in the Natural Sciences*, 45: 631–40.

Martini, Edwin A. 2012. *Agent Orange: History, Science, and the Politics of Uncertainty*. Amherst: University of Massachusetts Press.

Mauskopf, Seymour (ed.). 1993. *Chemical Sciences in the Modern World*. Philadelphia: University of Pennsylvania Press.
McDonald, Patrick D. 2008. "Waters Corporation: Fifty Years of Innovation in Analysis and Purification." *Chemical Heritage*, 26(2): 32–7.
McGrayne, Sharon B. 2001. *Prometheans in the Lab: Chemistry and the Making of the Modern World*. New York: McGraw-Hill.
McGucken, William. 1969. *Nineteenth-Century Spectroscopy: Development of the Understanding of Spectra, 1802–1897*. Baltimore, MD: Johns Hopkins Press.
McMillan, Frank M. 1979. *The Chain Straighteners. Fruitful Innovation: The Discovery of Linear and Stereoregular Synthetic Polymers*. London: Macmillan.
Meadows, Donella H.; Dennis L. Meadows, Jørgen Randers, and William W. Behrens III. 1972. *Limits to Growth: A Report for the Club of Rome's Project on the Predicament of Mankind*. Washington, DC: Universe Books.
Meyerson, Seymour. 1986. "Reminiscences of the Early Days of Mass Spectrometry in the Petroleum Industry." *Journal of Mass Spectrometry*, 21: 197–208.
Miller, Jon D., Rafael Pardo, and Fujio Niwa. 1997. *Public Perceptions of Science and Technology: a Comparative Study of the European Union, the United States, Japan, and Canada*. Madrid: BBV Foundation.
Miller, Samuel A. 1965 *Acetylene: Its Properties, Manufacture and Uses*, vol. 1. London: Benn.
Mody, Cyrus C.M. 2011. *Instrumental Community. Probe Microscopy and the Path to Nanotechnology*. Cambridge, MA: MIT Press.
Molony, Barbara. 1990. *Technology and Investment: The Prewar Japanese Chemical Industry*. Cambridge, MA: Harvard University Press.
Morange, Michel. 1998. *A History of Molecular Biology*. Cambridge, MA: Harvard University Press.
Morrell, Jack. 1993. "W. H. Perkin, Jr., at Manchester and Oxford: From Irwell to Isis." *Osiris*, 2nd Series, 8: 104–26.
Morris, Peter J.T. 1983. "The Industrial History of Acetylene: The Rise and Fall of a Chemical Feedstock." *Chemistry and Industry*, 120: 710–15.
Morris, Peter J.T. 1990. Review of Michael Crow, Barry Bozeman, Walter Meyer, and Ralph Shangraw, Jr. 1988. *Synthetic Fuel Technology Development in the United States: A Retrospective Assessment*. *Science*, 247(4950): 1593–4.
Morris, Peter J.T. 1994. "Synthetic Rubber: Autarky and War." In Peter J.T. Morris and Susan T.I. Mossman, *The Development of Plastics*. Cambridge: Royal Society of Chemistry.
Morris, Peter J.T. 1998. "Ambros, Reppe, and the Emergence of Heavy Organic Chemicals in Germany, 1925–1945." In Anthony S. Travis et al. (eds), *Determinants in the Evolution of the European Chemical Industry, 1900–1939*. Dordrecht: Kluwer.
Morris, Peter J.T. (ed.). 2002a. *From Classical to Modern Chemistry: The Instrumental Revolution*. Cambridge: Royal Society of Chemistry.
Morris, Peter. J.T. 2002b. "'Parts per trillion is a Fairy Tale': The Development of the Electron Capture Detector and Its Impact on the Monitoring of DDT." In Peter J.T. Morris (ed.), *From Classical to Modern Chemistry: The Instrumental Revolution*. Cambridge: Royal Society of Chemistry.
Morris, Peter J.T. 2004. "Abney, Sir William de Wiveleslie (1843–1920), civil servant and photographic scientist." *Oxford Dictionary of National Biography*. Oxford: Oxford University Press. Available online: http://www.oxforddnb.com/view/10.1093/ref:odnb/9780198614128.001.0001/odnb-9780198614128-e-30324.

Morris, Peter J.T. 2005. "Reppe Chemistry." In Colin Hempstead (ed.), *Encyclopedia of 20th-Century Technology*, vol. 2. M–Z. New York: Routledge.
Morris, Peter J.T. 2008. "Chemistry in the 21st Century: Death or Transformation?" In José R. Bertomeu-Sánchez, Duncan Thorburn Burns, and Brigitte van Tiggelen (eds), *Neighbours and Territories: The Evolving Identity of Chemistry*. Louvain-la-neuve: Mémosciences.
Morris, Peter J.T. 2009. "Regional Styles in Pesticide Analysis: Coulson, Lovelock and the Detection of Organochlorine Insecticides." In Peter J.T. Morris and Klaus Staubermann (eds), *Illuminating Instruments*. Washington, DC: Smithsonian Institution Press.
Morris, Peter J.T. 2015. *The Matter Factory: A History of the Chemistry Laboratory*. London: Reaktion Books.
Morris, Peter J.T. 2019. "A Tale of Two Nations: DDT in the USA and the UK." In Elisabeth Vaupel and Ernst Homburg (eds), *Hazardous Chemicals: Agents of Risk and Change (1800–2000)*. New York: Berghahn.
Morris, Peter J.T., and Anthony S. Travis. 1992. "A History of the International Dyestuff Industry." *American Dyestuff Reporter*, 81(November): 59–100.
Morris, Peter J.T., and Anthony S. Travis. 1997. "The Role of Physical Instrumentation in Structural Organic Chemistry in the Twentieth Century." In John Krige and Dominique Pestre (eds), *Science in the Twentieth Century*. Reading: Harwood.
Morris, Peter J.T., Anthony S. Travis, and Carsten Reinhardt. 2001. "Research Fields and Boundaries in Twentieth-Century Organic Chemistry." In Carsten Reinhardt (ed.), *Chemical Sciences in the 20th Century. Bridging Boundaries*. Weinheim: Wiley-VCH.
Morrison, Wayne M. 2018. "China's Economic Rise: History, Trends, Challenges, and Implications for the United States." Congressional Research Service 7-5700, RL33534. February 5. Available online: https://fas.org/sgp/crs/row/RL33534.pdf.
Mossman, Susan T.I., and Peter J.T. Morris. 1994. *The Development of Plastics*. Cambridge: Royal Society of Chemistry.
Muir, M.M. Pattison. 1907. *History of Chemical Theories and Laws*, 1st ed. New York: John Wiley and Sons.
Mukharji, Projit Bihari. 2016. "Parachemistries: Colonial Chemopolitics in a Zone of Contest." *History of Science*, 54, 362–82.
Münkler, Herfried. 2013. *Der Grosse Krieg: Die Welt 1914 bis 1918*. Berlin: Rowohlt.
Nakayama, Shigeru, Kunio Gotō, and Hitoshi Yoshioka (eds). 2001. *A Social History of Science and Technology in Contemporary Japan*, vol. 1, *The Occupation Period, 1945–1952*. Melbourne: Trans Pacific Press.
National Research Council. 1965. *Chemistry: Opportunities and Needs*. Washington, DC: National Academies Press.
National Research Council. 1987. *Opportunities in Chemistry: Today and Tomorrow*. Washington, DC: National Academies Press.
Ndiaye, Pap A. 2007. *Nylon and Bombs: DuPont and the March of Modern America*. Baltimore, MD: Johns Hopkins University Press.
Nelson, Donna J., and Diana C. Rogers. 2005. *A National Analysis of Diversity in Science and Engineering Faculties at Research Universities*. Available online: http://users.nber.org/~sewp/events/2005.01.14/Bios+Links/Krieger-rec4-Nelson+Rogers_Report.pdf.
Neswald, Elizabeth, David F. Smith, and Ulrike Thoms (eds). 2017. *Setting Nutritional Standards: Theory, Policies, Practices*. Rochester, NY: University of Rochester Press.

Newman, Richard S. 2016. *Love Canal: A Toxic History from Colonial Times to the Present*. New York: Oxford University Press.

New York State. 1981. *Love Canal. A Special Report to the Governor and Legislature*. Available online: https://www.health.ny.gov/environmental/investigations/love_canal/lcreport.htm.

New York Times. 1988. "Global Warming Has Begun, Expert Tells Senate" (June 24), 1, col. 1.

Nieto-Galan, Agustí. 2019. *The Politics of Chemistry: Science and Power in Twentieth-Century Spain*. Cambridge: Cambridge University Press.

Nobelprize.org. Available online: https://www.nobelprize.org/prizes/lists/all-nobel-prizes/.

NSF-NCSES (National Science Foundation. National Center for Science and Engineering Statistics). 1996. *Selected Data on Science and Engineering Doctorate Awards: 1995*. NSF 96-303 Arlington, VA: National Science Foundation. Available online: https://wayback.archive-it.org/5902/20160210155930/http://www.nsf.gov/statistics/s4095/.

NSF-NCSES (National Science Foundation. National Center for Science and Engineering Statistics). 2006. *S&E Doctorate Awards: 2005*. NSF 07-305. Arlington, VA: National Science Foundation. Available online: https://wayback.archive-it.org/5902/20160210153416/http://www.nsf.gov/statistics/nsf07305/.

NSF-NCSES (National Science Foundation. National Center for Science and Engineering Statistics). 2011. *Doctorate Recipients from United States Universities: 2010*. NSF 12-305. Arlington, VA: National Science Foundation. Available online: https://nsf.gov/statistics/archive-goodbye.cfm?p=/statistics/sed/2010/SED_2010.zip.

NSF-NCSES (National Science Foundation. National Center for Science and Engineering Statistics). 2017. *Doctorate Recipients from U.S. Universities: 2015*. NSF 17-306. Arlington, VA: National Science Foundation. Available online: https://nsf.gov/statistics/2017/nsf17306/data.cfm.

Nye, Mary Jo. 1993. *From Chemical Philosophy to Theoretical Chemistry: Dynamics of Matter and Dynamics of Disciplines 1800–1950*. Berkeley: University of California Press.

Nye, Mary Jo. 1996. *Before Big Science: The Pursuit of Modern Chemistry and Physics, 1800–1940*. Cambridge, MA: Harvard University Press.

Nye, Mary Jo. 2000a. "Physical and Biological Modes of Thought in the Chemistry of Linus Pauling." *Studies in History and Philosophy of Modern Physics*, 31B: 475–92.

Nye, Mary Jo. 2000b. "From Student to Teacher: Linus Pauling and the Reformulation of the Principles of Chemistry in the 1930s." In Anders Lundgren and Bernadette Bensaude-Vincent (eds), *Communicating Chemistry: Textbooks and their Audiences, 1789–1939*. Canton, MA: Science History Publications.

Nye, Mary Jo. 2001. "Paper Tools and Molecular Architecture in the Chemistry of Linus Pauling." In Ursula Klein (ed.), *Tools and Modes of Representation in the Laboratory Sciences*. Dordrecht: Kluwer.

Nye, Mary Jo. 2003. "Elements, Chemical." In John L. Heilbron (ed.), *The Oxford Companion to the History of Modern Science*. Oxford: Oxford University Press.

Nye, Mary Jo. 2012. *Michael Polanyi and His Generation: Origins of the Social Construction of Science*. Chicago, IL: University of Chicago Press.

Nye, Mary Jo. 2014. "Reaction Intermediates and Transition States: States of Matter or States of Mind." In Ursula Klein and Carsten Reinhardt (eds), *Objects of Chemical Inquiry*. Sagamore Beach, MA: Science History Publications.

Nye, Mary Jo. 2016. "The Republic vs. The Collective: Two Histories of Collaboration and Competition in Modern Science." *NTM*, 24: 169–94.

Nye, Mary Jo. 2018. "Boundaries, Transformations, Historiography: Physics in Chemistry from the 1920s to the 1960s." *Isis*, 109: 587–96.

O'Brien, Flann. 1964. *Dalkey Archive*. London: MacGibbon & Kee.

Olby, Robert C. 1974. *The Path to the Double Helix: The Discovery of DNA*. London: Macmillan.

Olby, Robert C. 1985. "The 'Mad Pursuit': X-Ray Crystallographers' Search for the Structure of Haemoglobin." *History and Philosophy of the Life Sciences*, 7: 171–93.

Olesko, Kathryn M. 1991. *Physics as a Calling: Discipline and Practice in the Königsberg Seminar for Physics*. Ithaca, NY: Cornell University Press.

Onion, Rebecca. 2016. *Innocent Experiments: Childhood and the Culture of Popular Science in the United States*. Chapel Hill: University of North Carolina Press.

Oreskes, Naomi, and Erik M. Conway. 2010. *Merchants of Doubt: How a Handful of Scientists Obscured the Truth on Issues from Tobacco Smoking to Global Warming*. London: Bloomsbury.

Orna, Mary Virginia (ed.). 2015. *Sputnik to Smartphones: A Half-Century of Chemistry Education*. Washington, DC: American Chemical Society.

Osterath, Brigitte. 2017. "Chemiegemeinschaft hinter der Mauer." *Nachrichten aus der Chemie*, 65: 1019–23.

Owen, Claude R. 1989. "All Quiet on the Western Front: Sixty Years Later." *Krieg und Literatur/War and Literature*, 1: 41–48.

Park Buhm Soon. 1999. "Chemical Translators: Pauling, Wheland and Their Strategies for Teaching the Theory of Resonance." *British Journal for the History of Science*, 32: 21–46.

Parker, Laura. 2018. "Plastics." *National Geographic*, June: 40–69.

Patterson, Gary. 2018. "A History of the Korean Chemical Society." In Seth C. Rasmussen (ed.), *Igniting the Chemical Ring of Fire: Historical Evolution of the Chemical Communities of the Pacific Rim*. London: World Scientific.

Pestre, Dominique (ed.). 1997. "Science, Medicine, and Industry: The Curie and Joliot-Curie Laboratories." Special issue of *History and Technology*, 13(4): 241–343.

Petri, Rolf. 1998. "Technical Change in the Italian Chemical Industry: Markets, Firms and State Intervention." In Anthony S. Travis, Harm G. Schröter, Ernst Homburg, and Peter J.T. Morris (eds), *Determinants in the Evolution of the European Chemical Industry, 1900–1939*. Dordrecht: Kluwer.

Pilcher, Richard B. 1917. "Chemistry in Wartime." *The Journal of Industrial and Engineering Chemistry*, 9: 411.

Pine, Lisa. 2010. *Education in Nazi Germany*. Oxford: Berg.

Pohl, Frederik and Cyril M. Kornbluth, 1953. *The Space Merchants*. New York: Ballantine.

Pohl, W. Gerhard, and Bernd B. Baumgartinger. 2005. *Links oder Rechts?: Wie Naturstoffe die Polarisationsebene des Lichtes drehen – Polarimetrie in der Zuckerindustrie – Geschichte einer Messmethode*. Linz: Rudolf Trauner.

Polanyi, Michael. 1936. "The Value of the Inexact." *Philosophy of Science*, 3: 233–4.

Pursell, Carroll W., Jr. 1969. "The Farm Chemurgic Council and the United States Department of Agriculture, 1935–1939." *Isis*, 60: 307–17.

Pye, Veronica I., and Ruth Patrick. 1983. "Ground Water Contamination in the United States." *Science*, 221(4612): 713–18.

Quirke, Viviane. 2005. "From Alkaloids to Gene Therapy: A Brief History of Drug Discovery in the 20th Century." In Stuart Anderson (ed.), *Making Medicines: A Brief History of Pharmacy and Pharmaceuticals*. London: Pharmaceutical Press.

Quirke, Viviane. 2006. "Putting Theory into Practice: James Black, Receptor Theory and the Development of the Beta-blockers at ICI, 1958–1978." *Medical History*, 50: 69–92.

Quirke, Viviane. 2013 "Thalidomide, Drug Safety Regulation, and the British Pharmaceutical Industry: The Case of Imperial Chemical Industries." In Jean-Paul Gaudillière and Volker Hess (eds), *Ways of Regulating Drugs in the 19th and 20th Centuries*. Basingstoke: Palgrave Macmillan.

Raabe, Wilhelm K. 1884. *Pfisters Mühle: Ein Sommerferienheft*. Leipzig: F.W. Grunow.

Rabkin, Yakov M. 1987. "Technological Innovation in Science: The Adoption of Infrared Spectroscopy by Chemists." *Isis*, 78: 31–54.

Radkau, Joachim. 2008. *Nature and Power: A Global History of the Environment*. Cambridge: Cambridge University Press.

Rae, Ian D. 2018. "Australian Chemists Crossing the Pacific to the Promised Land." In Seth C. Rasmussen (ed.), *Igniting the Chemical Ring of Fire: Historical Evolution of the Chemical Communities of the Pacific Rim*. London: World Scientific.

Raemaekers, Louis. 1919. *Raemaekers' Cartoon History of the War*. 3 volumes. London: Bodley Head.

Raitt, J. Gordon. 1966. *Modern Chemistry: Applied and Social Aspects*. London: Edward Arnold.

Ramberg, Peter J. 2003. *Chemical Structure, Spatial Arrangement: The Early History of Stereochemistry, 1874–1914*. Aldershot: Ashgate.

Ramirez, Francisco O., and Naejin Kwak. 2015. "Women's Enrollments in STEM in Higher Education: Cross-National Trends, 1970–2010." In Willie Pearson, Jr., Lisa M. Frehill, and Connie L. McNeely (eds), *Advancing Women in Science: An International Perspective*. Cham: Springer.

Rao, Chintamani Nagesa Ramachandra, and Suri Radhakrishna (eds). 1979. *Chemical Education in Developing Countries: Proceedings of a Conference Organized in Penang, Malaysia*. Madras: Committee on Science and Technology in Developing Countries of the ICSU.

Rasch, Manfred. 1989. *Geschichte des Kaiser-Wilhelms-Instituts für Kohlenforschung 1913–1943*. Weinheim: VCH.

Rasmussen, Nicholas. 2001. "Biotechnology before the 'Biotech Revolution': Life Scientists, Chemists and Product Development in 1930s–1940s America." In Carsten Reinhardt (ed.), *Chemical Sciences in the 20th Century. Bridging Boundaries*. Weinheim: Wiley-VCH.

Rasmussen, Nicholas. 2009. *On Speed. From Benzedrine to Adderall*. New York: New York University Press.

Rawls, Rebecca L. 2000. "Ahmed Zewail's Dynamic Chemistry." *Chemical and Engineering News*, 78(21): 35–9.

Rayner-Canham, Marlene, and Geoffrey Rayner-Canham. 1998. *Women in Chemistry: Their Changing Roles from Alchemical Times to Mid-Twentieth Century*. Washington, DC: American Chemical Society.

Rayner-Canham, Marelene F., and Geoff Rayner-Canham. 2008. *Chemistry was their Life: Pioneering British Women Chemists, 1880–1949*. London: Imperial College Press.

Reader, William J. 1970. *Imperial Chemical Industries: A History*, vol. 1, *The Forerunners, 1870–1926*. London: Oxford University Press.

Reader, William J. 1975. *Imperial Chemical Industries: A History*, vol. 2, *The First Quarter-century, 1926–1952*. London: Oxford University Press.

Reardon-Anderson, James. 1991. *The Study of Change: Chemistry in China, 1840–1949*. Cambridge: Cambridge University Press.

Reed, Peter. 2014. *Acid Rain and the Rise of the Environmental Chemist in Nineteenth-Century Britain: The Life and Work of Robert Angus Smith.* Aldershot: Ashgate.

Reed, Peter. 2015. "Making War Work for Industry: The United Alkali Company's Central Laboratory During World War One." *Ambix*, 62: 72–93.

Reinhardt, Carsten (ed.). 2001. *Chemical Sciences in the Twentieth Century: Bridging Boundaries.* Weinheim: Wiley-VCH.

Reinhardt, Carsten. 2002. "The Chemistry of an Instrument: Mass Spectrometry and Structural Organic Chemistry." In Peter J.T. Morris (ed.), *From Classical to Modern Chemistry: The Instrumental Revolution.* Cambridge: Royal Society of Chemistry.

Reinhardt, Carsten. 2004. "Chemistry in a Physical Mode. Molecular Spectroscopy and the Emergence of NMR." *Annals of Science*, 61: 1–32.

Reinhardt, Carsten. 2006a. "Wissenstransfer durch Zentrenbildung. Physikalische Methoden in der Chemie und den Biowissenschaften." *Berichte zur Wissenschaftsgeschichte*, 29: 224–42.

Reinhardt, Carsten. 2006b. *Shifting and Rearranging. Physical Methods and the Transformation of Modern Chemistry.* Sagamore Beach, MA: Science History Publications.

Reinhardt, Carsten. 2006c. "A Lead User of Instruments in Science. John D. Roberts and the Adaptation of Nuclear Magnetic Resonance to Organic Chemistry, 1955–1975." *Isis*, 97: 205–36.

Reinhardt, Carsten. 2010. "Zentrale einer Wissenschaft. Methoden, Hierarchie und die Organisation der chemischen Institute." In Rüdiger vom Bruch and Heinz-Elmar Tenorth (eds), *Geschichte der Universität Unter den Linden 1810–2010. Biographie einer Institution, Praxis ihrer Disziplinen*, vol. 5, *Transformation der Wissensordnung. Verwissenschaftlichung der Gesellschaft und Verstaatlichung der Wissenschaft.* Berlin: Akademie Verlag.

Reinhardt, Carsten. 2012. "Limit Values and the Boundaries of Science and Technology." *Comptes Rendus Chimie*, 15: 595–602.

Reinhardt, Carsten. 2014. "The Olfactory Object. Toward a History of Smell in the Twentieth Century." In Ursula Klein and Carsten Reinhardt (eds), *Objects of Chemical Inquiry.* Sagamore Beach, MA: Science History Publications.

Reinhardt, Carsten. 2017. "'This Other Method': The Dynamics of NMR in Biochemistry and Molecular Biology." *Historical Studies in the Natural Sciences*, 47: 389–422.

Reinhardt, Carsten. 2018. "What's in a Name? Chemistry as a Nonclassical Approach to the World." *Isis*, 109: 559–64.

Reinhardt, Carsten. 2020. "The Development of Research Methods as the Driving Force of Technosience." In Sabine Maasen, Sascha Dickel, and Christoph Schneider (eds), *TechnosScienceSociety. Technological Reconfigurations of Science and Society. Sociology of the Sciences Yearbook,* vol. 30. Berlin: Springer.

Reinhardt, Carsten, and Anthony S. Travis. 2004. "Wahrnehmungen und Realitäten der deutschen industriellen Forschung (1880–1925)." In Rolf Petri (ed.), *Technologietransfer aus der deutschen Chemieindustrie (1925–1960).* Berlin: Duncker und Humblot.

REM (Reichs- und Preussisches Ministerium für Wissenschaft, Erziehung und Volksbildung). 1938. "Neuordnung des höheren Schulwesens." *Deutsche Wissenschaft, Erziehung und Volksbildung*, 4: 46–56. Available online: https://goobiweb.bbf.dipf.de/viewer/image/991084217_0004/64/LOG_0073/.

Remane, Horst. 1987. "Zur Entwicklung der Massenspektroskopie von den Anfängen bis zur Strukturaufklärung organischer Verbindungen." *NTM*, 24: 93–106.

Remarque, Erich Maria. 1928. *Im Westen Nichts Neues.* Berlin: Propylaen.

Rhees, David J. 1993. "Corporate Advertising, Public Relations and Popular Exhibits: The Case of Du Pont." *History and Technology*, 10: 67–75.
Riegel, Léon. 1978. *Guerre et Littérature. Le bouleversement des consciences dans la littérature romanesque inspirée par la Grande Guerre (littératures française, anglo-saxonne et allemande) 1910–1930*. Nancy: Editions Klincksieck.
Ringer, Fritz K. 1979. *Education and Society in Modern Europe*. Bloomington: Indiana University Press.
Roberts, Elizabeth. 1988. *Women's Work 1840–1940*. London: Macmillan.
Roberts, Jody A. 2005. "Creating Green Chemistry: Discursive Strategies of a Scientific Movement." Ph.D. thesis, Virginia Polytechnic Institute, Blacksburg. Available online: https://vtechworks.lib.vt.edu/bitstream/handle/10919/27529/Roberts_Revised_Final_v2.pdf?sequence=1.
Roberts, John D. 1959. *Nuclear Magnetic Resonance: Applications to Organic Chemistry*. New York: McGraw-Hill.
Roberts, John D. 1990. *The Right Place at the Right Time*. Washington, DC: American Chemical Society.
Roberts, Lissa L., and Simon Werrett (eds). 2018. *Compound Histories: Materials, Governance and Production, 1760–1840*. Leiden: Brill.
Robinson, Ann E. 2018. "Chemistry Classrooms with Periodic Tables." Available online: https://www.pinterest.com/ann9robinson/chemistry-classrooms-with-periodic-tables/. Supplementary information provided by personal communication with Jeffrey A. Johnson, November 9, 2016.
Rocke, Alan J. 1984. *Chemical Atomism in the Nineteenth Century: From Dalton to Cannizzaro*. Columbus, OH: Ohio State University Press.
Rocke, Alan J. 1985. "Hypothesis and Experiment in the Early Development of Kekulé's Benzene Theory." *Annals of Science*, 42: 355–81.
Rocke, Alan J. 1993. *The Quiet Revolution: Hermann Kolbe and the Science of Organic Chemistry*. Berkeley: University of California Press.
Rocke, Alan J. 2010. *Image and Reality: Kekulé, Kopp, and the Scientific Imagination*. Chicago, IL: University of Chicago Press.
Roloff, Christine. 1988. *Von der Schmiegsamkeit zur Einmischung: Professionalisierung der Chemikerinnen und Informatikerinnen*. Pfaffenweiler: Centaurus.
Rooij, Arjan Van, and Ernst Homburg. 2002. *Building the Plant: A History of Engineering Contracting in the Netherlands*. Zutphen: Walburg.
Roqué, Xavier. 2001. "From Radiochemistry to Nuclear Chemistry and Cosmochemistry." In Carsten Reinhardt (ed.), *Chemical Sciences in the 20th Century: Bridging Boundaries*. Weinheim: Wiley-VCH.
Rosenbaum, Arthur Lewis. 2007. "Yenching University and Sino-American Interactions, 1919–1952." *The Journal of American–East Asian Relations*, 14: 11–60.
Rosenbloom, Al, and RuthAnn Althaus. 2010. "Degussa AG and its Holocaust Legacy." *Journal of Business Ethics*, 92: 183–94.
Rossiter, Margaret. 1995. *Women Scientists in America*, vol. 2, *Before Affirmative Action 1940–1972*. Baltimore, MD: Johns Hopkins University Press.
Rossiter, Margaret. 2012. *Women Scientists in America*, vol. 3, *Forging a New World since 1972*. Baltimore, MD: Johns Hopkins University Press.
Rovnyak, David, and Robert Stockland, Jr. (eds). 2007. *Modern NMR Spectroscopy in Education*. Washington, DC: American Chemical Society.
Rowland, F. Sherwood, and Mario J. Molina. 1994. "Ozone Depletion: 20 Years after the Alarm." *Chemical and Engineering News*, 72(33): 8–13.

Royal Society of Chemistry. 2015. *Public attitudes to chemistry*. Available online: www.rsc.org/campaigning-outreach/.../public-attitudes-chemistry/.

Rudolph, John L. 2002. *Scientists in the Classroom: The Cold War Reconstruction of American Science Education*. Basingstoke: Palgrave Macmillan.

Rudolph, John L. 2005a. "Epistemology for the Masses: The Origins of 'the Scientific Method' in American Schools." *History of Education*, 45: 341–76.

Rudolph, John L. 2005b. "Turning Science to Account: Chicago and the General Science Movement in Secondary Education 1905–1920." *Isis*, 96: 353–89.

Russell, Colin A. (ed.). 2000. *Chemistry, Society and Environment: A New History of the British Chemical Industry*. Cambridge: Royal Society of Chemistry.

Russell, Colin A., Noel G. Coley, and Gerrylynn K. Roberts. 1977. *Chemists by Profession: The Origins and Rise of the Royal Institute of Chemistry*. Milton Keynes: Open University Press.

Russell, Colin A., and John A. Hudson. 2012. *Early Railway Chemistry and its Legacy*. Cambridge: Royal Society of Chemistry.

Russell, Edmund. 2001. *War and Nature: Fighting Humans and Insects with Chemicals from World War I to Silent Spring*. Cambridge: Cambridge University Press.

Ruthenberg, Klaus. 2016. "About the Futile Dream of an Entirely Riskless and Fully Effective Remedy: Thalidomide." *Hyle: International Journal for Philosophy of Chemistry*, 22: 55–77. Available online: http://www.hyle.org/journal/issues/22-1/ruthenberg.pdf.

Saltzmann, Martin D. 1986. "The Development of Physical Organic Chemistry in the United States and the United Kingdom: 1919–1939." *Journal of Chemical Education*, 63: 588–93.

Sané, Krishna V. 1979. "Developing Countries." In Chintamani Nagesa Ramachandra Rao and Suri Radhakrishna (eds), *Chemical Education in Developing Countries: Proceedings of a Conference Organized in Penang, Malaysia*. Madras: Committee on Science and Technology in Developing Countries of the ICSU.

Scerri, Eric R. 2016. *A Tale of Seven Scientists and a New Philosophy of Science*. New York: Oxford University Press.

Schaffer, Simon. 2015. "Les cérémonies de la mesure." *Annales. Histoire, Sciences Sociales*, 70: 409–35.

Schneider, Tobias, and Theresa Lütkefend. 2019. *Nowhere to Hide: The Logic of Chemical Weapons Use in Syria*. Berlin: Global Public Policy Institute. Available online: https://www.gppi.net/media/GPPi_Schneider_Luetkefend_2019_Nowhere_to_Hide_Web.pdf

Schuder, Werner. 1956. "Vorwort." *Minerva: Jahrbuch der gelehrten Welt. Abteilung Universitäten und Fachhochschulen: 2: Aussereuropa*, 34: viii–xi.

Schummer, Joachim. 2006. "Historical Roots of the 'Mad Scientist': Chemists in 19th-century Literature." *Ambix*, 53: 99–127.

Schummer, Joachim. 2017. "Chemie als Teufelswerk? 2000 Jahre Chemiekritik." In Marc-Denis Weitze, Joachim Schummer, and Thomas Geelhaar (eds), *Zwischen Faszination und Verteufelung: Chemie in der Gesellschaft*. Berlin: Springer.

Schummer, Joachim. 2018. "Ethics of Chemical Weapons Research: Poison Gas in World War One." *Hyle: International Journal for Philosophy of Chemistry*, 24: 5–28. Available online: http://www.hyle.org/journal/issues/24-1/schummer.htm.

Schummer, Joachim. 2022. "Art and Representation: The Rise of the 'Mad Scientist.'" In Peter Ramberg (ed.), *A Cultural History of Chemistry in the Nineteenth Century*. London: Bloomsbury.

Schummer, Joachim, and Tom Børsen (eds). 2021. *Ethics of Chemistry: From Poison Gas to Climate Engineering*. Singapore: World Scientific.

Schummer, Joachim, Bernadette Bensaude-Vincent, and Brigitte Van Tiggelen (eds). 2007. *The Public Image of Chemistry*. Singapore: World Scientific.

Schwarzl, Sonja M. 2006. "Equal Opportunity in Chemistry in Germany." In Cecilia H. Marzabadi, Valerie J. Kuck, Susan A. Nolan, and Janine P. Buckner (eds), *Are Women Achieving Equity in Chemistry? Dissolving Disparity and Catalyzing Change*. Washington, DC: American Chemical Society.

Seeman, Jeffrey I. 2014. "R. B. Woodward, A Great Physical Organic Chemist." *Journal of Physical Organic Chemistry*, 27: 708–21.

Seeman, Jeffrey I. 2017. "R. B. Woodward: A Larger Than Life Chemistry Rock Star." *Angewandte Chemie International*, 56: 10228–45.

Segrè, Emilio. 1980. *From X-Rays to Quarks: Modern Physicists and Their Discoveries*. San Francisco, CA: W.H. Freeman.

Sella, Andrea. 2017. "Winkler's Bed." *Chemistry World*, 14(11): 70. Available online: https://www.chemistryworld.com/opinion/winklers-bed/3008164.article.

Sellers, Christopher, and Joseph Melling. 2012. *Dangerous Trade. Histories of Industrial Hazard across a Globalizing World*. Philadelphia, PA: Temple University Press.

Servos, John. 1990. *Physical Chemistry from Ostwald to Pauling: The Making of a Science in America*. Princeton, NJ: Princeton University Press.

Sime, Ruth Lewin. 2001. "The Search for Artificial Elements and the Discovery of Nuclear Fission." In Carsten Reinhardt (ed.), *Chemical Sciences in the Twentieth Century: Bridging Boundaries*. Weinheim: Wiley-VCH.

Simões, Ana, and Kostas Gavroglu. 2001. "Issues in the History of Theoretical and Quantum Chemistry, 1927–1960." In Carsten Reinhardt (ed.), *Chemical Sciences in the 20th Century. Bridging Boundaries*. Weinheim: Wiley-VCH.

Simon, Christian. 1999. *DDT: Kulturgeschichte einer chemischen Verbindung*. Basel: Christoph Merian.

Simon, Josep. 2011. *Communicating Physics. The Production, Circulation, and Appropriation of Ganot's Textbooks in France and England 1851–1887*. London: Pickering and Chatto.

Simon, Josep. 2019. "The Transnational Physical Science Study Committee: The Evolving Nation in the World of Science and Education (1945–1975)." In John Krige (ed.), *How Knowledge Moves: Writing the Transnational History of Science and Technology*. Chicago, IL: University of Chicago Press.

Simon, Josep, and Mar Cuenca-Lorente. 2012. "Science Education and the Material Culture of the Nineteenth-Century Classroom: Physics and Chemistry in Spanish Secondary Schools." *Science & Education*, 21: 227–44.

Singerman, David R. 2017. "The Limits of Chemical Control in the Caribbean Sugar Factory." *Radical History Review*, 1(127): 39–61.

Singh, M. Mahinder. 1989. "The Federation of Asian Chemical Societies: Its Formation, Administration and Activities." *Chemistry in Asia*, 1: 1–11.

SJR (SCImago Journal & Country Rank). 2018. Country Rankings in Chemistry (all subject categories), 1996–2017. Available online: https://www.scimagojr.com/countryrank.php?area=1600.

Skrebels, Paul. 2014. "A Poisonous Paradox: Representations of Gas Warfare in Post-Memory Films of the Great War." In Martin Löschnigg and Marzena Sokolowska-Paryz (eds), *The Great War in Post-Memory Literature and Film*. Berlin: de Gruyter.

Slater, John C. 1939. *Introduction to Chemical Physics*. New York: McGraw-Hill.

Slater, Leo B. 2001. "Woodward, Robinson, and Strychnine: Chemical Structure and Chemists' Challenge." *Ambix*, 48: 161–89.

Slater, Leo B. 2002. "Organic Chemistry and Instrumentation: R. B. Woodward and the Reification of Chemical Structures." In Peter J.T. Morris (ed.), *From Classical to Modern Chemistry: The Instrumental Revolution*. Cambridge: Royal Society of Chemistry.

Slater, Leo B. 2004. "Malaria Chemotherapy and the 'Kaleidoscopic' Organisation of Biomedical Research during World War II." *Ambix*, 51: 107–34.

Slobodkin, Gregory, and Miles Pickering. 1988. "An Inside View of Soviet Chemical Education." *Journal of Chemical Education*, 65: 3–5.

Southwold, Stephen ("Miles"). 1931. *The Gas War of 1940*. London: Scholartis Press.

Smil, Vaclav. 2001. *Enriching the Earth: Fritz Haber, Carl Bosch, and the Transformation of World Food Production*. Cambridge, MA: MIT Press.

Smith, John K., Jr. 1985. "The Ten-Year Invention: Neoprene and Du Pont Research, 1930–1939." *Technology and Culture*, 26: 34–55.

Smith, Leonard V., Stéphane Audoin-Rouzeau, and Annette Becker. 2003. *France and the Great War, 1914–1918*. Cambridge: Cambridge University Press.

Spear, Hilda D., and Sonya A. Summersgill. 1991. "Poison Gas and the Poetry of War." *Essays in Criticism*, 41: 308–22.

Spitz, Peter H. 1988. *Petrochemicals: The Rise of an Industry*. New York: Wiley.

Spitz, Peter H. (ed.) 2003. *The Chemical Industry at the Millennium: Maturity, Restructuring, and Globalization*. Philadelphia, PA: Chemical Heritage Press.

Spitz, Peter H. 2019. *Primed for Success: The Story of Scientific Design Company: How Chemical Engineers Created the Petrochemical Industry*. Cham: Springer.

Steen, Kathryn. 2014. *The American Synthetic Organic Chemicals Industry: War and Politics, 1910–1930*. Chapel Hill: University of North Carolina Press.

Steinhauser, Thomas. 2014. *Zukunftsmaschinen in der Chemie: Kernmagnetische Resonanz bis 1980*. Frankfurt am Main: Peter Lang.

Stephens, Trent, and Rock Brynner. 2001. *Dark Remedy: The Impact of Thalidomide and its Revival as a Vital Medicine*. Cambridge, MA: Perseus.

Stephenson, Neal. 1988. *Zodiac*. New York: Grove/Atlantic.

Stocker, Jack H. (ed.). 1998. *Chemistry and Science Fiction*. Washington, DC: American Chemical Society.

Stoff, Heiko. 2012. *Wirkstoffe. Eine Wissenschaftsgeschichte der Hormone, Vitamine und Enzyme, 1920–1970*. Stuttgart: Steiner.

Stoff, Heiko, and Anthony S. Travis. 2019. "Discovering Chemical Carcinogenesis. The Case of Aromatic Amines." In Ernst Homburg and Elisabeth Vaupel (eds), *Hazardous Chemicals. Agents of Risk and Change, 1800–2000*. New York: Berghahn.

Stokes, Raymond G. 1985. "The Oil Industry in Nazi Germany, 1936–1945." *Business History Review*, 59: 254–77.

Stokes, Raymond G. 1988. *Divide and Prosper: The Heirs of I.G. Farben under Allied Authority, 1945–1951*. Berkeley: University of California Press.

Stokes, Raymond G. 1994. *Opting for Oil: The Political Economy of Technological Change in the West German Chemical Industry, 1945–1961*. Cambridge: Cambridge University Press.

Stokes, Raymond G. 2000. *Constructing Socialism: Technology and Change in East Germany, 1945–1990*. Baltimore, MD: Johns Hopkins University Press.

Stranges, Anthony N. 1982. *Electrons and Valence: Development of the Theory, 1900–1925*. College Station: Texas A&M University Press.

Stranges, Anthony N. 1984. "Friedrich Bergius and the Rise of the German Synthetic Fuel Industry." *Isis*, 75: 643–67.

Stranges, Anthony N. 1985. "From Birmingham to Billingham: High-pressure Coal Hydrogenation in Great Britain." *Technology and Culture*, 26: 726–57.

Stranges, Anthony N. 2000. "Germany's Synthetic Fuel Industry, 1927–1945." In John E. Lesch (ed.), *The German Chemical Industry in the Twentieth Century*. Dordrecht: Kluwer.

Strasser, Bruno. 2006. "A World in One Dimension: Linus Pauling, Francis Crick, and the Central Dogma of Molecular Biology." *History and Philosophy of the Life Sciences*, 28: 491–512.

Sugden, Samuel. 1924. "CXLIL: A Relation between Surface Tension, Density, and Chemical Composition." *Journal of the Chemical Society, Transactions*, 125: 1177–89.

Szabadváry, Ferenc. 1966. *History of Analytical Chemistry*. Oxford: Pergamon Press.

Szinicz, Ladislaus. 2005. "History of Chemical and Biological Warfare Agents." *Toxicology*, 214: 167–81.

Tarbell, Dean S., and Ann T. Tarbell. 1986. *Essays on the History of Organic Chemistry in the United States, 1875–1955*. Nashville, TN: Folio.

Tarbell, Stanley. 1976. "Organic Chemistry. Centennial American Chemical Society 1876 to 1976." *Chemical and Engineering News*, 54(15): 110–27.

Tarr, Joel A. 1996. *The Search for the Ultimate Sink: Urban Pollution in Historical Perspective*. Akron, OH: University of Akron Press.

Tarr, Joel A. 2002. "Industrial Waste Disposal in the United States as a Historical Problem." *Ambix*, 49: 4–20.

Teissier, Pierre. 2014. *Une histoire de la chimie du solide: Synthèses, formes, identités*. Paris: Hermann.

Thackray, Arnold, David Brock, and Rachel Jones. 2015. *Moore's Law: The Life of Gordon Moore, Silicon Valley's Quiet Revolutionary*. New York: Basic Books.

Thackray, Arnold, and Minor Myers, Jr. 2000. *Arnold O. Beckman: One Hundred Years of Excellence*. Philadelphia, PA: Chemical Heritage Foundation.

Thackray, Arnold, Jeffrey L. Sturchio, P. Thomas Carroll, and Robert F. Bud. 1985. *Chemistry in America, 1876–1976*. Dordrecht: D. Reidel.

Timmermans, Stefan, and Steven Epstein. 2010. "A World of Standards, but not a Standard World: Toward a Sociology of Standards and Standardization." *Annual Review of Sociology*, 36: 69–89.

Tomkins, Judith. 1996. "Instruments for Water Quality Management." In John Hassan, Paul Nunn, Judith Tomkins, and Iain Fraser (eds), *The European Water Environment in a Period of Transformation*. Manchester: Manchester University Press.

Toumey, Christopher P. 1992. "The Moral Character of Mad Scientists: A Cultural Critique of Science." *Science, Technology, and Human Values*, 17: 411–37.

Trager, Rebecca. 2018. "Bigger but Slimmer." *Chemistry World*, 15(1): 26–7. Available online: https://www.chemistryworld.com/news/chemical-industry-roundup-2017/3008406.article.

Travis, Anthony S. 1989. "Science as Receptor of Technology: Paul Ehrlich and the Synthetic Dyestuffs Industry." *Science in Context*, 3: 383–408.

Travis, Anthony S. 1993. *The Rainbow Makers: The Origins of the Synthetic Dyestuffs Industry in Western Europe*. Bethlehem, PA: Lehigh University Press.

Travis, Anthony S. 2002a. "Contaminated Earth and Water: A Legacy of the Synthetic Dyestuffs Industry." *Ambix*, 49: 21–50.

Travis, Anthony S. 2002b. "Instrumentation in Environmental Analysis, 1935–1975." In Peter J.T. Morris (ed.), *From Classical to Modern Chemistry: The Instrumental Revolution*. Cambridge: Royal Society of Chemistry.

Travis, Anthony S. 2004. *Dyes Made in America, 1915–1980: The Calco Chemical Company, American Cyanamid and the Raritan River*. Jerusalem: Edelstein Center.

Travis, Anthony S. 2012. "*Silent Spring* at 50: Earth, Water, and Air." *Ambix*, 59: 83–7.
Travis, Anthony S. 2018. *Nitrogen Capture: The Growth of an International Industry (1900–1940)*. Cham: Springer.
Travis, Anthony S., Harm G. Schröter, Ernst Homburg, and Peter J.T. Morris (eds). 1998. *Determinants in the Evolution of the European Chemical Industry, 1900–1939: New Technologies, Political Frameworks, Markets and Companies*. Dordrecht: Kluwer.
Tremblay, Jean-François. 2017. "Chinese Chemistry Ph.D. Grads are Forgoing U.S. Postdocs." *Chemical and Engineering News*, 95(41): 21–3.
Trischler, Helmuth. 2016. "The Anthropocene: A Challenge for the History of Science, Technology, and the Environment." *NTM*, 24: 309–35.
Tsaparlis, Giorgios. 2003. "Globalisation in Chemistry Education Research and Practice." *Chemistry Education: Research and Practice*, 4: 3–10.
Tu, Anthony T. 1999. "Overview of Sarin Terrorist Attacks in Japan." In Anthony T. Tu and William Gaffield (eds), *Natural and Selected Synthetic Toxins: Biological Implications*. Washington, DC: American Chemical Society.
Tullo, Alexander H. 2017. "C&EN's Global Top 50: Strong Industry Performance Continues as Profits Rise Despite Declining Sales." *Chemical and Engineering News*, 95(30): 30–5. Available online: https://cen.acs.org/articles/95/i30/CENs-Global-Top-50.html?utm_source=Old50&utm_medium=InLine&utm_campaign=CEN.
Tuttle, William M. 1981. "The Birth of an Industry: The Synthetic Rubber 'Mess' in World War II." *Technology and Culture*, 22: 35–67.
Ulitz, Arnold. 1920. *Ararat*. Munich: A. Langen
US Bureau of Labor Statistics. 2000, May 2010, May 2017. *Occupational Employment Statistics: National* (data for occupational group 19-2031: Chemists). Available online: https://www.bls.gov/oes/tables.htm.
US Bureau of the Census. 1981, 1992. *Statistical Abstract of the United States*. Washington: [US] GPO. Available online: https://www.census.gov/library/publications/time-series/statistical_abstracts.html.
Verband der Laboratoriumsvorstände an Deutschen Hochschulen. 1914–1939. *Berichte des Verbandes der Laboratoriumsvorstände an Deutschen Hochschulen*, 16–31.
Verkade, Pieter E. 1985. *A History of the Nomenclature of Organic Chemistry*. Dordrecht: D. Reidel.
Virta, Robert L. 2006. *Circular 1298: Worldwide Asbestos Supply and Consumption Trends from 1900 through 2003*. Reston, VA: US Geological Survey.
Vogel, Sarah A. 2013. *Is it Safe? BPA and the Struggle to Define the Safety of Chemicals*. Berkeley: University of California Press.
Vonnegut, Kurt. 1963. *Cat's Cradle*. New York: Delta.
Waddington, David J. (ed.). 1984. *Teaching School Chemistry*. Paris: UNESCO.
Waddington, David J., and Henry W. Heikkinen. 2015. "Developments in Chemical Education." In Mary Virginia Orna (ed.), *Sputnik to Smartphones: A Half-Century of Chemistry Education*. Washington, DC: American Chemical Society.
Wagner, Bernd C. 2000. *IG Auschwitz: Zwangarbeit und Vernichtung von Häftlingen des Lagers Monowitz 1941–1945*. Munich: K.G. Saur.
Walker, Mark. 1989. *German National Socialism and the Quest for Nuclear Power, 1939–1949*. Cambridge: Cambridge University Press.
Wang, Zuoyue. 2010. "Transnational Science during the Cold War: The Case of Chinese/American Scientists." *Isis*, 101: 367–77.

Warner, Deborah Jean. 2007. "How Sweet It Is: Sugar, Science, and the State." *Annals of Science*, 64: 147–70.

Watch, Daniel. 2008. *Building Type Basics for Research Laboratories*, 2nd ed. Hoboken, NJ: Wiley.

Warwick, Andrew. 2003. *Masters of Theory: Cambridge and the Rise of Mathematical Physics*. Chicago, IL: University of Chicago Press.

Wedeen, Richard P. 1993. "The Politics of Lead." In Helen E. Sheehan and Richard P. Wedeen (eds), *Toxic Circles: Environmental Hazards from the Workplace into the Community*. New Brunswick, NJ: Rutgers University Press.

Weil, John A., and James R. Bolton. 2007. *Electron Paramagnetic Resonance: Elementary Theory and Practical Applications*, 2nd ed. Hoboken, NJ: Wiley-Interscience.

Weingart, Peter. 2007. "Chemists and their Craft in Fiction Film." In Joachim Schummer, Bernadette Bensaude-Vincent, and Brigitte Van Tiggelen (eds), *The Public Image of Chemistry*. Singapore: World Scientific.

Weininger, Stephen J. 1984. "The Molecular Structure Conundrum: Can Classical Chemistry Be Reduced to Quantum Chemistry?" *Journal of Chemical Education*, 61: 939–44.

Weininger, Stephen J. 2000. "'What's in a Name?' From Designation to Denunciation. The Nonclassical Cation Controversy." *Bulletin for the History of Chemistry*, 25: 123–31.

Weininger, Stephen J. 2018. "Delayed Reaction: The Tardy Embrace of Physical Organic Chemistry by the German Chemical Community." *Ambix*, 65: 52–75.

Weir, Ronald. 1995. *The History of the Distillers Company, 1887–1939: Diversification and Growth in Whisky and Chemicals*. Oxford: Clarendon Press.

Wells, H.G. 1914. *The World Set Free*. London: Macmillan.

Wells, H.G. 1933. *The Shape of Things to Come*. London: Hutchinson.

WHO [World Health Organization]. 2017. "Asthma Fact Sheet." Available online: http://www.who.int/mediacentre/factsheets/fs307/en/.

WHO Regional Office of Europe. 2010. *WHO Guidelines for Indoor Air Quality: Selected Pollutants*. Available online: http://www.euro.who.int/__data/assets/pdf_file/0009/128169/e94535.pdf.

Wik, Reynold Millard. 1962. "Henry Ford's Science and Technology for Rural America." *Technology and Culture*, 3: 247–58.

Wilhelm, Kate and Theodore L. Thomas. 1970. *Year of the Cloud*. New York: Doubleday.

Wilson, Gordon D. 1994. "Polythene: the early years." In Peter J.T. Morris and Susan Mossman, *The Development of Plastics*. Cambridge: Royal Society of Chemistry.

Wilson, Jennifer. 2015. "Dame Kathleen Lonsdale (1903–1971): Her Early Career in X-ray Crystallography." *Interdisciplinary Science Reviews*, 40: 265–78.

World Bank. 2016. *Poverty and Shared Prosperity 2016*. Available online: http://www.worldbank.org/en/topic/poverty/overview.

Wotiz, John H. 1971. "Higher Education and Research in Chemistry in the USSR." *Journal of Chemical Education*, 48: 60–8.

Yergin, Daniel. 1991. *The Prize: The Epic Quest for Oil, Money, and Power*. London: Simon and Schuster.

Zalasiewicz, Jan, Colin Waters, Mark Williams, Colin Summerhayes, and Phil Gibbard. 2017–18. "Where are we with the 'Anthropocene'?" *Geoscientist*, 27(11): 17–19.

Zhao Kuanghua (ed.). 2003. *Zhongguo huaxue shi jinxiandai juan* [*The History of Chemistry in China, Modern and Contemporary Periods*]. Nanning: Guangxi Educational Press.

Ziolkowski, Theodore. 2015. *The Alchemist in Literature: From Dante to the Present*. Oxford: Oxford University Press.

CONTRIBUTORS

José Ramón Bertomeu-Sánchez, University of Valencia, Spain

Yasu Furukawa, The Graduate University for Advanced Studies, SOKENDAI, Japan

Antonio García-Belmar, University of Alicante, Spain

Lijing Jiang, Johns Hopkins University, USA

Jeffrey Allan Johnson, Villanova University, USA

Peter J.T. Morris, Science Museum, London and University College London, UK

Mary Jo Nye, Oregon State University, USA

Peter Reed is an Independent Scholar, in Carmichael, California, USA

Carsten Reinhardt, Bielefeld University, Germany

Joachim Schummer is an Independent Scholar in Berlin, Germany

Anthony S. Travis, Edelstein Center, Hebrew University, Israel

INDEX

ab initio calculations 35
Abbott Laboratories (company) 19
Abelson, Philip H. 32
Abney, William de W. 75
absent-minded scientist trope 205–10
Academy of Sciences, Beijing 194–5
accidents involving toxic chemicals, *see also
    disasters, industrial* 124
acetone 154–5
acetylene 16, 154–5, 157–8
Acid Precipitation Act (1980) (US) 130
acid rain 23, 129–30
Afghanistan and Iraq, invasion of 196
Africa 174, 185, 188, 195
Agent Orange defoliant 115, 169
Agilent Technologies (company) 95
air pollution 127–31, 139–40, 220
AkzoNobel (company) 171
Alkali Inspectorate (UK) 128
Allied Chemical and Dye Corp. (company)
    152
Almy, Charles 166
American Chemical Society (ACS) 25, 27,
    143, 174, 197, 199, 202
American Institute of Chemical Engineers
    (AIChE) 19
American Petroleum Institute 54
American Viscose Corporation (AVC)
    (company) 158
ammonia synthesis, *see nitrogen fixation*
Anderson, Philip W. 42–3

Anglo-Persian Oil Co. (later BP) (company)
    155
Anthropocene era 119, 147–8
antibiotics 5, 18–19, 108, 111, 162–4
antifreeze 154
antimalarial drugs, *see under
    pharmaceuticals*
apocalyptic events in fiction 214–16
apparatus, chemical 53, 74, 90–1, 94–5
Arab–Islamic nations 196
Armstrong, Edward Frankland 61–2
Arndt, Fritz 34, 40
Arrhenius, Svante 45, 144
Arribas Jimeno, Siro 70
"Art of the First World War" (1998),
    online exhibition 213–4
asbestos 110, 136–7
Asia, chemistry in 26, 174, 188–95,
    199–202
Associated Electrical Industries (AEI)
    (company) 87
asthma 139
Aston, Francis 7, 32, 46, 86
AstraZeneca (company) 96
Atlantic Richfield (ARCO) (company) 166
Atlas Powder Co. (company) 152
atomic bomb 110, 215–16
atomic fission 32, 110, 224, 219
atomic force microscopy (AFM) 8, 12, 46
atomic nucleus 33–4, 82
atomic number 31–2

atomic theory 30–2, 39
Auden, W.H. 176–7
Australia 174
automatic control of industrial processes 20–1
automobile manufacturing 153–4
Aventis (company) 172

Baekeland, Leo H. 11
Baeyer Society 176
Ball, Philip 121
Ballard, J.G. 219–20
Bartlett, Neil 32
BASF (company) 3, 16, 20, 103–4, 150, 152, 154, 156, 164–5, 171
  Oppau ammonia factory 150
  Oppau experimental station 94
  Oppau disasters (1921 & 1948) 150, 216
Bayer (company) 3, 18, 102, 108–9, 151–2, 164, 171–2
Becher, Johannes R. 215
Bechgaard, Klaus 44
Beckman Arnold O. (company) 85
Beckman, Arnold O. (person) 91
Beckman instruments (company) 85, 91
Beevers, C. Arnold 77
Beilsteins Handbuch 59, 65
Bell Telephone Laboratories (company) 91
Bensaude-Vincent, Bernadette 119
Bergius process for sugar 154–5
Bergius process for synthetic gasoline 156, 161
Bergius, Friedrich K.R. 16, 154–6
Berlin University 99
Bernal, John Desmond 76–7
Berthelot, P.E. Marcellin 74
Beuys Joseph 218–19
Bevan, Edward J. 126
Bhopal disaster (1984) 20, 138–9, 169, 216
biblical themes 219, 222
Biemann, Klaus 57, 88
Big Bang (cosmology) 14
Big Pharma 114
Big Science 112
Binning, Gerd 8
biochemical engineering 19–20
biochemistry 13–14, 18, 27, 37, 46–7, 65–6, 88, 107, 118

biology 37, 45, 112–121
  synthetic biology 99, 119
bio-organic chemistry 118
Black, James W. 19, 167
Blakely, Calvin R. 95
Bloch, Felix 82
blockbuster drugs 19, 167
Bodenstein, Max E. A. 41
Bohr, Niels 31–4
Bologna Declaration (1999) and Bologna Process 181–2, 196, 198
Bonino, Giovanni Battista 55–6
Book of Enoch 222
Born, Max 33
Bosch, Carl 16, 156
Bosch, Hieronymus 215
Boyle, T.C. 221
Bragg, W. Lawrence 5, 11, 76
Bragg, William H. 5, 11, 76
brain drain from Europe to America 26, 185
Brattain, R. Robert 85
Brazil 188
Briggs, Mitzi 93
Britain, *see United Kingdom*
British Antarctic Survey 23
British Dyestuffs Corporation (company) 152
British Industrial Solvents (company) 155
British Nylon Spinners (company) 160
British Petroleum (BP) (company) 89
Brock, William H. 47
Broglie, Louis V.P.R. de 33
Brookhaven National Laboratory (New York State) 132
Bruker (company) 95
Brunner, John (author) 220
Brunner, Mond & Co. (company) 152
buckminsterfullerene 12, 39
Buna, *see rubber, synthetic*
Bunsen, Robert 9, 53, 80
Burbidge, E. Margaret 15
Burbidge, Geoffrey R. 15

Cain, Gordon A. 170
calcium carbide 155
calcium cyanamide 150–2, 155
California Air Resources Board (CARB) 129

INDEX

California Institute of Technology (Caltech) 34, 86, 91
Callendar, Guy S. 144
Cambridge University 75, 184
carbon disulfide 137
carbon nanotubes 12, 44
carbon-14 dating 14
Caro, Nikodem 150
Carothers, Wallace Hume 11, 158
Carr, Emma Perry 141–3
Carson, Rachel L. 18, 89, 110, 115, 132, 169, 217–20
Carter, President James E., Jr. 169
Carver, George Washington 155–6
Casale, Luigi 156
Cassella (company) 152
Cat's Cradle (1963), novel (Vonnegut) 220
catalysis 32, 39, 41
Celanese (company) 107, 158, 171
cellophane 158, 159
Chadwick, James 32
Chain, Ernst B. 162, 164
Chan, Garnet K. 45
Chemical Abstracts 23, 59, 65
chemical analysis 8–9, 22–3, 53, 70–1, 177
Chemical Bond Approach (CBA) Project 63
chemical bonding 5, 32–6
chemical century, twentieth century as the 26, 113
Chemical Education Material (CHEM) Study Project 63
Chemical Education Research and Practice in Europe (journal) 197
chemical engineering 19–21, 198, 193
  as an academic discipline 166
  as a separate field 104
chemical engineering firms 166
chemical industries 16, 20, 104–9, 160, 167–72, 176, 189, 204
  decline since 1965 170–2
  fundamental changes after 1980 170–2
  in interwar period 151–3
  not uniformly successful in making profits 168
  response to environmental and health issues 217–18
chemical reactions 4–5
Chemical Society (of London) 25
Chemical Society of Japan (CSJ) 189–90

chemical structures, determination of 5–10, 36–8, 47–8, 54–7, 75–88, 118–19, 141–2
chemical weapons 103, 124–5, 151, 184, 194, 196, 210–15, 221–2
  nerve agents 3, 125, 161
Chemical Weapons Convention (1992) 125
chemistry
  attractiveness of 188
  branches of 4, 13, 32, 72, 104, 106, 113, 117, 204
  characteristic features of 119–20
  concerned with resolving problems rather than memorizing data 63–5
  confidence in 149
  core concepts of 30
  disciplinary status and centrality of 30, 101, 113–15, 119, 194, 202
  diversification and growth in the twentieth century 204
  duality as a science and an industry 100, 112
  ethics of 223
  existential crisis in 27
  global spread of 25–6, 53, 195–9, 202
  image of 4, 62, 71–2, 115–16, 121, 174, 189, 194, 202, 203–23
  new sub-disciplines in 11
  quality of facilities for study of 197
  reputation of and role in society 100, 112, 116, 196
chemistry clubs for children 62
Chemistry in Context (CiC) 64
Chemistry in the Community (ChemCom) 64
chemistry sets 62, 69
chemists
  challenges for 101, 115
  dispensability of 98
  diversity of the profession 173–4, 187, 197–8
  employment of 145–6
  fundamental task of 113–14
  knowledge and influence of 120–1
  mostly disinterested in the history of their subject 223
  national and international organizations of 25, 27, 65, 143, 148, 174–6, 184–5, 188–90, 195, 197–9, 202

number of 24, 26, 101, 175–6, 182–5, 187, 189, 191, 197, 199
  role of 116–17, 124, 149, 183
  self-image of 112, 114, 120, 223
  social isolation of 223
  types of work 23–4, 173–4
  use of the term 4
  women as 24–5, 182–4, 187, 191, 197–200
Chemours (company) 170
chemurgic movement 155–6, 162–3
Chicago World's Fair (1933) 152–3
Chilean nitrate 102, 150–1
China 26, 167, 174, 186, 188, 192–5, 199–201
  Cultural Revolution 194
  Great Leap Forward 194
Chinese Chemical Society 195
Chinese Communist Party 195
chlorofluorocarbons (CFCs) 22–3, 41, 130–1, 154, 172
Christian tradition 115–16
chromatography 9–10, 80, 88–91
CIBA (company) 152
City and Guilds of London Institute 61–2
Claassen, Howard 32
Clariant (company) 167
Clark, Birge M. 93
Claude, Georges 156
Clean Air Act (UK, 1956) 129
Clean Air Acts (US) 127
Clean Seas campaign 136
Clean Water Acts (US) 134
climate change 23, 129, 144–8, 219–20
Coblentz, William W. 54
Cold War 110, 112, 185, 207–9
collaboratories 100
comics and comic strips 204–10
Commercial Solvents Corp. (company) 154
computers, quantum 44
computers, use of 23, 77, 196, 204
Conant, James B. 211
concentration camps 108–9, 161–2
Conoco (company) 170
Consolidated Electrodynamics Corporation (CEC) (company) 87, 91
consumer products 123, 126
Convention on Bacteriological Weapons (1975) 125

Corey, Elias J. 23, 38
Corey, Robert B. 37
Corteva (company) 171
cosmochemistry 14–5, 53, 117–8
Coulson, Charles A. 35, 60
Coulson, Dale M. 133
Courtaulds (company) 158, 160
Covestro (company) 172
Crab Nebula (Messier 1) supernova remnant 15
Cram, Donald J. 38
Crichton, Michael 221
Crick, Francis 14, 37, 47, 116
Cross, Charles F. 126
crown ethers 12–13, 38
Crutzen, Paul J. 147–8
crystallization 88
Curie, Irène 32
Curie, Marie 32, 114
Curl, Robert F., Jr. 12
cyclohexane 138

dangerous substances 20, 74, 110, 120, 123–4, 134, 137–9, 140, 216–7
Das letzte Kapitel (1930, The Last Chapter), poem (Kästner) 215
data mining 23
DDT (dichlorodiphenyltrichloroethane) 16–18, 22, 89, 131–2, 162–3, 169, 217–18
Dead Sea Scrolls, dating of 14
Debye, Peter 46
DECHEMA (German Society for Chemical Apparatus) 19
decolonization 185
Degesch (company) 109
degradative (or ultimate) analysis 8–9, 81, 87, 118–9
degree qualifications; *see also doctoral students* 143, 187, 190
DeLillo, Don 121
demographic changes 123, 143–6
demonstrations, chemical 61, 177
Department of Scientific and Industrial Research (UK) 151
Der Krieg (1932, The War) painting (Dix) 215
Deutsche Chemische Gesellschaft (DChG) 25, 176, 199, 202

developing countries 185
Dewar, Michael J.S. 35
Dewey & Almy (company) 166
Dewey, Bradley 166
Die Ballade des Vergessens (1926, The Ballad of Forgetting), poem (Klabund) 212
Die letzten Tage der Menschheit (1915–1922, The Last Days of Humanity), play (Kraus) 215
differential thermal analysis 11
dioxins 22, 139, 169, 217
Diplom-Chemiker 181, 187
disaster accounts 214–22
disasters, industrial 20, 115, 138–9, 169, 216–7
Discol motor fuel 155
Disney, Walt(er) E. 209
Distillers (company) 155
Dix, Otto 215
Djerassi, Carl 54, 57, 88, 93
DNA (deoxyribonucleic acid) 14, 37, 47, 112, 116
Dobson, Gordon M.B. 22–3
doctoral students 181–7, 192–8
Domagk, Gerhard 160
Donegani, Guido 152, 156, 164
Donora event (1948) 127–8
Doomsday fiction 218–21
Dow (company) 164, 169, 171
DowDuPont (company) 171
Doyle, Arthur Conan 206
Dresselhaus, Mildred 44
Drexler, Eric K. 57
Dreyfus, Henri and Camille 158
Du Pont (company) 22, 86, 119, 130–1, 137, 151–4, 157–8, 165, 168, 170–1
  move into synthetic fibers 158–160
  patent agreement with ICI voided 164
  merger with Dow 171
Du Pont, Pierre S. 154
Dubbs, Carbon Petroleum 166
Dubson, Michail I. 215
Duisberg, Carl 102
Dulce et Decorum est (1917), poem (Owen) 212
dyes, synthetic 77, 102–3, 134, 150–1, 160, 170, 172
Dyson Perrins, Charles William 79
DyStar (company) 172

earth system sciences 119
East Texas oilfields 154
Education in Chemistry (journal) 197
education, chemical *see also teaching practices* 61–3, 173–84, 186–8, 192–202
  general science programs 62
  heuristic method of teaching 61–2
  industrial funding of 176
  school laboratories 62, 75
  state funding of 63, 191, 195, 200
Egbert, Robert 166
Einstein, Albert 114
electron diffraction 37, 46
electron paramagnetic resonance, *see spectroscopy*
electron theory 4, 31–4, 40, 48
electrophoresis 53
elements, chemical 31–2
  abundance of 14
elites 176–8, 185–6, 190, 192
Elmer, Charles W. 85
Endo, Akira 19
England, *see also United Kingdom* 75, 116, 135–6, 176–8
Enraf-Nonius (company) 79
environmental chemistry 22–2, 223
environmental movement 169, 216, 218
environmental problems 3, 18, 22, 121, 127–40, 168, 172, 189
Environmental Protection Agency (EPA) (US) 21, 129, 132, 135–8
Ernst, Richard R. 56, 95
ersatz materials 104, 110
Ertl, Gerhard 41
Eschenmoser, Albert 38
Esso (company) 164
ethanol (ethyl alcohol) 22, 155, 162
ethnic diversity 24–5, 198
European Union (EU) 132
everyday life, chemistry in 126–7
experimental practice 52–4
  interaction with theoretical knowledge 54
explosives 102–3, 124, 140, 150–2, 156, 211
ExxonMobil (company) 171
Eyring, Henry 41

fairy tales 210
Falkland Islands (1914), battle of 150

false positives and false negatives 68
famine 105, 219
Farm Chemurgic Council 155
Fauser, Giacomo 156
Federation of Asian Chemical Societies
    (FACS) 188–9, 195, 202
Federation of European Chemical Societies
    (FECS) 188, 197, 199
Fenn, John B. 95
fertilizers 16, 104–6, 151
fibers, synthetic 110, 126–7, 158–160, 168
    nylon 22, 126–7, 158–160
    rayon 126, 136–7, 158
films and the film industry 203, 206–7
Fischer, Emil 99, 102–3, 108, 119
Fischer–Tropsch process 156, 161, 169
Flash Gordon, fictional character 207, 209–10
Fleck, Ludwik 68
Fleming, Alexander 108, 164
Flint, Michigan 135
Flixborough disaster (1974) 138, 169
Florey, Howard W. 162, 164
Flory, Paul J. 11, 94
Food and Drug Administration (US) 132
food labelling 66–7
Ford (company) 154
Ford, Henry 22, 155–6
Formosa Plastics (company) 171
formulae, chemical 59
Fowler, William A. 15
Foxboro (company) 20
France 25, 40, 151–2, 178
Franco, Francisco 184
Francoeur, Eric 60
Frank, Adolph 150
Frankau, Gilbert 212
Frank-Caro process 150
Franklin, Rosalind E. 14, 37
Freeman, Ray 95
Friedrich, Bretislav 42
Fryer, John 192
Fukui, Kenichi 26, 35, 188, 190
Fulbright scholarships 191
fullerenes 12, 39, 44
fume hood (fume cupboard) 10, 74, 97
funding for research, military 92
funding for research, state 92, 112–13

Ganot, Adolphe 61
Gao Chongxi 194
GARIOA (Government Appropriation for
    Relief in Occupied Areas) program
    191
gas chromatography (GC) 9, 22, 88–91
    electron-capture detector 9, 22, 89
    flame ionization detectors 9, 89
    gas chromatography-mass spectrometry
    (GC-MS) 89
gas masks 212–14
Gasangriff (1916, Gas Assault), poem
    (Ulitz) 215
gasoline (petrol) 135, 154, 157
gasoline, synthetic 20, 104, 156, 161, 169
Gavroglu, Kostas 35, 46
Geim, Andre K. 12
gender equality 144
General Motors (company) 154
genetics and genetic engineering 47, 114, 170
Geneva Convention on air pollution 130
Geneva Protocol (1925) 125, 213
geochemistry 14, 117–18
Gérard, Jean 184
Gerber, Christoph 8
German Democratic Republic 164
Germany 13, 19, 26, 40–1, 55, 60, 74,
    91, 93, 96, 99–104, 150–1, 174–8,
    181–7, 190, 197, 199, 202, 208,
    213, 220
    primacy in chemistry of 26, 102, 185, 202
Gesellschaft Deutscher Chemiker (GDCh) 25
Gibbs, Josiah Willard 11
GlaxoSmithKline (company) 96
global hazards, chemical 207
global science 195–9
global warming *see climate change*
glove box for manipulating chemicals 10, 97
Goldschmidt, Victor M. 14
Goodrich, B.F. (company) 157
Grace, W.R. & Co (company) 166
graphene 12
Graves, Robert 213
Green Bank radio telescope (West Virginia) 15

green chemistry 21–2, 170
Griesheim-Elektron (company) 155
Grignard, Victor 39–40
groundwater 134–5
Grünenthal (company) 217
Gutowsky, Herbert S. 56, 83
Gyro Gearloose, fictional character 209–10

Haagen-Smit, Arie Jan 129
Haber, Fritz 3, 16, 41, 103, 150–1, 210–11
Haber–Bosch process 3, 16, 104, 150–1, 155–6
Hague Conventions (1899 and 1907) 211
Hahn, Otto 32, 110, 214
Halabja chemical weapons attack (1988) 125
Halcon (later Halcon SD) (company) 166–7
Hale, William J. 155
Handler, Philip 121
Hansen, James E. 147
Hartree, Douglas R. 60
Hassel, Odd 36
Haszeldine, Robert N. 94
Hauptman, Herbert A 77
Hauptmann, Gerhart 211
Haworth, Walter N. 36
hazardous substances, *see dangerous substances*
Health and Safety Executive (HSE) (UK) 138
Heatley, Norman G. 162–4
Heisenberg, Werner 33–4
Heitler, Walter 33–4
herbicides *see pesticides*
Hercules Powder Co. (company) 152
Herschbach, Dudley R. 30, 42, 48
Herty, Charles H. Sr. 155
HIAG (Holzverkohlungs-Industrie AG) (company) 154
Hieber, Walter 41
higher education 177, 186–9
    in Asia 189
    global expansion of 186
    women in 187
high-pressure chemistry 16, 103–4, 154, 156, 158

Hilger, Adam (company) 83
Hill, Julian W. 158
Hinselwood, Cyril N. 41, 79
Hirota, Noboru 31, 41
Hirs, Christophe H.W. 114
Hitler, Adolf 3–4
Hodgkin, Dorothy Crowfoot 5–6, 37–8, 77–8, 142
Hoechst (company) 3, 18, 151–2, 154–5, 164, 168, 171–2
Hoechst Marion Roussel (company) 172
Hoffmann, Roald 35, 38, 48
Hofmann, August Wilhelm 1
home environment and asthma 140
Hong Kong 168
Hoover, Herbert Jr. 91
Hopkins, Frederick Gowland 13–14
Horace (Roman poet) 212
Hörlein, Heinrich 18
Horrocks, Sally M. 140
Horváth, Csaba 90
Hoyle, Fred 14–15
HPLC (high-performance liquid chromatography) 9–10, 90–1, 96
Huber, Josef F.K. 90
Hückel, Erich 34–5
Hückel, Walther 34
human immunodeficiency virus (HIV) 19
Humphreys and Glasgow (company) 166
Hund, Friedrich 34
Hunter, Norman 209
Huntsman, John Sr. 170
Hussein, Saddam 125
hydrogen bonding 37
hydrogen, metallic form of 16

IBM (company) 8, 12
IG Farben (company) 3–4, 11, 16–18, 102, 104, 108, 151, 154–64, 176
    Auschwitz factory 108–9, 161
    broken up by Allies 164
    Elberfeld factory 108, 161
    Leuna ammonia factory 104, 151–2, 156
    Ludwigshafen factory 104, 161
    factories seized by Soviet Union 164
    trial of IG Farben executives 164
    use of slave labor 108–9, 161, 164

Im Westen Nichts Neues (1928, All Quiet on the Western Front), novel (Remarque) 213
Imperial Chemical Industries (ICI) (company) 16, 19, 21–2, 152, 154–5, 156–8, 160, 165, 167–8, 171
　Billingham factory 104, 156–7
　Grangemouth factory 160
　ICI Pharmaceuticals 17, 167
　patent agreement with Du Pont voided 164
　polyethylene, development of at 151, 158
India 199, 202
indoor environments 139–40
industrial waste 220
Ineos (company) 171
infrared spectroscopy, *see* spectroscopy
Ingold, Christopher Kelk 34, 40
insecticides *see* pesticides
Inspec (company) 171
Institut Pasteur, Paris 18, 160
Institution of Chemical Engineers 19
instrument makers, relationship with chemists 74, 92
instrument manufacture 83–4
instrumental community 57
instrumental practice 3, 10, 23, 53–7, 118–19
　interaction with theory 57–60
instrumental revolution 53–4, 70–3, 80–92
　first use of the term 80
interdisciplinarity 11, 223
internal combustion engine 127, 129
International Association of Chemical Societies 184
International Chemistry Olympiad 186–7
International Congress of Applied Chemistry 184
International Hydrogenation Patents Co. (company) 157
International Labour Organization (ILO) 67
International Program of Chemical Security 68–9
International Research Council 184
International Symposium on University Chemical Education (1969) 188

International Union of Pure and Applied Chemistry (IUPAC) 25, 65, 184, 186, 188, 195
International Year of Chemistry (IYC, 2011) 202
invisible practices 64–77
isotopes 32
Italy 55–6, 177–8, 184

James Bond films 215
James, Anthony T. 9, 89, 132
Janssen-Cilag (company) 19
Japan 20, 26, 168, 174, 186, 188–93, 195, 199–200
Johnson, William S. 86, 93
Joliot, J. Frédéric 32, 46
Jones, David E.H. ("Daedalus") 12
Journal of Chemical Education 174
journals 176, 189–90, 197, 200
Jupiter, metallic hydrogen core of 16

Kagaku kyoiku (Chemical Education) (journal) 189
Kagaku kyoiku sinpojiumu (Chemical Education Symposium) (journal) 189
Kaiser Wilhelm Institute for Coal Research 156
Kaiser Wilhelm Institute for Physical Chemistry and Electrochemistry 103, 151
Kaiser Wilhelm Institute of Chemistry 110
Kaisers, Georg 215
Karle, Jerome 77
Kästner, Erich 215
Kawai, Maki 200
Keeling, Charles D. 23
Kekulé, F. August 36, 46
Kellogg, M.W. (company) 166–7
Kendrew, John C. 38, 77
Kirchhoff, Gustav R. 9, 53, 80
Kirchner, Justus G. 9
klystron 85
Kohn, Walter 48
Korea 186
Kornbluth, Cyril M. 220
Kraus, Karl 212, 215
Krausnick, Michail 220
Krebs (citric acid) cycle 14
Krebs, Hans A. 13–14

Kroto, Harold W. 12
Kuhlmann (company) 152
Kunming University 193
Kyoto Imperial University 190
Kyoto Protocol (1997) 147

Labinger, Jay A. 32, 40
laboratories 1–2, 10, 52, 61–2, 74–5, 79–80, 83, 92–9, 178–9
  benches in 74–5, 79, 93–5, 97
  classical type 52, 92–3, 96, 99
  design of 74–5, 92–8
  electricity in 75
  health & safety, influence on design of laboratories 96
  industry, in 75
  robotics in 97–8
  schools, in 62, 75
laboratory practices 31, 36, 46, 69, 177–8, 194, 196–7, 200
lacquer, artificial 153
Landau, Ralph 166
Langmuir, Irving 11, 33, 40, 220
language issues 65, 199–200
Lanxess (company) 172
Lapworth, Arthur 40
Latin America 174, 188, 195
Laukien, Günther 95
Lavoisier, Antoine 31
Lawrence Livermore National Laboratory (California) 16
Le Bel, Joseph Achille 36
lead poisoning 135
lectures 61, 179–82
Lee, Yuan T. 42, 188
Lehn, Jean-Marie 13, 38
Lenin, Vladimir I. 176
Lennard-Jones, John 35, 60
Leslie, May Sybil 140
Lewis, Gilbert N. 33, 59
Lewis, Warren K. 166
Libby, Willard F. 14
Liebig Society 176
Liebig, Justus 61, 65, 176
Limits to Growth (1972) 220
Lipson, Henry 77
London, Fritz W. 33–4, 41
Longuet-Higgins, H. Christopher 35

Lonsdale, Kathleen 5, 76–9
Los Angeles 129
Love Canal (New York State) 134, 169, 172, 218
Lovelock, James E. 9, 22, 133, 148
Löwdin, Per-Olov 43
Lowry, T. Martin 40
Lucas, Howard J. 40
Lukyanovich, V.M. 44
Luther, Martin 218–19
LyondellBasell (company) 171

mad alchemist trope 205–7, 222
mad scientist trope 204–10, 219–23
malaria 131–2, 162
Mao Zedong 186
Marion Merrell Dow (company) 171
Mark, Herman F. 11, 37
Martin, Archer J.P. 9, 88–90, 132
Marvel Comics 208
Mason, Clesson E. 20
mass media 203–5, 222
mass spectrometry (MS) 7–10, 57, 79–80, 85–8, 95–7
  HPLC-MS 9–10, 96
  time of flight MS 96
Massachusetts Institute of Technology (MIT) 19, 91, 166
materials science 11–12, 27, 42–5, 111–12, 115, 119, 171
mathematical chemistry 35
mathematical concepts and techniques 117
matter, theory of 40
Mauna Loa (Hawaii) 23
May & Baker (company) 18, 160
McCall, David W. 83
McCrae, John 211
McGowan, Harry D. 154
McLafferty, Fred W. 57
McMillan, Edwin M. 32
McMillen, Wheeler 156
McWilliam, Ian 9
medicine, *also see pharmaceuticals* 24, 27, 45, 53, 56, 92, 104, 108, 118–19, 188
Meerwein, Hans 40
Meitner, Lise 32, 110
Mendeleev, Dmitrii I. 180

Merck (US company) 19, 163
metallurgy 11–12
methanol synthesis (Pier process) 16, 104, 154–5
methyl isocyanate 139
metric system 65
Metropolitan-Vickers (company) 79
Meyer, Kurt H. 11, 37
Mickey Mouse, series of Disney films 209
microchips 21–2
Miller, Stanley L. 116
Minamata Bay disaster 20, 216–17
mineralogy 14
minority groups 124
Mitsubishi Chemicals (company) 171
models, *see molecular models*
molecular biology 14, 26, 114
molecular models 37–9, 56, 58–60, 180
molecular orbital theory 34–5, 43
molecules 36–9, 46–7, 48
    found in outer space 16
Molina, Mario 41
Monastral Fast Blue dye 160
monoclonal antibodies as drugs 19
Monsanto (company) 170, 172
Montecatini (company) 152, 164
Montedison (company) 171
Montreal Protocol (1987) 130–1
Moore, Gordon E. 91
Moore, John C. 90
Moore, Stanford 90
Moseley, Henry G.J. 31
Moulton, John Fletcher Moulton, Baron 151
Moureu, Charles 210
Müller, Paul H. 16, 162
Mulliken, Robert S. 34–5, 60
munitions factories 140–1
Murphy, Walter J. 141

nanotechnology 12, 44, 57, 115, 119
National Academy of Science (US) 113
National Institute for Medical Research (UK) 89
National Institutes of Health (US) 92
National Research Council (US) 111, 115, 151
National Science Foundation (US) 63, 92
national styles of research 56, 60

Natta, Giulio 168
Natural Environment Research Council (UK) 131
natural materials replaced by synthetic materials 126, 157
natural products, preference for 168–9, 217
Naturwissenschaften im Unterricht Chemie (Sciences in Instruction – Chemistry) (journal) 197
Nazi regime in Germany, 26, 55, 108–9, 156–7, 208–9
    mass murder by 109
Negishi, Ei-ichi 192
Neoprene, *see rubber, synthetic*
Nernst, H. Walther 45, 103
nerve agents *see under chemical weapons*
neutron 32
New Zealand 174
Nichitsu (company) 152
Nichols, Mike 112
Nier, Alfred O.C. 87
nitrogen fixation 41, 103, 150, 152, 156, 167
Nixon, President Richard M. 129
Nobel Industries (company) 152, 154
Nobel prizes 27, 45–7, 192
noble gases 32–3
nomenclature, chemical, *see terminology*
non-fiction writing 204, 216, 220
Nordenholt's Million (1923), novel (Connington) 219
Norrish, Ronald G.W. 42
Norsk Hydro (company) 171
Northern Regional Research Laboratory (Illinois) 18, 163
Novoselov, Konstantin S. 12
Noyori, Ryoji 192
nuclear chemistry 32, 53, 117
nuclear magnetic resonance (NMR) 6–10, 40, 46, 56, 82–6, 95–6, 196, 200
    FT-NMR 95
Nuclear Magnetic Resonance (Roberts) 86
nuclear technology 119
nuclear war 215
Nuffield Foundation 61
Nuffield, William Morris, Viscount 79
numerical methods 60
nutritional standards 65–6

Nye, Mary Jo 117
nylon, *see under synthetic fibers*

O'Brien, Flann 220
oceans, pollution of 135–6, 172
Ogg, Richard A., Jr. 86
oil crises (1973 and 1979) 169
oil prices 149, 165, 169–72
Olah, George A. 40
Oppau disasters (1921 and 1948) 150, 216
Oppenheimer, Frank 63
oral contraceptives 19, 111, 167
Osawa, Eiji 12
Oswald, Wilhelm 45, 179
Owen, Wilfred 212
Oxford University 184
    chemistry degree, Part II 1
    Chemistry Research Laboratory (CRL) 96–7
    Dyson Perrins (DP) Laboratory 1–2
    penicillin research 18, 162–3
    Physical Chemistry Laboratory (PCL) 79
ozone hole 22–3, 130

Pacific garbage patch 172
Packard, Martin E. 82, 85–6
Panoramix (Getafix), fictional character 209–10
paper tools 30, 58–9
parachor 75–6
Parson, Alfred L. 59
Pasteur, Louis 36
Patterson, Arthur L. 77
Pauling, Linus 34, 37, 46–7, 59–60, 86, 180–1
Pederson, Charles J. 12–13, 38
Peking University 194
penicillin 5, 18–19, 37, 108, 162–4
periodic table 31–2, 180–1, 194–5
    year of the (2019) 195
Perkin, Richard Scott 85
Perkin, William Henry, Jr. 1–2, 79, 102
Perkin-Elmer (company) 85, 91
Perutz, Max 38, 77
pesticides 16–18, 104–6, 131, 167, 169, 172, 217–18
    daminozide (Alar) 18
    glyphosates 18, 170

neonicotinoids 18
    also see DDT
petrochemicals 3, 20, 162, 164–70
petroleum companies 89, 154–5, 157, 162, 164, 166, 170–1
Pfisters Mühle (1884, Pfister's Mill) novel (Raabe) 218
Pfizer (company) 163
pH meter 75, 85
pharmaceuticals, *also see medicine* 3, 5, 10, 18–19, 26, 67, 75, 88, 90, 96, 98, 108, 111, 134, 160, 162, 167, 170–1
    adalimumab 19
    antimalarial drugs 16, 108, 162
    beta-blockers 19, 167
    chloroquine 18, 162
    cimetidine 19, 167
    M&B 693 160
    mepacrine 162
    Prontosil 18, 160
    proton inhibitor drugs 19, 167
    propranolol 19, 167
    simvastatin 19, 22
    statins 19, 22
    sulfonamide drugs 18, 108, 160
    also see penicillin and oral contraceptives
pharmacology 18–19, 167
Phillips, David C. 77
philosophy of chemistry 47, 223
physical chemistry 179–80
physical organic chemistry 38, 86, 118
physics 5, 45–8, 73, 86, 112–19, 179–80
Pilcher, Richard B. 124
Pimentel Report (1987) 115–16, 121, 172
plastics 11, 22, 37, 106, 110, 116, 133–6, 156–8, 165, 168–9, 172
    failure of Ardil and Corfam 168
    plastic bottles 168
    Plexiglas (Perspex) 158
    polyethylene (Polythene) 16, 155, 158, 168
    polyurethanes 110, 168
    polyvinyl chloride (PVC) 20, 157, 168
    stereospecific polymers 168
poetry 211–15
Pohl, Frederik 220
Poison (1919), poem (Frankau) 212

poison gas, *see chemical weapons*
Polanyi, John C. 42, 44
Polanyi, Michael 41, 43
pollution 132–46, 172, 194, 205, 216–23
  measurement of 22
polymer chemistry 11, 37, 106
polymers, *also see plastics* 37, 43, 56–7
Pope, William Jackson 210–11
Porter, George 42
postdoctoral fellows in chemistry 199
poverty 123
Power-Gas Corporation (company) 166
Pregl, Fritz 211
Pretorius, Victor 8
Proctor, Warren G 83
Professor Balthazar (1967–1978) TV series 210
Professor Branestawm, fictional character 209
proteins, structure of 7, 14, 24, 35–7, 46–7, 77–8, 83, 96, 141
publications, chemical 200–2
Purcell, Edward M. 82
Pye, W.G. (company) 85

quality control 67, 70, 129
quantum chemistry 4–5, 34, 35, 41, 43–4, 47, 60, 114, 117, 190
quantum mechanics 5, 40
quantum theory 12, 33–5, 47
Quate, Calvin F. 8

Raabe, Wilhelm 218
Radushkevich, Leonid Viktorovich 44
Raemaekers, Louis 213
Raemaekers' Cartoon History of the War (1919), drawings (Raemaekers) 213
Raskob, John J. 154
Ratcliffe, James A. 170
Rathenau, Walther 151
raw materials for the chemical industry 154–5
rayon, *see under synthetic fibers*
reaction mechanisms 4, 39–42
Reed, R. Ivor 88
Reeve, Arthur Benjamin 206
regulatory concepts and processes 67, 127, 137
Rehnberg, Harry 166

Reinhardt, Carsten 52, 71, 120
Reiterlied (1914, Cavalry Song), poem (Hauptmann) 211
religion 116, 222
Remarque, Erich Maria 213
Reppe, J. Walter 16, 158
retrosynthetic analysis of targets 23
Revelle, Roger 147
Rhône-Poulenc (company) 171, 172
Roberts, John D. 86
Robertson, Alex 87
Robertson, John M. 77
Robinson, Robert 12, 38, 40, 46
Rockefeller Foundation 13, 192
rocks, age of 14
Rohm & Haas (US company) 158
Röhm (German company) 158
Rohrer, Heinrich 8
Rose, Frank W. 54
Rossiter, Margaret 141
Rowland, Sherwood 41
Royal Institute of Chemistry (RIC) 25
Royal Institution, London 76–7
Royal Society of Chemistry (RSC) 25, 27, 148, 197
rubber, synthetic 20, 85, 108–9, 157, 161–2, 165–6, 168
Rudolph, John L. 60
Runge, Friedlieb Ferdinand 9
Russia *see Soviet Union*
Rutherford, Ernest 31–2, 46, 101

Sabatier, Paul 39–40
Sabic (company) 167, 171
saccharimeters 53
safety regulations 126
Salisbury incident (2018) 3, 125
Salters Chemistry project 64
San Jose, California 91
Sandoz disaster (1986) 217
Sanger, Frederick 47
Sanofi (company) 172
scanning tunnelling microscope (STM) 7–8, 12, 42
Schlack, Paul 160
Schlenk, Wilhelm J. 39–40
Schrödinger, Erwin 33
Science and Art Department (UK) 75
science fiction 204, 207, 210, 214–15, 220

Scientific Design (company) 166
scientific method 61
Scottish Dyes (company) 160
Senderens, Jean-Baptiste 39–40
September 11 attacks (2001) 196, 198
Seveso disaster (1976) 20, 115, 138–9, 169, 217
Shaik, Sason S. 34
Shell (company) 85, 89, 154–5, 157, 164
Shelley, Mary 206
Shimomura, Osamu 192
Shirakawa, Hideki 192
Shoolery, James N. 86
Shoppee, Charles W. 87
Sidgwick, Nevil V. 33, 36–7
Siegel, Jay S. 201
Silent Spring (Carson) 132, 217
Silicon Valley 85, 91
Simões, Ana 35, 46
Sinopec (company) 171
Slater, John Clarke 34, 117
Slichter, Charles P. 83
Smalley, Richard E. 12
Smith, Kline & French (company) 19, 167
smog 127–8
Snelson, Kenneth D. 59
Snyder, Lloyd R. 90
Société chimique de France (SCF) 25
Society of Chemical Industry 25
Soddy, Frederick 32
Soderbergh, Steven 221
solid-state science 11, 42–5
Solvay (company) 152, 171
Sommerfeld, Arnold 33
South Africa 169
Southwold, Stephen 215
Soviet Union 25–6, 92, 162, 164, 176–7, 184–6, 193, 202
   collapse of 195
spectroscopy 5–9, 53–6, 80–9, 91–2, 95–7, 118, 142
   atomic absorption 9, 80
   electron paramagnetic resonance (EPR) 56–7
   Fourier-transform infrared (FTIR) 89
   infrared 5–7, 23, 42, 53, 81–2, 85, 89
   Raman spectroscopy 56
   ultraviolet 5–6, 22–3, 38, 76, 81–3, 85, 91

x-ray fluorescence spectroscopy 9, 71
x-ray spectroscopy 9, 14, 31
also see mass spectrometry and nuclear magnetic resonance
Sputnik crisis (1957) 63, 92, 186
St. Gobain (company) 152
Stalin, Josef 176
Stalinist planning 185
Standard Oil of New Jersey (company) 154–5, 157, 161
standardization 52, 65–7
Stanford Research Park 86, 93
Stanford University 82–3, 85–6, 91
   Stauffer buildings 93–4
State of Fear (2004), novel (Crichton) 221
Staudinger, Hermann 11, 37
Stauffer, John Jr. 93
Stein, William H. 90
stellar nucleosynthesis 14–5
Stephenson, Neal 221
stereochemistry 59
steroid chemistry 167
Stewart, Alfred Walter ("J.J. Connington") 219
Stockholm Convention (2001) 132
STOE (company) 79
Stoermer, Eugene F. 147
Strassmann, Fritz 32, 110, 214
Stubbens, Sidney W.J. 85
Sugden, Samuel 75
superconductivity 43
supercritical fluid extraction 21
Superman and Spider-Man, fictional characters 208
supervillain trope 205–9
supramolecular chemistry 12–13
surface chemistry 11, 41
Svedberg, Theodor 37
Swedish Academy of Sciences 45
Sydney University 87
Synge, Richard L.M. 9, 88–9
synthesis, organic 10, 23, 38–9, 53–4, 80–1
Syria 125, 196

Taiwan 188, 195
Taniguchi, Norio 44
Tarbell, Ann Tracy 80
Tarbell, D. Stanley 80
Tati, Jacques 112

Taube, Henry 94
teaching practices 52, 59–64, 71
Terman, Frederick E. 93
terminology, chemical 59, 65
terrorism 125
tetraethyl lead gasoline additive 127, 135, 154
textbooks of chemistry 61–2, 180, 192
thalidomide disaster (1961) 67, 169, 217
The Adventures of Tintin (1929 onwards), comic strip novels (Hergé) 209
The Burning World (1964), novel (Ballard) 219–20
The Food of the Gods and How It Came to Earth (1904), novel (Wells) 219
The Mad Doctor (1933), film (Disney) 209
The Sheep Look Up (1972), novel (Brunner) 220
The World Set Free (1914), novel (Wells) 214, 219
theoretical practices and knowledge 52–4, 58
Thiele, F. K. Johannes 40
Thompson, Harold W. 91
Thomson, Joseph John 40, 86
threshold limit value (TLV) 67, 110
Thudichum, Ludwig 13
Tiananmen Square incident (1989) 195
tobacco smoke, health risks of 110
Tokyo subway incident (1995) 125
Townes, Charles H. 15
toxic substances, *see dangerous substances*
traditional chemical techniques, survival of 72
tropinone, synthesis of 12
Tswett (or Tsvet), Mikhail 9, 89
Tuskegee syphilis study (1932–1972) 67
Tyndall, John 144

Ulitz, Arnold 215
ultimate analysis, *see degradative analysis*
ultracentrifuge 37, 53
ultraviolet spectroscopy, *see spectroscopy*
UMIST (University of Manchester Institute of Science and Technology) 94–5
Unicam Instruments (company) 85
Unilever (company) 171
Union Carbide (company) 139, 152, 154–5, 157, 164, 171, 216

United Alkali Co. (company) 152
United Kingdom, *see also England* 13, 19, 25–6, 40, 54, 62, 65, 91, 94, 102, 125, 128–9, 131, 133, 140, 151, 153, 157–8, 160–1, 168–9, 184, 186, 188, 197, 199
United Nations Economic Commission for Europe (UNECE) 130
United Nations Framework Convention on Climate Change (UNFCCC) 147
United Nations Security Council 125
United States 26, 100, 150, 177–82, 186–7, 200
  Affirmative Action legislation 187
  higher education in 177–86, 191, 197, 202
  instrument manufacture, leading force in 91–2
  world leadership in chemistry 74, 91, 185, 202
United States Department of Agriculture (USDA) 18, 132, 156
United States Supreme Court 152
Universal Oil Products (UOP) (company) 166
universities 177–82, 186–7, 200
University Chemistry Education (journal) 197
Urey, Harold C. 116
US Steel (company) 127–8

valence bond theory 34–5
van 't Hoff, Jacobus Henricus 36, 45
Varian Associates (company) 85–6, 91, 95
Varian, Russell H. 85, 91
Varian, Sigurd F. 85
Verband der Laboratoriumsvorstände an deutschen Hochschulen (Association of Laboratory Directors at German Colleges and Universities) 176, 179, 181
Verein Deutscher Chemiker (VDCh) 25, 176
Verne, Jules 209
Vestal, Marvin L. 95
Vietnam War 169, 186
vinyl chloride monomer 20, 157
volatile organic compounds (VOCs) 140
Vonnegut, Kurt 220

Wacker (company) 155
Wallace, Don and Fran 71
Warburg, Otto 13
Warwick, Andrew 58
Washburn, Edward W. 54
Washburn, Harold W. 91
Wasserman test 68
Waterhouse, Alfred 79
Waterhouse, Paul 79
Waters Associates (company) 90–1
Waters, James L. 90–1
Watson, James D. 14, 37, 47, 116
Weichselfelder, Theodor 39–40
Weiler-ter Meer (company) 152
Weininger, Stephen J. 40–1
Weizmann process 154
Wells, H.G. 214–15, 219
westernization 174
Westheimer, Frank H. 113–16
Wheland, George W. 34
Wilkins, Maurice H.F. 14, 47
Winkler, Fritz 20
Winthrop Chemical (company) 162
Wisconsin, University of 194
women
    discrimination against 140 (*see also* chemists, as women)
    employment in the chemical industry 140–1
    in higher education 187
    roles and status of 24–5, 124
Wonder Woman, fictional character 208
wood distillation 154
Woodward-Hoffmann rules 35, 38
Woodward, Robert Burns 35, 38, 44, 76, 88, 91
Woodward rules 38, 76
working environments 136–8

World Health Organization (WHO) 139
World War I 17, 19, 29, 54, 102–4, 116, 123–6, 134, 140, 144, 149–51, 154, 164, 172–4, 184, 205–7, 210–15, 219–22
    British naval blockade 4, 102, 150
    chemists' war, as the 1, 29, 124, 183
    Hindenburg program 151
World War II 25, 56–7, 108, 110, 125, 131, 141, 149, 157–8, 161, 164, 172, 183–4, 190, 208

x-ray crystallography 5, 7, 11, 14, 37, 46, 76–83, 141
    "direct methods" 77
    computers used in 77
    heavy atom method 77
    Lipson-Beever strips 77
    Patterson function 77
x-ray diffraction 11–12, 37, 40
x-ray fluorescence spectroscopy *see spectroscopy*
x-ray spectroscopy *see spectroscopy*
Xu Shou 192

Yale University 178–9
yam, Mexican (Dioscorea mexicana) 167
Yara (company) 171
Yenching University 192–3
Ytterby mine, Sweden 14
Yu, Fu Chun 83

Zavoisky, Yevgeny K. 56
Zeneca (company) 171
Zewail, Ahmed H. 42
Ziegler, Karl W. 168
Zodiac (1988), novel (Stephenson) 221
Zyklon B 109